Blockchain Technology for Water and Environmental Systems

In an era defined by climate uncertainty, resource scarcity, and the urgent need for sustainable innovation, blockchain technology emerges as a transformative force across water and environmental systems. This groundbreaking volume brings together global experts to explore how decentralized technologies are reshaping the way we manage water, protect ecosystems, and build climate-resilient infrastructure.

Spanning foundational concepts to cutting-edge applications, the book offers a comprehensive journey through blockchain's evolution—from its cryptographic roots to its integration with smart cities, IoT, and ecological governance. Readers will discover how blockchain enhances transparency in water trading, revolutionizes flood risk management, supports sustainable agriculture, and drives innovation in environmental monitoring and policy.

Structured across eight thematic parts, the chapters delve into intelligent water systems, consensus mechanisms, disaster resilience, and industrial transformation—offering both theoretical insights and practical case studies. Whether you're a researcher, policymaker, technologist, or sustainability advocate, this volume provides the tools and vision to harness blockchain for a greener, smarter future.

Edited by leading scholars and practitioners, *Blockchain Technology for Water and Environmental Systems* is a timely and essential resource for anyone seeking to understand and apply blockchain in service of global sustainability.

With contributions from global thought leaders and case studies spanning continents, *Blockchain Technology for Water and Environmental Systems* is a vital resource for academics, practitioners, and innovators seeking to harness the power of decentralization for a more resilient and sustainable future.

Blockchain Technology for Water and Environmental Systems

Edited by
Saeid Eslamian
Faezeh Eslamian

CRC Press
Taylor & Francis Group
Boca Raton London New York

CRC Press is an imprint of the
Taylor & Francis Group, an **informa** business

Designed cover image: Shutterstock

First edition published 2026
by CRC Press
2385 NW Executive Center Drive, Suite 320, Boca Raton FL 33431

and by CRC Press
4 Park Square, Milton Park, Abingdon, Oxon, OX14 4RN

CRC Press is an imprint of Taylor & Francis Group, LLC

ISBN: 978-1-032-98612-8 (hbk)
ISBN: 978-1-032-98614-2 (pbk)
ISBN: 978-1-003-59959-3 (ebk)

DOI: 10.1201/9781003599593

Typeset in Times
by SPi Technologies India Pvt Ltd (Straive)

Dedication

To all those who believe in the quiet strength of kindness, the enduring light of peace, and the shared humanity that binds us across borders and generations.

> *"Peace begins where kindness is planted, and humanity blooms when we choose compassion over conflict."*

Saeid Eslamian, 2026

Contents

Part IV Blockchain Application in Ecological Protection and Water Quality

Part V Blockchain Technology for Environmental Innovation, Conservation and Sustainability

Part VI Blockchain for Sustainable Practices in Irrigation and Crop Systems

Part VII Blockchain Application in Disaster Management

Part VIII Blockchain for Hydrosystem and Industrial Transformations in Smart Cities: A Forward-Looking Approach

Preface

In the face of mounting environmental challenges and the urgent need for sustainable development, the convergence of emerging technologies with ecological stewardship has never been more critical. Among these innovations, blockchain technology stands out—not merely as a tool for financial systems, but as a transformative framework for transparency, accountability, and resilience across water and environmental domains.

This book was born from a shared vision: to explore how blockchain can be harnessed to address some of the most pressing issues of our time—water scarcity, climate change, ecological degradation, and the need for inclusive, data-driven governance. It brings together a diverse group of researchers, practitioners, and thought leaders from around the world, each contributing unique insights into how decentralized systems can support smarter water management, sustainable agriculture, disaster resilience, and environmental innovation.

Structured across eight thematic parts, the chapters span foundational theory, real-world applications, and forward-looking perspectives. From integrating blockchain with IoT in smart water systems to revolutionizing supply chains and hydrosystems in smart cities, this volume offers both depth and breadth—designed to inspire academics, engineers, policymakers, and sustainability advocates alike.

As editor, we are deeply grateful to the contributors whose expertise and dedication have shaped this work. Their interdisciplinary collaboration reflects the spirit of innovation and global cooperation that this book seeks to promote.

May this volume serve not only as a scholarly resource, but also as a catalyst for dialogue, action, and hope—for a future where technology and humanity work hand in hand to build a more peaceful, equitable, and sustainable world.

Saeid Eslamian and Faezeh Eslamian
Isfahan University of Technology and McGill University

Editor Biographies

Saeid Eslamian received his PhD in civil and environmental engineering from University of New South Wales, Australia, in 1998. Saeid was a visiting professor at Princeton University and ETH Zurich in 2005 and 2008, respectively. He has contributed to more than 1,000 publications in journals, conferences and books. Currently, he is a full professor of Resilient Climate and Water Systems and the Director of the Excellence Center in Risk Management and Natural Hazards in Isfahan University of Technology. His scientific interests are floods, droughts, water reuse, climate change adaptation, sustainability and resilience. Prof. Eslamian is now recognized as Extraordinary Professor in Education and Science by the Government of the United States of America.

Faezeh Eslamian holds a PhD in bioresource engineering from McGill University, Canada. Her research focuses on the development of a novel lime-based product to mitigate phosphorus loss from agricultural fields. Faezeh completed her bachelor's and master's degrees in civil and environmental engineering from Isfahan University of Technology, Iran, where she evaluated natural and low-cost absorbents for the removal of pollutants such as textile dyes and heavy metals. Furthermore, she has conducted research on worldwide water quality standards and wastewater reuse guidelines. Faezeh is an experienced multidisciplinary researcher with an interest in soil and water quality, environmental remediation, water reuse and drought management.

Contributors

Vaibhav Aggarwal
United Nations University FLORES
Dresden, Germany

Monika Agrawal
Jaipuria Institute of Management
Indore
Madhya Pradesh, India

Burak Arslan
Faculty of Economics and Administrative Sciences
Anadolu University
Eskişehir, Türkiye

Gautam Bapat
Dr Vishwanath Karad MIT World Peace University
Pune, Maharashtra, India

Salvatore Cannella
Department of Civil Engineering and Architecture
University of Catania
Italy

João Casqueira Cardoso
Director of Master of Humanitarian Action, Cooperation and Development
Center of Studies for the Population, Economy and Society (CEPESE)
University Fernando Pessoa
Portugal

Harpreet Kaur Channi
Department of Electrical Engineering
Guru Nanak Dev Engineering College
Ludhiana, Punjab, India

Faezeh Eslamian
Department of Bioresource Engineering
McGill University
Montréal, QC, Canada

Saeid Eslamian
Department of Water Science and Engineering
College of Agriculture, Isfahan University of Technology
Isfahan, Iran

Simon Fernandez-Vazquez
Faculty of Economics
University of Oviedo
Spain

Asghar Akbari Foroud
Electrical and Computer Engineering Faculty
Semnan University
Semnan, Iran

Tulsi Pawan Fowdur
Electrical and Electronic Engineering Department
University of Mauritius
Moka, Mauritius

Omar Alexander León García
Faculty of Economics
University of Oviedo
Spain

Mahendra Gooroochurn
Mechanical and Production Engineering Department
Faculty of Engineering
University of Mauritius
Moka, Mauritius

Elena Gurgu
Economic Sciences Department
Faculty of Economic Sciences
Spiru Haret University of Bucharest
Bucharest, Romania

Nguyen Thi Hang
Faculty of Economic Information System
University of Information and Communication Technology
Can Tho City, Viet Nam

Farkhondeh Jabari
Power Systems Operation and Planning Research Department
Niroo Research Institute
Shahrak Ghods, Tehran, Iran

Irena Jindrichovska
Department of Financial Management
Metropolitan University Prague
Prague, Czech Republic

Karkuzhali S
Department of Computer Science and Engineering
Mepco Schlenk Engineering College
Sivakasi, Tamilnadu, India

Mohammad Karrabi
Electrical and Computer Engineering Faculty
Semnan University
Semnan, Iran

Anuj Kumar
Head of Research
Rushford Business School
Switzerland

Pulkit Kumar
Department of Electrical Engineering
Chandigarh University
Gharuan, Mohali, Punjab

C. Beulah Christalin Latha
Karunya Institute of Technology and Sciences
Coimbatore, India

Ritika Mahajan
Department of Management Studies
Malaviya National Institute of Technology Jaipur
Jaipur, Rajasthan, India

Mousa Maleki
Department of Water Science and Engineering
College of Agriculture, Isfahan University of Technology
Isfahan, Iran

Prateek Pandey
Department of Computer Science and Engineering
Jaypee University of Engineering and Technology
Madhya Pradesh, India

Sujni Paul
Faculty of Computer Sciences
Higher Colleges of Technology
Dubai

Bharat Ramdas Pawar
Department of Electronics & Computer Engineering
CSMSS Chh Shahu College of Engineering
Chh Sambhaji Nagar, Maharashtra, India

Shraddha Phansalkar
Computer Science and Engineering
School of Computing
MIT Art, Design, and Technology University
Pune, India

Purvi Pujari
Vijay Patil School of Management
DY Patil University
Navi Mumbai, India

Julián Ramírez
Faculty of Economic Sciences
National University of Colombia
Colombia

Ana Rodriguez
Health Sciences, Human Rights and Humanitarian Studies
Center of Studies for the Population, Economy and Society
Portugal

Yaser Sabzevari
Department of Water Science and Engineering
College of Agriculture, Isfahan University of Technology
Isfahan, Iran

Esmaeil Sajadi
Water Resources Department
Tarbiat Modares University
Tehran, Iran

Aanandha Saravanan
Department of ECE, Vel Tech Rangarajan
Dr. Sagunthala R&D Institute of Science and Technology
Chennai, India

G. Sasikala
Department of ECE
Vel Tech Rangarajan Dr. Sagunthala R&D Institute of Science and Technology
Chennai, India

B. Sathyasri
Department of Electronics and Communication Engineering
Vel Tech. Rangarajan Dr. Sagunthala R&D Institute of Science and Technology
Chennai, India

Senthilkumar S
Department of Chemistry
Ayya Nadar Janaki Ammal College
Sivakasi,Tamilnadu, India

Pathan Mohd. Shafi
Computer Science and Engineering
School of Computing
MIT Art, Design, and Technology University
Pune, India

Swati Dattatray Shirke
Computer Science and Engineering
School of Engineering & Technology
Pimpri Chinchwad University
Pune, India

Parminder Singh
Department of Electrical Engineering
Chandigarh University
Gharuan, Mohali, Punjab, India

S.V. Evangeline Sonia
Computer Science and Engineering
Karunya Institute of Technology and Sciences
Coimbatore, India

Rahul Ganpatrao Sonkamble
Computer Science and Engineering
School of Engineering & Technology
Pimpri Chinchwad University
Pune, India

Tanuj Surve
Haas School of Business
University of California
Berkeley, USA

Mehdi Talaie
Electrical and Computer Engineering Faculty
Semnan University
Semnan, Iran

Amit Kumar Tyagi
Department of Fashion Technology
National Institute of Fashion Technology
New Delhi, India

Erginbay Uğurlu
Department of Economics and Finance
Faculty of Economics and Administrative Sciences
Istanbul Aydın University
Istanbul, Türkiye

Part I

An Introduction to Blockchain Technology

1 Blockchain Technology for Intelligent Water Management and Sustainable Development

Mehdi Talaie, Farkhondeh Jabari, and Asghar Akbari Foroud

1.1 INTRODUCTION

According to the report of the Third World Water Forum, clean, freshwater is a precious and limited resource for achieving sustainable development, economic growth, social stability and poverty reduction. According to the principles of sustainability and in order to maintain sustainable development, available water resources must be preserved not only for the present generation but also for future generations. Achieving such goals requires the protection of natural resources and their optimal use.

Global demand for water is continuously growing due to population growth and economic development, climate change and other factors. Also, the non-optimal use of water resources and reserves, excessive extraction of resources, improper management and changing the climatic conditions have disrupted the natural balance of water resources. Therefore, on the one hand, existing water resources are decreasing and becoming polluted, and the demand for water is increasing. Due to the limited water resources, sustainable use of these resources are inevitable, and adopting appropriate policies, protecting water resources, water sustainability and optimal management of water resources have also attracted the attention of many researchers and policymakers.

Water resources management is considered one of the greatest challenges of the current century. Studies conducted on the issue of water, as one of the most central components of sustainable development, showed that improper management is the cause of many existing problems. The current management system faces fundamental problems. Current water management systems are decentralized, and data sharing is challenging. Also, data storage in this system lacks proper security and there is the possibility of data forgery or distortion. Accessing data is time consuming. Accurate monitoring is also difficult due to the extensiveness of the water network.

Blockchain technology has many advantages. It can help in the effective and efficient management of water resources and create sustainable infrastructure to ensure water security and management. Blockchain is able to store generated data in a decentralized manner with high security. Blockchain enables the collection, processing and analysis of data from water distribution networks to optimize resources and improve decision-making. Therefore, it is essential to examine the performance of blockchain technology in optimizing water management systems and advancing the achievement of the UN Sustainable Development Goals (SDGs) (Mattila et al., 2022).

1.1.1 Definition of Blockchain Technology

Centralized systems rely on third parties or a central entity. The data in this system is stored, controlled and processed by third parties or the central entity, and data security is also provided by them. Therefore, the cost of operating the system is high due to the existence of a central intermediary.

DOI: 10.1201/9781003599593-2

Also, this system suffers from issues such as scalability, data accessibility, reliability, security, lack of privacy and anonymity of participants. In order to solve the problems of centralized systems, such as the issue of security and privacy, decentralized systems such as blockchain technology have been able to attract wide attention in the global community (Bordel et al., 2019).

Blockchain technology is a different approach to managing and sharing data. Blockchain is a peer-to-peer (P2P) network that stores all information in a distributed database. Blockchain is a combination of two words: block and chain. This technology is actually a chain of blocks. All the data stored in this database are stored in a chain of blocks according to the time sequence. The block chain grows continuously with the addition of new blocks. This chain is called blockchain (Patil & Sangeetha, 2024). In each block, it is possible to store any type of data, such as a person's medical records or the details of a financial network's transactions. This data is shared among all or a group of nodes. The data stored in this system will never be deleted. It is also not possible to change the time or data (Eslamian & Motevallian, 2014).

Blockchain is a free network and every user can join it and store their data in it or create a new block in the chain by following the technical standards. For this reason, it is called public blockchain. Of course, in blockchain, the levels of cooperation and access to system data may be different for different nodes, and different nodes have different levels of access or viewing of data. Also, as previously stated, this network is a decentralized network and does not depend on a central controller (Sundaresan et al., 2021). This system allows users to communicate with each other without the need for trusted third parties or a central institution. Therefore, there is no need for a central entity to ensure the security, management, processing and storage of data. The decentralization of blockchain technology and the removal of a trusted third party or central institution will reduce the cost of operation and increase the efficiency of the system (Zheng et al., 2018).

The task of validating and confirming each new block is the responsibility of network users, and each block must be validated and confirmed by the majority of users of this system, which is called consensus. Therefore, it is not possible to change the data or delete it, unless the majority of nodes agree with it (Vranken, 2017). This will cause transparency in the operation of the system, because each member of the network participates in the validation of data and blocks by comparing their version of the ledger with others.

Each block is distinguished from other blocks by encryption. Encryption prevents fraud and changes to data stored in blockchain technology. The main application of encryption is to protect privacy and increase security and reliability in the blockchain system. In other words, the security and accuracy of the data stored in this database is maintained by means of digital signatures and encryption (Khorasany et al., 2019). In fact, blockchain technology is a chain of different blocks that are connected to each other at regular intervals. The data in the blocks depends on the blockchain network, but the time stamp and the password key are present in all types of blockchain. Each block is encrypted and each block refers to the block before it. Therefore, blockchain includes the history of all blocks. As a result, it is possible to trace all the previous blocks based on the last stored block. Therefore, the possibility of hacking, changing and manipulating or deleting the data stored in this system is almost impossible (Vranken, 2017), and this will increase the security and reliability of this system. Increase the number of blocks also improves the security and reliability of the system (Fernández-Caramés & Fraga-Lamas, 2018).

One of the main features of blockchain technology is the recording of all changes in blocks, so that no block can be deleted or modified. It is almost impossible to change the data in the blockchain, unless all the blocks are rewritten. As a result, this technology has high security. Decentralization, transparency, high security, anonymity of users' identities and automatic execution of contracts are among the advantages of this technology. In fact, blockchain technology acts as a bridge to pass from a centralized system to a decentralized system. Blockchain technology was firstly used for financial system of an encrypted currency called Bitcoin (Patil & Sangeetha, 2024), where users could trade with each other without any intermediary. In fact, decentralized digital currencies are the first, most mature and most important and successful application of blockchain technology. But the scope of application of this technology is very wide. It can be used in many different fields such as the development of the Internet of Things (IoT), to process intelligence, and for financial services, insurance, supply chain and many other cases (Atlam et al., 2020).

In recent years, new applications called smart contracts have been added to the main function of blockchain technology. These programs enable the decentralized storage of data by using mechanisms that provide the possibility of performing real operations in a decentralized manner. Also, many programs can be run automatically through smart contracts. Blockchain has attracted a lot of attention due to its prominent advantages such as distributed structure, cost reduction, immutable nature and security benefits and privacy protection. Blockchain technology has not yet been widely used in the real world. But the provision of the necessary infrastructure, such as the development of artificial intelligence or the IofT, has created hope for increasing the applications of this technology. In fact, blockchain technology is a general tool for creating P2P, decentralized and secure applications. It is also possible to benefit from this technology in systems that need space to store data and eliminate lack of trust. For example, manufacturing companies can record their product information on the database from the beginning of production to the time of delivery to the customer. This guarantees the authenticity of the goods and the quality of raw materials for the customer (Atlam et al., 2020).

1.1.1.1 Contents of a Block

Each block can be considered as a page of a ledger. Blocks consist of several components. Blocks are distinguished from each other by header block and block body. In general, each block consists of two parts, header block and block body (Zheng et al., 2018). The data of the header block section can be different based on their application. But in general, the block header section includes data such as block number, time stamp, protocol type, previous block hash, difficulty, block hash value and Merkel root value. Also, the most important data of the block body section includes the list of completed transactions and data information. In the block chain, each block (except the first block) in addition to referring to the previous block or its parent, also stores the hash value of the previous block (Merkel root value of the previous block).

The first block in the blockchain is defined as the Genesis block. This block has no parent and subsequent blocks will be connected to the Genesis block. Therefore, in order to change a block, all the previous blocks must also be changed, and this makes it very difficult or impossible to hack the system. The hash value of the previous block is obtained using the hash encryption function, that is, the hash algorithm of the previous block. The block creation time is indicated by the time stamp. In blockchain technology, difficulty is defined in order to limit the time spent to create a new block. Difficulty is a measure of how difficult it is to create a block in the blockchain. The task of the Merkel tree root is to analyze and summarize the data stored in the block body through the hash function. This process causes data integration in blockchain technology. Merkel tree ensures the integrity and validity of the data in a block. Merkel tree summarizes all the data of a block in a hashed value. It also makes it possible to check the presence or absence of a specific data in the block. It can be said that Merkel tree is a binary tree. In each block, each leaf of the Merkel tree is equal to the hash value of a specific data, and each non-leaf node is equal to the sum of the hash value of its two children.

The root node of the Merkle tree is defined as the Merkle Root or Root Hash. The Merkel tree construction process for a block with four transactions is shown in Figure 1.1. At first, after calculating the hash value of each of the transactions A, B, C and D, its value is placed in the leaves of the tree. Then, after summing the hash value of both leaves together, the resulting hash value is placed in the parent of those two leaves. This process continues until the root of the Merkel tree is built. Creating a Merkel tree has three general advantages:

- It does not require a lot of memory to store data.
- The hash value of a node is small, which reduces network traffic.
- It guarantees the integrity and validity of the data in a block.

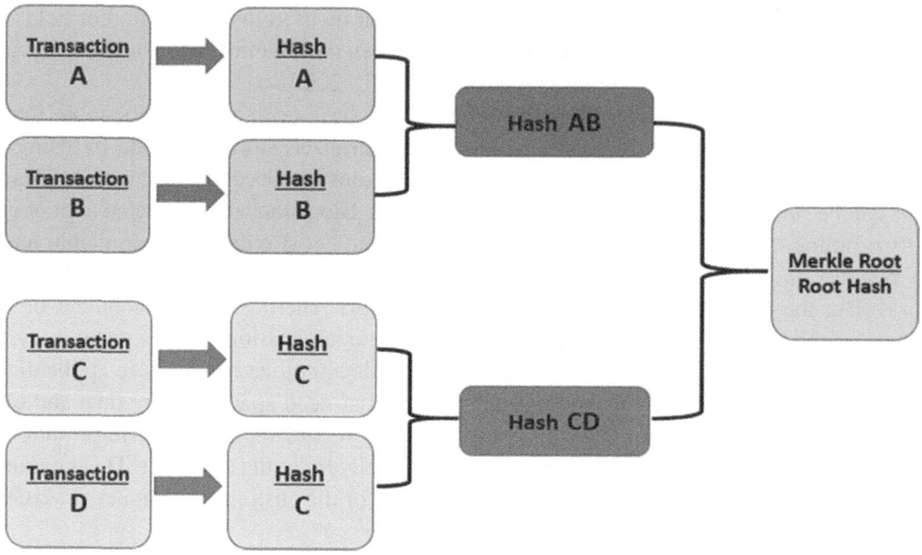

FIGURE 1.1 An example of building a Merkel tree for a block with four transactions.

1.1.1.2 Blockchain Performance

Blockchain technology is based on three technologies: encrypted keys, distributed network with a common ledger and service protocols. Blockchain needs the participation of users to survive and maintain its security. Therefore, there must be an incentive to attract users' participation. Service protocols will motivate users to participate in confirming blocks, saving records and maintaining network security. For example, in Bitcoin, nodes compete to obtain the answer to a mathematical problem, and the winning node is awarded a reward. These three technologies are not new, but their coordination and cooperation have created a new technology. Because blockchain is a distributed network, data is distributed to all nodes of the network. So all nodes have a single copy of the data. Each node has a private key and a public key. By combining private and public encryption keys, a digital identity is created. The main purpose of this combination in blockchain technology is to create a secure digital identity.

Digital signature enables ownership control. But to establish security in digital communication, only control over ownership is not enough. In addition to security, activity in the blockchain must be free and every user can create and approve blocks. Every user is allowed to use encryption keys and can join the network. This issue is realized by a distributed network. The decentralized performance of blockchain technology is achieved due to the benefit of other technologies such as hashing, digital signature and consensus mechanisms. By combining cryptography and distributed network, digital interactions become possible.

Another technology used in blockchain is consensus protocols. As mentioned, in blockchain technology, data will be added to the chain in the form of a block by some network nodes. The block acceptance process by network users is based on a process called consensus. After the block is approved by network users, the approved block will become part of the blockchain and will be added to the chain of blocks. The consensus process can be different depending on the type, application, purpose, protocol and rules of a blockchain. All users will decide on the new block with their participation, and the block approved by the majority of participants will be added to the chain (Soliman & Leon-Garcia, 2014).

1.1.2 Blockchain Technology Principles (Component Technologies)

Blockchain is not a fundamental and innovative technology. Rather, it is a set of processes and technologies such as encryption and hashing, P2P distributed network and consensus protocols. A description of each of the aforementioned technologies is provided as follows.

1.1.2.1 Distributed Ledger

A distributed ledger is actually a distributed and immutable database of records. In this database, the recorded data is updated and maintained by each node. For this reason, it is almost impossible to change or delete data by a person or a central institution. Therefore, the distributed database has high security and reliability. In a distributed database, data is stored independently by each node and is not registered and updated by a central organization. Each node in the network processes each data and then votes on whether the data is registered or not. If the majority approves, that data is stored in the database. Due to the consensus, the distributed ledger is always up to date and all nodes have a complete copy of the database. Therefore, the distributed ledger has many advantages over a simple database. In this technology, additional costs and intermediaries have been avoided in order to provide security and communication.

1.1.2.2 Consensus Protocols

Blockchains are distributed databases that allow those who do not trust each other to exchange with each other. In this system, there is no need to trust a person, third parties, or a central institution such as a bank. Blockchain nodes reach agreement and consensus about rules, status and value (Dinh et al., 2018). As mentioned, the blockchain is a decentralized system and every node is able to add blocks. Therefore, in order to ensure reliability, it is very important to confirm the data that is going to be added to the blockchain. This data must be confirmed by all blockchain nodes, which is called general agreement or consensus. Consensus means creating an agreement between the nodes of a blockchain system. Consensus in blockchain is established around three key aspects: network rules, network status (ensuring that fraud is not possible) and the value of the thing exchanged in the network. Blockchains use different algorithms to reach consensus. Different algorithms can play an effective role in blockchain performance, including network scalability, speed of reaching consensus, energy consumption and security. Therefore, the most important and widely used consensus protocols, including proof of work (PoW), proof of stake (PoS), practical byzantine fault tolerant (PBFT) and authority of proof (PoA), are described next (Sukhwani et al., 2017).

1.1.2.2.1 Proof of Work (PoW) Protocol

Proof of work is a protocol. This protocol rewards users who participate in validating and creating new blocks. The main purpose of this protocol is to prevent data manipulation and cyberattacks. Cynthia Dwork and Naor Moni proposed the main idea of the proof-of-work protocol in 1993. The term PoW was also published by Markus Jakobsson and Juels Ari in a 1999 article. This idea was used in 2008 by the cryptocurrency Bitcoin (Gramoli, 2020). In this type of consensus, nodes compete with each other to add a new block. This competition is designed in such a way that it is not easy to win, but it is easy to prove to others. Also, all nodes in this algorithm have the same score. Therefore, in general, this algorithm has three characteristics: difficult solution, fairness and simple proof.

This consensus has a complex calculation process. In this consensus, each network node changes a part of the block header called nonce and calculates the hash value. The calculated hash value must be smaller than a number determined by the network called the difficulty level. All the nodes of the network continuously change the nonce value and calculate the hash value, and this process continues until the hash value is smaller than the difficulty level. When a node succeeds in calculating the hash of a block whose value is smaller than the difficulty level of the network, it informs other nodes. Other nodes must verify the validity of this claim. After confirmation, the data of the new block is checked for fraud and if confirmed, the new block will be added to the block chain. Nodes that compete with each other to calculate the hash value in order to receive rewards are called miners. The PoW process is also called mining. This consensus requires a complex calculation process and a lot of time. Also, in the PoW process, users with higher processing capabilities can be more successful. PoW consensus is designed in such a way that with the passage of time, the difficulty level of its calculations increases automatically.

1.1.2.2.2 Proof of Stake (PoS) Protocol

As previously stated, PoW consensus has a complex calculation process and therefore requires a lot of electrical energy. So the PoW consensus is very inefficient in terms of energy consumption. Therefore, in order to optimize this process, another consensus called proof of stake was introduced, which consumes much less electricity. In the proof-of-stake protocol, the reward amount of the nodes is determined based on the amount of their contribution in the validation process. A proof-of-stake protocol is a method of validating data based on collective agreement. Although the process of the proof-of-stake protocol is completely different from the proof-of-work protocol, the goals of both protocols are similar. Unlike the proof-of-work algorithm, in which nodes participate in the process of confirming data and creating a new block by solving mathematical problems in order to receive rewards, in the proof-of-stake algorithm, the amount of participation of each node in this process is proportional to the amount of its shares or capital. In the PoS algorithm, the selected node to create the next block is selected through a pseudo-random process. This choice depends on the amount of shares or capital of that node. In this case, it is not possible to predict which node will be selected. The main idea of PoS was proposed in 2011. Blackcoin Nxt, Peercoin, NuBit Qora and Nav coin were the first digital currencies that used this algorithm.

In the proof-of-stake protocol, instead of searching for an amount of nonce in an unlimited space, it asks users to prove the amount of their share, and a node that has more assets and shares is more successful. Those with higher processing ability have more assets. In this protocol, it is suggested that each node, instead of spending money to increase the processing power, spend the same money to buy shares and increase its success by the same amount (Zheng et al., 2018). In PoS consensus, the winner is selected based on the amount of capital and shares, and the nodes with more shares will dominate the network. Therefore, there will be no justice in this consensus. Proof-of-stake protocol has advantages over proof-of-work protocol. This protocol consumes less energy compared to the working protocol. Therefore, in terms of energy consumption, this method is faster and more optimal than the PoW protocol and has good security. One of the weaknesses of PoS is that a node can accompany both branches when creating a branch. This can prevent the formation of consensus among nodes (Sriyono, 2020).

1.1.2.2.3 Authority of Proof (PoA)

In order to enjoy the advantages of PoS and PoW consensus, another consensus protocol called proof of validity was proposed. In this algorithm, reliable people have the ability to confirm the data and create a new block. This consensus has a higher processing speed. In PoA, a block will be added to the blockchain when it is confirmed by N nodes.

1.1.2.2.4 Proof of Byzantine Fault Tolerance (PBFT)

This algorithm refers to the consensus between two nodes in a distributed network with the presence of malicious nodes. In fact, PBFT is able to tolerate Byzantine faults. This algorithm performs the consensus process in the best way by assuming that a number of network nodes are defective and relying on a set of trusted nodes. In the PBFT protocol, one of the trusted nodes is selected as the leader and other nodes are considered as substitutes (Cachin, 2016). When the leader node receives a request, it sends that request to the alternate nodes and the consensus process will begin. After the consensus process, the leader node and alternate nodes will announce the result of the process to the requester. In order to carry out the consensus process in PBFT, a vote will be done by all the trusted nodes. The result is acceptable when the majority of nodes vote positively. Each time the PBFT consensus process is performed, a new block will be added to the blockchain. There is no competition in the PBFT consensus protocol. This protocol is suitable for a block chain whose users trust each other to some extent.

1.1.2.2.5 Failure of Consensus Protocol

When the nodes cannot agree with each other, a phenomenon called consensus protocol failure will occur. The consensus algorithm should be designed in such a way that system users can reach an agreement in any situation. Even if some nodes become defective or malfunction, this agreement must be reached. There are two types of consensus protocol failures:

- **Sudden failure**: When the consensus process suddenly stops, that process has suffered a sudden failure.
- **Byzantium failure**: Any threat that is not in the sudden failure group is defined as Byzantine failure. This failure is often caused by malicious actions of a hacker. In order to prevent Byzantine failure, the algorithm should be designed in such a way that the selection of nodes to create a block in the network is random.

1.1.2.2.6 Some characteristics of consensus algorithms

Some characteristics of consensus algorithms are described below.

- **Complex calculation**: Many consensus algorithms used in blockchain technology such as PoW or PoS have a complex calculation process and therefore require a lot of energy (electricity).
- **Scalability**: In blockchain technology, the process of consensus and confirmation of blocks is done by network users. This causes the scalability feature. Therefore, the time and energy required for data processing increases exponentially with the increase in the number of network users.
- **Operational power**: In blockchain technology, the number of data that can be stored is defined as operational power (efficiency).
- **Delay**: Ensuring that a node is confirmed by participating nodes in the blockchain requires an unconditional delay.

1.1.2.3 Data Encryption

In order to establish digital security, data encryption is considered an essential technology. Data encryption involves hiding the original data through a mathematical algorithm, so that the data can only be recognized by the receivers. There are two types of data encryption processes in blockchain technology. The first process is defined as encryption. This process involves one-to-one translation of data from one data set to another. Due to the fact that in this process the data is encoded with a mathematical formula, by knowing the desired formula, that data can be decoded. The second encryption process in the blockchain system is defined as hashing.

1.1.2.3.1 Hash Function in Blockchain

In some situations, data confidentiality is unimportant. But it is very important to ensure the correctness of the data and ensure that the data is not changed or manipulated. In order to ensure the integrity of the data and prevent fraud and change of the data recorded on the blockchain, cryptographic hash functions are used. In fact, the hash determines the identity of the block and its contents. In general, hash is a practical tool to detect changes in the block and it is always unique like a fingerprint. The encryption hash function receives bit strings of different length as input and produces a bit string of fixed length as output. The output of cryptographic hash function is called code or hash value. It is a text string hash and a mathematical function is used to create it; its most important use is to prevent fraud in the system. The hash of a data is always fixed and unique and the hash value of different data is not the same. Hash functions are widely used in computer science. The fundamental difference between a cryptographic hash function and a standard hash function is that the cryptographic hash function must at least have a one-way feature. As mentioned, the first block of the blockchain is called the Genesis block or the initial block. In the block chain, the hash of the next blocks, in addition to the address of the previous block, also stores the hash data of the previous block. Due to the existence of the previous hash block in the new hash block, the blockchain blocks are all connected to each other. Therefore, any change in the data of a block will change its hash. As a result, the next blocks will be invalid. This process increases the security and reliability of blockchains. With the increase of blocks, the security and reliability in the block chain also increases (Joshi et al., 2023).

1.1.2.4 Smart Contracts

Smart contracts are computer programs that include all provisions related to the terms of the contract and can automatically implement the terms of the contract and all targeted actions. Also, this contract will not be implemented until the conditions mentioned in the smart contract are met. In fact, a smart contract is a protocol for participating, facilitating, validating, or enforcing the terms of a contract. The smart contract carries out transactions and rules and processes related to an agreement in a fully guaranteed, automatic, independent manner and without the need for the intervention of third parties or a central institution. In 1994, Nick Szabo introduced the distributed application program (smart contracts). In fact, the phenomenon of the distributed application program was a great idea and a new tool for the automatic implementation of contracts between the contract parties. This idea was not considered until the creation of programmable payments and cryptocurrencies. After the introduction of crypto currencies, the blockchain program and the smart contract were able to make payments with preplanned conditions. Also, the use of smart contracts on the Ethereum platform increased the use of this technology in various fields. Smart contracts are stored in the blockchain distributed database and the correct execution of these contracts is done by consensus algorithms. Since the operation of smart contracts is based on blockchain technology, they have many advantages, including

- **High security**: Due to the registration of smart contracts in distributed databases, the activities and registrations of smart contracts are traceable and irreversible. Therefore, it is not possible to delete, change, or manipulate these contracts.
- **Distributed**: Due to the decentralized nature of the database in blockchain technology, it is possible to view the stages or results of the smart contract for all network nodes.
- **Affordable**: Due to the execution of these contracts on the blockchain platform, it is possible to perform them automatically. Therefore, there is no need for intermediaries and third parties or central institutions. As a result, the cost of its implementation is reduced and the speed of the processes is also increased.

Smart contracts, despite many advantages, like any other technology have disadvantages, such as:

- **The human factor**: Programs related to a contract are written by humans. So there is a possibility of error.
- **Uncertain legal status**: Standard and international rules for smart contracts have not yet been established. Therefore, the legal status of these types of contracts may change with the preparation of a series of regulations by some countries.
- **Writing fee**: Programs related to smart contracts require programming. Therefore, an expert programmer should be used to write these types of contracts.

1.2 BLOCKCHAIN HISTORY

The history of blockchain can be divided into three different generations.

1.2.1 First Generation Blockchain

The main idea of blockchain goes back to an article from 1976. This article has proposed the decentralization of data storage. In 1991, Stuart Haber and Scott Stornetta described a secure cryptographic blockchain. In 1992, Bayer, Haber and Stornetta added the hash tree to the scheme. In addition to improving the efficiency of the blockchain, this idea made it possible to collect several documents in one block (Narayanan et al., 2018). A blockchain database can be managed autonomously. This system benefits from a P2P network and a distributed scheduling server. The first blockchain technology was introduced to the world through an article by Satoshi Nakamoto (pseudonym of a person or persons whose identity remains unknown) in 2008 under the title "Bitcoin: A P2P Electronic Cash System" (Altaf et al., 2023). This article introduces electronic cash that can be transferred directly and online from one person to another without the need of a third party or central institution.

1.2.2 Second Generation of Blockchain

The Ethereum platform was unveiled as the second generation of blockchain in late 2013, with the aim of creating and implementing distributed programs and smart contracts. The founder of the second generation blockchain is Vitalik Buterin. He solves many of the limitations of the blockchain in the Ethereum platform. Ethereum is also based on the public blockchain, which has the ability to implement and execute smart contracts. The Ethereum platform is designed as the second generation of blockchain to run the programming code of decentralized applications. It allows programmers to run their programs on this platform. Ethereum is an accessible software based on blockchain technology. This platform allows users to build and develop decentralized applications. The probability of failure, fraud, or data modification, censorship and third-party interference in the programs that run on the Ethereum platform is very low. It has very good flexibility, security, trust and speed. The second generation of blockchain technology has been very effective in evolving this technology and transforming it from a support database to a public platform for running decentralized applications and smart contracts. The second generation of blockchain uses the proof-of-stake protocol instead of the proof-of-work protocol. Use of PoS reduces energy consumption.

1.2.3 Third Generation of Blockchain

The first generation of blockchain technology gained attention with the introduction of Bitcoin and cryptocurrencies. The second generation of blockchain technology was completed by developing the capabilities of this technology and implementing decentralized programs and smart contracts. But the third generation blockchain has taken into account goals such as high scalability, interoperability, stability, privacy and self-governance. Considering that it is not possible to combine these four features in the first and second generation of blockchain technology, the third generation is emerging to have and combine these features. In the third generation of blockchain technology, communication between different blockchains has been proposed in order to exchange data flow. Also, different layers have been defined for managing agreements and contracts, conducting transactions and communicating with other blockchains. Cardano, ARK, Lisk, Bitshares and Steem can be mentioned from the third-generation blockchain technology, which, while having the ability to collaborate and scale, also increases the ability of self-governance. Cardano benefits from the Ouroboros PoS consensus process. Ouroboros is a proof-of-stake protocol with mathematically proven security. Cardano has a calculation layer and a consensus layer. The consensus layer acts as the balance of the ledger and aims to improve scalability. This layer uses a proof-of-stake algorithm to create new blocks and verify data. The calculation layer contains data about the calculations and how to perform the transaction. This layer is responsible for smart contracts and legal control. The separation of two layers of consensus and calculation increases privacy, allows greater flexibility of smart contracts and enables customization of contracts.

1.3 CLASSIFICATION OF BLOCKCHAIN SYSTEMS

Blockchain systems can be classified into three categories based on the method of acceptance and consensus: public blockchain, private blockchain and consortium blockchain (Vranken, 2017):

1.3.1 Public Blockchain

In public blockchain, anyone can join and all users can also participate in the validation process (consensus process). This type of blockchain is completely decentralized. Also, all the records are exposed to the public and it is almost impossible to manipulate and distort these records. This type of blockchain is often used when the opinion of all users and participants in that system is important. Due to public participation in this type of blockchain, a lot of time will be spent to distribute data and blocks. Therefore, this type of blockchain has less speed, efficiency and flexibility compared to other types. It is also very difficult to provide security in the public blockchain, where important data can be recorded and stored encrypted. Bitcoin is an example of public blockchain (Vranken, 2017).

1.3.2 Private Blockchain

This type of blockchain is accessible only to selected and approved individuals or nodes. Access to this type of blockchain is limited and the data and records stored are visible only to selected individuals or nodes (Vranken, 2017). In addition, private blockchain may have different access levels, where nodes have varying degrees of authority to access and view data. Therefore, anonymous users in private blockchain cannot write or view data. Also, the selected nodes are only from a specific organization and the consensus process will be controlled by them. So a private blockchain is a centralized network. It is also possible to manipulate or change data in the private blockchain if the majority of users agree with it. Therefore, this type of blockchain is less secure compared to the public blockchain (Zheng et al., 2018).

1.3.3 Consortium Blockchain

Consortium blockchain is a combination of private blockchain and public blockchain and is also known as hybrid blockchain. This type of blockchain can be created by companies, organizations, or a group of people who cooperate with each other. But only a group of nodes are selected to confirm the activities. Therefore, carrying out the consensus process is the responsibility of this preselected group and it should be done with their participation (Vranken, 2017).

Consortium blockchain is similar to private blockchain. But in this type of blockchain, instead of a specific organization, a consortium will monitor the system. Also, the process of consensus and production of blocks will be done by a selected group with their participation. Other nodes can exchange with each other, but they do not have the right to participate in the consensus process or create new blocks (Vranken, 2017). This type of blockchain is made by several organizations and it is semi-public. It can also be considered decentralized to some extent (Zheng et al., 2018) because only a small number of nodes are selected to participate in the consensus process.

In a consortium blockchain, all data and records can be visible to the public (selected and unselected nodes), but some specific data can be protected.

1.3.4 Comparison of Effective Factors in Different Types of Blockchain

The effective factors in different types of blockchain have been examined and compared as follows:

- **Consensus process**: In the public blockchain, all nodes and users can participate in the consensus process. In the consortium blockchain, the process of consensus and production of blocks will be done only by a selected group and with their participation. In the case of private blockchain, the responsibility of controlling the consensus process is the responsibility of only one organization. In a public blockchain, all users can join the consensus process. But in private and consortium blockchains, nodes need permission to join the consensus process.
- **Distributed network**: The public blockchain is decentralized and consortium blockchain is relatively decentralized. But the private blockchain is completely centralized, because it is controlled by selected nodes only from a specific organization.
- **Read access**: All nodes in the public blockchain are allowed to read data and records. While in private blockchain, data is visible only to selected individuals or nodes, and in consortium blockchain, access to data and records depends on certain conditions.
- **Immutability**: In public blockchain, data and records are stored in all nodes. Therefore, it is almost impossible to manipulate or change data in this type of blockchain. Data in a private or consortium blockchain is stored only in a limited number of selected nodes. Therefore, it is less secure and the data can be easily changed or manipulated.
- **Efficiency**: Considering that in the public blockchain, all users can participate in the consensus process, it will take a lot of time to distribute data and produce blocks. Therefore, the throughput of transactions is limited and the delay time is high. But due to the existence of fewer nodes, private and consortium blockchains have more optimal efficiency than public blockchains.

The characteristics of blockchain types are presented in Table 1.1. Blockchain technology can also be classified into two categories, permissioned (private) and permissionless (public), based on the level of access of individuals or nodes to stored data and records. In the permissionless (public) blockchain, the validation process is not limited. Also, any person or group has access to the blockchain and is able to add blocks to it. In the permissioned blockchain, which includes private blockchain and consortium blockchain, only limited and approved people or selected nodes will be able to add new blocks. A copy of all data can be available to a limited number of users.

TABLE 1.1
Features of Blockchain Types

		Types of Blockchain		
	Property	**Public Blockchain**	**Private Blockchain**	**Consortium Blockchain**
Attributes	Determination of Consensus	All nodes	A specific organization	Selected nodes
	Consensus process	Anyone can join the consensus process.	Requires a permission	Requires a permission
	Distributed network	Decentralized	Centralized	Somewhat decentralized
	Read access	Public	It can be public or limited	It can be public or limited
	Immutability	Manipulation is almost impossible.	Manipulation is possible.	Manipulation is possible.
	Efficiency	Low	High	High

1.4 ANALYSIS OF STRENGTHS, WEAKNESSES, OPPORTUNITIES AND THREATS (SWOT)

The strengths, weaknesses, opportunities and threats of blockchain technology will be examined next.

1.4.1 Features and Strengths of Blockchain Technology

Blockchain technology has attracted a lot of attention due to its prominent features such as distributed structure, immutable nature and privacy. Some of the features of blockchain technology and its advantages over traditional databases are discussed in the following (Tian, 2016).

1.4.1.1 Distributed (Decentralization)

Centralized systems depend on a central institution or third parties. In this system, every transaction is controlled and validated by the central institution or third parties. The dependence of this system on a central institution increases the cost. It is also possible to make changes in the data or delete it in this system. But blockchain is a distributed and decentralized network. This technology enables communication between users without the need for third parties or a central institution. This system is formed by connecting the nodes to each other, which will also control the consensus process. Therefore, the development and operation costs of this system are significantly reduced (Zheng et al., 2018). Another advantage of this system is that it is difficult to manipulate data or hack the system. Hackers have to hack at least 51% of the nodes to break into this system, which is almost impossible.

1.4.1.2 Stability

In most centralized systems, data records are stored for a limited period of time. Also, if a person or people access the data records or if there is a problem with the server, the data records may be deleted. But in blockchain technology, all data and transactions in the network are stored in blocks and the blocks are distributed in the network. Therefore, it is impossible or very difficult to manipulate data in such a system. Also, considering that every block in the network is validated by other users, any change or manipulation of data can be easily recognized (Zheng et al., 2018). Blockchain technology is able to store data in an unlimited and immutable way.

1.4.1.3 Anonymity

The existence of a central entity or a third party causes the lack of privacy and anonymity of users in centralized systems. In decentralized blockchain technology, each user can interact with others using a generated address (public key). It is also able to add a new block to the chain or call a block from the chain using the permissions it has. In this technology, the identity of the user who registered the report is not clear (Misra et al., 2023).

In addition, each user can generate several addresses in order to prevent their real identity from being revealed. In blockchain technology, there are no more third parties or central entities that want to store users' private information. This also preserves privacy in blockchain technology (Zheng et al., 2018). Confidentiality of transactions and reports using cryptographic keys is the most important feature of this technology after the issue of data protection and security.

1.4.1.4 Transparency and Traceability

Each block in blockchain decentralized technology is first validated and then stored with a time stamp (Mattila et al., 2022). The time stamp shows the creation time of a block in the blockchain. Also, each block is cryptographically linked to two adjacent blocks. Therefore, all users can easily validate and track all the records related to a block, including its verification, storage and creation time through any accessible node in this decentralized system. This helps the traceability as well as the transparency of the data stored in the block chain (Zheng et al., 2018).

1.4.1.5 Security and Non-falsification of Data

The distributed database of blockchain technology creates a completely different digital support by storing data throughout its network (Mattila et al., 2022). This system is able to remove the vulnerability points that reduce network security in a centralized database. This technology provides an innovation in data storage and distribution, and this is another feature of blockchain technology.

The data stored in the distributed database can be edited, copied and transferred from all over the world. This feature can reduce network and data security. In order to deal with this problem, a copy of this data is distributed in the network and there is no centralized copy. Therefore, this process increases the security and reliability of blockchains.

1.4.1.6 Elimination of Third Parties and Cost Reduction

The performance of blockchain technology leads to fundamental changes such as decentralization and elimination of third parties or central institutions (Vranken, 2017). This technology allows users and participants to communicate directly with each other without the need for third parties or central institutions. Therefore, the costs of this system are reduced due to the absence of third parties or central institutions.

1.4.1.7 Programmable

Users can ensure the automatic execution of their programs by creating smart contracts.

1.4.2 Risks, Weaknesses and Threats of Blockchain

Blockchain technology has attracted attention due to its outstanding advantages such as increasing accuracy in performance by eliminating human labor in the verification process, reducing costs by eliminating intermediaries or third parties and decentralization which increases data security. As a new technology, this has many challenges. In order to benefit more from this technology and ease its implementation, it is very valuable to identify and solve these challenges. Therefore, in this section, the challenges of blockchain technology have been studied from technical, individual-organizational, legal, financial and environmental aspects.

1.4.2.1 Technical Challenges

1. **Programming and web development skills**
   Blockchain technology developers should have enough skills for web development and programming. They should also be able to analyze and compare blockchain systems, their algorithms and protocols. They should have sufficient knowledge about how data structures work and the role of encryption, graphs and linked lists (Skiba, 2020).
2. **Quantum computers**
   Due to the appearance of quantum computers and the increase in computing power in the coming years, it is predicted that the failure of the basic security of the blockchain (public key encryption, private key and digital signature) will be possible for hackers reasonably soon. The design of the keys can be made more powerful so that they cannot be easily broken. In the future, nodes should be equipped with artificial intelligence and machine learning capabilities (Salah et al., 2019).
3. **Security restrictions**
   Trust is one of the most important principles for communication between two users. This principle is possible in computer networks by means of authentication and permission

checking. Sharing valuable data between network users and maintaining its security is one of the goals of blockchain technology. Security restrictions consist of two parts: cyber-attacks and the availability of sensitive data to everyone.

a. Types of cyber-attacks include hacking, double spending, Parity wallet attacks, network-level attacks, flexibility attacks, true DoS attacks against Ethereum, scale attacks, SYBIL attacks, DDoS attacks and flexibility attacks and other threats to data integrity (Jimoh et al., 2019).
b. Availability of sensitive data for everyone: In blockchain technology, all data and records are publicly available to all users. If decentralized blockchain is used in sensitive environments such as patients' medical records, these data can be misused.

4. **Save data**
   In the decentralized blockchain system, there is no need for a central server to store data and its records. In this technology, data is stored on the user's computer. As time passes and data increases, a computer with a larger memory is needed. Therefore, the storage process can be considered a technical challenge for some users with low storage capacity.
5. **The immaturity of blockchain technology**
   The emerging blockchain technology is still developing and the main focus of this technology is on improving issues such as security, privacy and scalability. Therefore, this technology is very vulnerable to other problems such as capacity, system failure, and other issues.
6. **Integration problems with other new technologies**
   Integrating blockchain with other new technologies such as the IoT and AI can be challenging. IoT devices have little computing resources or storage capacity. Databases in the IoT usually operate centrally, while in blockchain technology, each node stores records independently. Therefore, integrating the IoT with blockchain can be somewhat challenging. If these problems are not solved, we will face a big challenge in using blockchain technology.
7. **Size, operational capacity and bandwidth**
   If the number of nodes in a system increases continuously, the blockchain will expand to such an extent that it will not be able to store a copy of all the ledgers related to the data. This issue is of particular importance, as such limitation can severely affect system operational power.
8. **Constraints of big data processing**
   Proof-of-work protocol requires complex mathematical calculations to confirm transactions. Also, it is challenging to manage, investigate and optimally analyze a large amount of data.
9. **Shortage or lack of proper infrastructure**
   There are many differences between developed and developing countries regarding access to new technologies and the Internet. For example, studies show that more than half of Nigerians do not have access to the Internet.
10. **Absence of international standard**
    Due to the lack of standards in blockchain technology architectures, it is difficult to establish communication between organizations and institutions, and systems are often created based on different parameters, including privacy, consensus protocols and programming languages. If there is no international standard and if this problem is not solved, the use of blockchain technology will face a big challenge.
11. **Time and speed**
    Although blockchain is very efficient in transactions and log storage systems, its speed is lower than centralized databases. In blockchain technology, all nodes must participate in validating and confirming a block. As a result, with the increase in the number of users, the review and validation time will be longer. Due to the fact that the number of blockchain users can continuously increase, the duration of implementation of blockchain-based programs can be challenging.

1.4.2.2 Environmental Challenges

a. **Energy loss**
In the PoW protocol, only the first node that succeeds in solving the problem is able to add a new block. Therefore, considerable energy will be spent in order to find a solution to the problem. The time and energy of other nodes that did not find the solution have been effectively wasted.
b. **Compliance with environmental laws**
The use of fossil fuels in energy production can have harmful effects on the environment. Therefore, in order to prevent the irreparable damage of fossil fuels to nature, attention should be given to replacing these sources of energy production with clean and renewable energy sources. The high energy consumption of this technology can cause an imbalance in energy consumption. Therefore, actual energy usage will be different from expected levels, which can cause challenges.

1.4.2.3 Financial Challenges

Currently, in blockchain systems, data validation and verification are done by software and with high energy consumption. The significant increase in energy prices in recent years has caused an increase in the operating costs of this system. Therefore, this issue should be taken into consideration. Other costs may include the expenses associated with creating a suitable infrastructure, providing hardware and software, ongoing maintenance and upgrades of the blockchain system and the cost of recovering the system in the event of cyberattacks.

1.4.2.4 Individual, Social and Organizational Challenges

Individual, social and organizational challenges in implementing blockchain technology are discussed as follows:

a. **Trust in new technologies**
There are always oppositions to changes. Many people trust less in the use and capabilities of new systems such as blockchain technology, and some organizations may avoid giving their information to this system due to lack of trust. There are complex concepts and processes in blockchain technology that can be difficult for ordinary people to understand.
b. **Organizational readiness**
Managers of organizations must accept that in the blockchain system there is no need for the presence of a third authority. It also has good security and safety. The level of preparation of organizations from managers to employees can play a very effective role in accepting this technology in the organizational framework. Some government organizations have concerns about the dangers of using these technologies. Additionally the implementation of a blockchain platform requires the cooperation of multiple institutions and stakeholders. Therefore, a new governance model is needed, and this means organizational transformation, which often leads to adverse reactions within the organization.
c. **Lack of skills and skilled labor**
Since blockchain technology is new, there are very few experts who have sufficient skills. Due to the limited number of experts in this technology, hiring the technical expert needed to manage and implement this system can be challenging.
d. **Opposition of the central institution or third parties**
Because the use of blockchain technology can reduce or eliminate the privileged and effective role of the central institution or trusted third parties, it is expected that most third parties or trusted institutions such as banks will oppose this technology, causing challenges in its implementation.
e. **Participation of all stakeholders**
The implementation of a system based on blockchain technology is successful when all stakeholders cooperate with each other. The success of the blockchain technology system depends on the participation of all stakeholders.

1.4.2.5 Legal Challenges

- **Legal and compliance issues**
  Some legal and compliance issues are:
  - **Verification of real identity**: In order to complete official documents, including real estate transfer systems, electronic elections, and so on, it is necessary to confirm the identity of people.
  - **Controlling the legality and effectiveness of the contract**: In blockchain technology, it is not possible to control the legality and effectiveness of the contract.
  - **Tax**: Uncertainty of financial transactions such as selling consumer goods and doing activities that can be done by blockchain technology, it is not possible to check and determine taxes.
  - **Impossibility of modifying the block chain**: Due to the impossibility of changing data in the blockchain system, it is not possible to modify documents. However, reversibility of transactions or change of property ownership can be legal.
- **Network Administration**
  Creating and managing the blockchain system with the participation of various organizations and stakeholders can be challenging. Determining factors such as the type of blockchain, the location of nodes, responsibility for writing smart contracts, dispute resolution mechanisms, applicable regulations and standards and other related considerations can cause challenges.
- **Lack of specific laws and courts to resolve disputes**
  To help define the rights and obligations of the parties, manage agreements and resolve disputes, the existence of laws and regulatory frameworks is necessary, which has not yet been achieved in blockchain technology.

1.5 APPLICATION OF BLOCKCHAIN

Blockchain technology can be used to provide various services due to features such as the elimination of intermediaries, independence from third parties or other entities, low operating costs, easy integration and the possibility of registering documents in a public and decentralized manner. This technology was originally created for the scheduling of digital documents and is a completely distributed and accessible information recording system. This technology is now used in many financial and non-financial fields and very diverse and wide areas such as the global economy, banking, financial markets, creating the infrastructure of basic crypto currencies, voting systems, medical services, intelligent transportation, government services, databases, digital identity creation, IoT development and many other applications. Also, this technology has the ability to use applications such as smart contracts and distributed applications. The use of blockchain has a great impact on data security. This technology can be used in management of a wide range of service and goods supply chains.

1.5.1 Banking

Financial institutions can increase the speed and reduce the time and cost of transactions with the help of blockchain technology. It also can be used in existing financial infrastructures such as stocks, bonds and many other fields. Using this technology, data related to inventory and assets can be encrypted and produced, managed and stored with high security.

1.5.2 Medical and Health Services

Information about patients is of great importance. Protection of these data is one of the priorities of every medical center. Using blockchain technology, patient records can be stored with high security. Also, if necessary, they can provide medical records to other hospitals and medical centers.

1.5.3 Government Services

Blockchain technology has attracted the attention of various countries and governments due to the digital storage of data and the provision of immutable history.

1.5.4 Copyright Law

One of the problems of software producers is noncompliance with copyright law, which blockchain technology could be used to resolve.

1.5.5 Electronic Elections

Electronic election systems can be subjected to cyberattacks or hacker penetration. Among other things, the US presidential electronic election system was attacked by hackers in 2016. Blockchain technology can be used to validate and prevent fraud or cyberattacks in electronic elections.

1.5.6 Digital Identity Recognition

Cryptographic keys grant ownership rights to users and enable the establishment of digital relationships. Blockchain technology can provide an opportunity to create a powerful digital identity identification system.

1.5.7 Use of Blockchain in Internet of Things

The IoT has new opportunities and challenges, and if this technology is used correctly and optimally, life in the future can undergo a huge transformation. One of the most important applications of blockchain is the integration of this technology with the IoT, which is often used in smart networks. The IoT is a network for managing certain electronic devices connected to the Internet. The IoT has many advantages and can simplify system monitoring and reduce system operating costs by combining software and sensors and creating an information exchange network. The integration of blockchain technology with the IoT increases the security and reliability of the network. The development of IoT devices requires both communication infrastructure and effective device management, which is also possible with the integration of blockchain technology.

1.5.8 Application of Blockchain in Artificial Intelligence

Artificial intelligence is one of the newest technologies that can be integrated with blockchain. By using blockchain technology, a huge amount of data can be stored in a completely stable manner, and with the help of artificial intelligence, it is possible to recognize or predict patterns in the created chain data.

1.5.9 Blockchain Applications in Energy Exchanges

Blockchain, which is developed on a P2P network, can facilitate decentralized energy management and save time, cost and complexity of energy exchange. In smart grids, unlike traditional markets, energy can be exchanged directly between consumers and producers, which can lead to complex financial transactions. Meanwhile, time sensitivity is a critical factor in both energy exchanges and financial transactions, the latter of which must be done online. In addition to facilitating these energy and financial exchanges, blockchain technology can also guarantee the security of transactions (Rahmadika et al., 2018). Also, energy exchange between smart grids can cause many challenges such as fraud and data manipulation, false data entry, high costs and unreliable transactions. Blockchain technology, without the presence of intermediaries such as a third party or a central institution, can create transparency and facilitate financial processes and data flow between producers and consumers. In energy markets, blocks can be used to store all data related to energy exchanges and all terminals can track and monitor exchanges (Khaqqi et al., 2018).

Blockchain technology is able to guarantee the security of transactions made in energy exchanges by using digital currencies and smart contracts to facilitate financial transactions in the energy market. Each block in the blockchain can store the information about multiple financial transactions encoded with hash functions. Therefore, all financial transactions can be followed and traced by the stakeholders of the market, and this increases the transparency of financial transactions and the energy market. The P2P structure of decentralized blockchain technology can improve the security of financial transactions compared to centralized transactions. Dynamic pricing in blockchain technology occurs in the structure of financial proposals, and the process of making transactions in the energy exchange market is done automatically by smart contracts (Shrestha & Nam, 2019).

1.5.10 Application of Blockchain in Cyber Security

Due to the increase of cyberattacks such as injection of false data and manipulation of data that can affect network performance, network security has become essential. In smart networks, many data such as the amount of energy requested by the consumer, the price offered by producers and the amount of energy produced by each producer are exchanged between producers and consumers. Therefore, data security is vital in smart grids. Due to the ability to track and store data in the decentralized blockchain database, this technology can be used as one of the solutions to increase security and prevent fraud, cyberattacks and incorrect data entry in smart networks (Kong et al., 2019).

1.5.11 Blockchain and Environment

Blockchain technology in smart grids affects the environment in multiple ways, such as carbon tax collection, recycling management, supply chain transparency, energy distribution and incentive mechanisms, which will be discussed in the following.

- **Electrical energy**: Blockchain technology can lead to the development and increase of renewable energy sources. Because the blockchain structure can facilitate the exchange of renewable energy resources with the network, the development of renewable energy sources is able to reduce the environmental problems caused by the use of fossil fuels. Therefore, the amount of environmental pollution caused by fossil fuels in power plants is reduced. The blockchain mining process causes an increase in the consumption of electrical energy. Due to the fact that many power plants use fossil fuels to generate electricity, they can increase the emission of environmental pollution. Therefore, it is very important to study the advantages and disadvantages of using blockchain in order to preserve the environment and prevent the emission of polluting gases.
- **Supply chain**: The structure of the blockchain is such that it can cause transparency in the processes of production, transportation and delivery.
- **Incentives**: Blockchain technology is able to encourage society to produce less greenhouse gases by creating incentives based on environmental conditions.
- **Environmental treaties**: The security of data exchanges in blockchain technology can contribute to the transparency of the work done in the field of environmental treaties. Therefore, security against fraud and data manipulation is reduced.
- **Carbon tax**: Due to the secure nature of data exchanges and digital financial transactions in blockchain technology, carbon taxes associated with digital transactions can be easily calculated and making cheating or manipulation virtually impossible.

1.6 APPLICATION OF BLOCKCHAIN IN WATER INDUSTRY AND ITS SUSTAINABLE DEVELOPMENT

Rapid population growth and urbanization and economic development have increased the global demand for water (Sundaresan et al., 2021). Also, climate change is another important factor that can cause increased water demand. The lack of reliable water resources (Parkinson & Mark, 2005), disproportionate policies in water management, poor control and monitoring of water quality, damage to the water infrastructure network, lack of proper education in the way of water consumption, excessive wastage of water and uncontrolled losses have left countless people exposed to water shortages (Lin et al., 2018). Benefiting from safe, clean and sustainable water is one of the fundamental rights of human societies. Therefore, water management is a major societal concern and one of the important priorities of SDG 6 in order to preserve water resources, improve operation of resources and ensure the safe quality of drinking water.

The third World Water Forum introduced fresh and clean water as a precious and limited resource for sustainable development, economic growth, social stability and poverty reduction. Therefore, the correct water management system is very important and human societies are currently forced to adopt sustainable technologies and prudent solutions, and to build stable infrastructure for smart water management and the safe, high-quality supply of water. Currently, most water systems are managed in a traditional and centralized way, which can cause problems such as increasing costs, reducing efficiency and resources, and data storage insecurity (Dogo et al., 2019).

In traditional and centralized management systems, data sharing is very challenging: the stored data lacks sufficient security and it can be deleted, manipulated or changed. In this system, a lot of time must be spent to access the data. Other challenges of the traditional management system include a difficult monitoring and control process, which can increase the costs of data verification.

In the management of water resources, many departments and organizations cooperate with each other. However, the lack of trust between them can create island databases, which is very effective for the coordination between these organizations and departments in order to manage the water system, although it can also reduce the efficiency of this process (Sundaresan et al., 2021).

Therefore, in order to manage water resources efficiently, the best and most appropriate measures should be taken. Today, the collection, storage, processing and analysis of water network data in order to control and monitor in real time, secure data management, optimize resources, improve decision-making and create smart water management with the help of new and growing technologies such as blockchain AI and IoT are possible. The ability to use smart management technologies holds significant potential for enhancing sustainable water management practices. Smart water management can include issues such as resource management and water supply, water distribution networks, water quality, optimal water use, monitoring of the security and infrastructure of the water network, wastewater management and so on (Dogo et al., 2019).

Smart water management can control and monitor the water network continuously and remotely and fix it if there is a problem with or damage to the network. Also, smart water management is able to preventatively manage network maintenance. Therefore, studying the role and impact of new technologies for water management is essential, such as blockchain technology, which in recent years has been widely discussed in academic literature and research papers on the effective role of this technology in water management systems. The features and characteristics of blockchain technology, such as immutability, auditability, transparency, traceability of stored transactions and so on, bring many benefits to the integration of blockchain technology with water management systems. This also makes the water management system smarter and more useful than the traditional system. Studies also show that blockchain-based decentralized water management system can be useful for the entire system, including the processes of water supply, distribution, consumption and wastewater management (Dogo et al., 2019).

Blockchain technology can play an effective role in improving water management systems. In a water management system, the use of blockchain technology can help establish trust and reliability among stakeholders. The encryption process in this technology improves data security and makes it almost impossible to manipulate, change or delete data and information. Blockchain technology can reduce corruption in water management systems due to its ability to record and store records of each transaction without the possibility of changing, deleting or manipulating the recorded data. Also, by creating transparency in the records of each transaction among the stakeholders, it is able to improve the way of managing and trading water resources (Zhang et al., 2020).

Also, this technology creates a good perspective for stronger water market mechanisms and develops a P2P water trading market using smart contracts. Blockchain technology is able to monitor the process of market regulation and clearing more effectively (Poberezhna, 2018). The blockchain-based water management system can be used to manage the water supply chain and monitor water quality. Implementation of blockchain technology in water supply management systems can improve water supply. It also helps to track water quality and increase public awareness about water quality (Pérez Ortiz, 2018). In the blockchain-based water management system, all stakeholders have access to the same data about water quality or quantity in order to make more appropriate decisions. A number of researchers have recommended implementing wastewater management and recycling systems based on blockchain technology, using wastewater reuse rates and quality indicators as key parameters (Hakak et al., 2020).

In the blockchain-based water management system, all monitoring data is encrypted using blockchain, then distributed among stakeholders through a P2P network. In addition, the blockchain-based water management system changes from a centralized state to a decentralized state, and a part of the management work is assigned to each node. Also, blockchain-based water management improves the process of traditional water data storage and the process of issuing and monitoring water withdrawal permits. It is possible to use these data to solve the challenges related to the water management system. Monitoring records in blockchain technology can reduce the cost of maintaining records. Also, blockchain technology allows secure data sharing at a low cost. Decentralized data management of blockchain technology improves water resource management.

Accurate monitoring of the water network is very difficult due to its multiple links. The blockchain-based water management system is capable of remotely managing, controlling and optimizing all aspects of water distribution. This system pursues goals such as implementing intelligent measurement and monitoring throughout the water distribution network, increasing security, improving data analysis, and increasing revenue and efficiency in the network. Therefore, blockchain technology can play a very effective role in optimizing water management systems. Also, the use of blockchain can be considered as a solution for the sustainable development of water resources.

1.6.1 Benefits of Blockchain for Smart Water Management

Blockchain-based decentralized intelligent water management has many advantages compared to traditional centralized management. The following are some of the key benefits.

- New technologies such as blockchain and IoT are the main solutions for modern water management, enabling intelligent systems and supporting the achievement of Sustainable Development Goal 6 (SDG 6)(clean water).
- Consumers receive a clear and accurate report on the quality and quantity of water consumed and its related costs. This report can be effective in making informed decisions. Therefore, consumer satisfaction is maintained in this management system.
- Due to the decentralization of the blockchain-based water management system, the central entity or a third party is removed from the intelligent water management and the cost of management and operation is reduced.
- Using blockchain technology in intelligent water management can create visibility and transparency in the system.

- Decentralized blockchain systems, by providing transparency in transaction records between parties, can improve water management and trading practices, increase transaction rates, and create trust between stakeholders.
- In a blockchain-based water management system, data cannot be altered or deleted, ensuring maximum security within the intelligent water management framework.
- The blockchain-based water management system brings reliability and increased management efficiency, as well as secure data storage, data security and high data transmission efficiency to the management system.
- Blockchain technology is able to provide accessibility, auditability, high data traceability and immutability in the water management system.

1.6.2 Blockchain Framework for Sustainable Water Management

Blockchain technology can have a great impact on optimizing smart water management and advancing the achievement of UN-SDGs. Water resource management based on blockchain can include four layers: the data collection and transmission layer, application layer, network layer and interaction layer. The data collection layer, with the help of the monitoring system, plays an effective role in collecting and transferring water-related data in real time. The use of the network layer and the application layer is to store distributed data and share data between nodes. Water sources, water supply chain, water consumers and sewage treatment plants can be considered as nodes of the system. Data is stored by nodes in blocks. Blocks are connected in series with time stamps and a P2P network is created between nodes. Data exchange between nodes and development of smart contract is done by the network layer. The application layer consists of four application modules for water resources management, water supply management, water consumption management and wastewater management. After creating the system, the interactive layer creates an integrated platform and an information portal. Due to the establishment of trust between nodes, monitoring costs are reduced. The smart contracts enable the automatic execution of water management functions at each node, thus increasing the efficiency of the system.

The water resource management module is used to ensure the proper quantity and quality of water resources. The amount of potable water stored in reservoirs is very important for supplying water to customers. The water level of the reservoirs is measured in real time by sensors and transferred to the water source node block with a time stamp. The water level information of the reservoirs is compared with the control indicators by a data analysis program and the results are stored by the blockchain. The obtained data and results of real-time evaluation are used in optimization algorithms to optimize water supply. Water quality is very important, so it can be tracked in different nodes by smart water management based on blockchain. Since the data in this technology has a time stamp, it is nearly impossible to change the data. Therefore, the data can be traced while remaining highly secure. By monitoring and tracking water quality, it is possible to quickly determine the location of a pollutant or poor water quality. It is possible to benefit from blockchain technology to monitor the quality of water sources, water supply network pipes, treatment plants and other sectors. The water supply management module consists of two parts: water management and maintenance management of infrastructure facilities of the water distribution network. By using smart contracts, the resulting data can be compared with the permitted amounts of water withdrawal. Also, water quality monitoring data is used for network management in real time. The water supply management module also controls the leakage of pipes. The water management module focuses on customer behavior and aims to increase water efficiency and reduce water consumption. All users can access their consumption and quota information, allowing them to effectively monitor and manage their water consumption. Water consumption data is compared with programmed values by smart contracts. In case of consumption exceeding the allowed limit, gradual pricing of water and incentives will be applied in order to reduce water consumption. In the management module of the wastewater treatment system, the volume of incoming and outgoing wastewater is measured and evaluated against quality standards using smart contracts, and the results are then stored in the blockchain Therefore, the quality of water coming out of the sewage treatment plant will also be controlled. Issuance, renewal and change of water withdrawal permits can be managed by this system.

1.7 CONCLUSION

As a recent technology, blockchain has created a different approach to data management and sharing. This P2P network stores all information in a distributed database. Blockchain is a free network, and any user can join it and store their information in it or create a new block in the chain. Of course, in blockchain, the levels of cooperation and access to system data may be different for different nodes. This network is a decentralized network and is not dependent on a central controller. The decentralization of blockchain technology will reduce operating costs and increase system efficiency. It is also the responsibility of the network users to validate each new block. Each block must be validated and confirmed by the majority of the users of this system, which is called consensus. This will bring transparency to the operation of the system. Each block is distinguished from other blocks by encryption. Encryption prevents fraud and alteration of stored data, preserves the privacy and identity of users, and increases security and reliability in the blockchain system. It is possible to trace all the previous blocks based on the last stored block. Therefore, the possibility of hacking, changing and manipulating or deleting the data stored in this system is almost impossible, and this will increase the security and reliability of this system.

Each block consists of two parts, header and block body. The block header section includes data such as block number, time stamp, protocol type, previous block hash, difficulty, block hash value and Merkel root value. The most important data of the block body section includes the list of completed transactions and data information.

As stated, blockchain is a set of processes and technologies such as encryption and hashing, P2P distributed network and consensus protocols. The consensus process can vary depending on the type, application, purpose, protocol and rules of blockchain. The most important and widely used consensus protocols are proof-of-work, proof-of-stake, practical Byzantine fault tolerant (PBFT) and authority of proof (PoA).

Proof-of-work (PoW) protocol has three characteristics: difficult solution, fairness and simple proof. This consensus requires a complex computational process and a lot of time. The PoS protocol consumes less energy compared to the working protocol. Therefore, in terms of energy consumption, this method is faster and more optimal than the proof-of-work protocol and has good security. The PoA protocol was proposed based on the advantages of the two consensus protocols PoS and PoW. This consensus has a higher processing speed. When the nodes cannot agree with each other, a phenomenon called consensus protocol failure will occur.

The history of blockchain can be divided into three different generations. The first-generation blockchain has proposed the decentralization of data storage. The Ethereum platform was unveiled as the second generation of blockchain with the aim of creating and implementing distributed programs and smart contracts. The probability of failure, fraud or data modification, censorship and third-party interference in the programs that run on the Ethereum platform is very low and it has very good flexibility, security, trust and speed. The third generation blockchain has taken into account goals such as high scalability, interoperability, stability, privacy and self-governance.

Blockchain systems can be classified into three categories based on the method of acceptance and consensus: public blockchain, private blockchain, and consortium (combined) blockchain. In a public blockchain, all users can participate in the consensus process. But in private and consortium blockchains, nodes need permission to join the consensus process. Public blockchain is decentralized and consortium blockchain is relatively decentralized. But the private blockchain is completely centralized. All nodes in the public blockchain are allowed to read data and records. While in private blockchain, data is visible only to selected individuals or nodes, and in consortium blockchain, access to data and records depends on certain conditions.

In public blockchain it is almost impossible to manipulate or change data. Private or consortium blockchains are less secure. In the public blockchain, the throughput of transactions is limited and the delay time is high. But private and consortium blockchains have more optimal efficiency than public blockchains. Blockchain technology has outstanding advantages over traditional systems, including increased accuracy in performance, reduced costs, decentralization that increases data security and reliability, transparency that increases efficiency and productivity, programmability and privacy, blockchain faces numerous challenges from technical, individual-organizational, legal, financial and environmental aspects.

The scope of application of blockchain technology is very wide due to features such as elimination of intermediaries and lack of dependence on third parties or entities, low operating costs, integration and the ability to store data in a public and decentralized manner. This technology is now used in many financial and non-financial fields and very diverse and wide areas such as the global economy, banking and financial markets, creating the infrastructure of basic cryptocurrencies, voting systems, medical services, intelligent transportation, government services, databases, digital identity creation, IoT development and many other applications.

Additionally, given that water management for conserving water resources, exploiting resources and ensuring appropriate water quality is an important priority for SDG 6, blockchain technology can be used to optimize water resource management and sustainable development systems.

Today, the collection, storage, processing and analysis of water network data in order to control and monitor in real time, secure data management, optimize resources, improve decision-making and create smart water management with the help of new and growing technologies such as blockchain artificial intelligence and IoT are possible.

Blockchain-based decentralized water management systems can be useful through the entire system, including the process of water supply, distribution, consumption and wastewater management. Also, this technology creates a good perspective for stronger water market mechanisms. Using the blockchain technology in smart water management can bring visibility and transparency to the system, improve the way water is managed and traded, increase transaction rates, build trust between stakeholders, increase the security of the smart water management system, increase system efficiency, and secure data storage and security to the management system. Therefore, the use of blockchain can be considered as a solution for the sustainable development of water resources.

REFERENCES

Altaf, A., Iqbal, F., Latif, R., Yakubu, B. M., Latif, S., & Samiullah, H. (2023). A survey of blockchain technology: Architecture, applied domains, platforms, and security threats. *Social Science Computer Review, 41*(5), 1941–1962.

Atlam, H. F., Azad, M. A., Alzahrani, A. G., & Wills, G. (2020). A review of blockchain in Internet of Things and AI. *Big Data and Cognitive Computing, 4*(4), 28.

Bordel, B., Martin, D., Alcarria, R., & Robles, T. (2019). A blockchain-based water control system for the automatic management of irrigation communities. Paper presented at the *2019 IEEE International Conference on Consumer Electronics (ICCE)*, Las Vegas, NV USA.

Cachin, C. (2016). Architecture of the hyperledger blockchain fabric. Paper presented at the *Workshop on Distributed Cryptocurrencies and Consensus Ledgers*, Zurich, Switzerland.

Dinh, T. T. A., Liu, R., Zhang, M., Chen, G., Ooi, B. C., & Wang, J. (2018). Untangling blockchain: A data processing view of blockchain systems. *IEEE Transactions on Knowledge and Data Engineering, 30*(7), 1366–1385.

Dogo, E. M., Salami, A. F., Nwulu, N. I., & Aigbavboa, C. O. (2019). Blockchain and internet of things-based technologies for intelligent water management system. In *Artificial Intelligence in IoT* (pp. 129–150). Springer International Publishing.

Eslamian, S., & Motevallian, S. S. (2014). Sustainability in Urban Water System. In *Handbook of Engineering Hydrology, Ch. 27, Vol. 1: Fundamentals and Applications*, Ed. By Eslamian, S., Taylor & Francis Group, CRC Group, USA, 549–562.

Fernández-Caramés, T. M., & Fraga-Lamas, P. (2018). A review on the use of Blockchain for the Internet of Things. *IEEE Access, 6*, 32979–33001.

Gramoli, V. (2020). From blockchain consensus back to Byzantine consensus. *Future Generation Computer Systems, 107*, 760–769.

Hakak, S., Khan, W. Z., Gilkar, G. A., Haider, N., Imran, M., & Alkatheiri, M. S. (2020). Industrial wastewater management using blockchain technology: Architecture, requirements, and future directions. *IEEE Internet of Things Magazine, 3*(2), 38–43.

Jimoh, F. O., Abdullahi, U. G., & Ibrahim, I. A. (2019). An overview of blockchain technology adoption. *Journal of Computer Science, 7*(2), 26–36.

Joshi, P., Tewari, V., Kumar, S., & Singh, A. (2023). Blockchain technology for sustainable development: A systematic literature review. *Journal of Global Operations and Strategic Sourcing, 16*(3), 683–717.

Khaqqi, K. N., Sikorski, J. J., Hadinoto, K., & Kraft, M. (2018). Incorporating seller/buyer reputation-based system in blockchain-enabled emission trading application. *Applied Energy, 209*, 8–19.

Khorasany, M., Mishra, Y., & Ledwich, G. (2019). A decentralized bilateral energy trading system for peer-to-peer electricity markets. *IEEE Transactions on Industrial Electronics, 67*(6), 4646–4657.

Kong, X., Zhang, J., Wang, H., & Shu, J. (2019). Framework of decentralized multi-chain data management for power systems. *CSEE Journal of Power and Energy Systems, 6*(2), 458–468.

Lin, Y.-P., Petway, J. R., Lien, W.-Y., & Settele, J. (2018). Blockchain with artificial intelligence to efficiently manage water use under climate change. *Environments 5*, 34.

Mattila, V., Dwivedi, P., Gauri, P., & Ahbab, M. (2022). Blockchain for environmentally sustainable economies: Case study on 5irechain. *International Journal of Social Sciences and Management Review, 5*(2), 50–62.

Misra, D. K., Kumar, P., & Khari, M. (2023). Achieving the Security in Cyber World through Blockchain: A Taxonomy. Paper presented at the *2023 10th International Conference on Computing for Sustainable Global Development (INDIACom)*, New Delhi, India.

Narayanan, A., Bonneau, J., Felten, E., Miller, A., & Goldfeder, S. (2018). *Bitcoin and cryptocurrency technologies: A comprehensive introduction*. Princeton University Press.

Parkinson, J., & Mark, O. (2005). *Urban stormwater management in developing countries*: IWA Publishing.

Patil, P. B., & Sangeetha, M. (2024). A comprehensive performance analysis of a hyperledger fabric-powered blockchain network for cross-border fund transfers. *Procedia Computer Science, 233*, 723–732.

Pérez Ortiz, Y. (2018). How blockchain technology could improve the quality of drinking water in Puerto Rico. Available at SSRN: https://ssrn.com/abstract=3266166

Poberezhna, A. (2018). Addressing water sustainability with blockchain technology and green finance. In *Transforming climate finance and green investment with blockchains* (pp. 189–196): Elsevier.

Rahmadika, S., Ramdania, D. R., & Harika, M. (2018). Security analysis on the decentralized energy trading system using blockchain technology. *Jurnal Online Informatika, 3*(1), 44–47.

Salah, K., Rehman, M. H. U., Nizamuddin, N., & Al-Fuqaha, A. (2019). Blockchain for AI: Review and open research challenges. *IEEE Access, 7*, 10127–10149.

Shrestha, R., & Nam, S. Y. (2019). Regional blockchain for vehicular networks to prevent 51% attacks. *IEEE Access, 7*, 95033–95045.

Skiba, R. (2020). Blockchain technology as a health and safety contributor in the transport and logistics industry–human resource requirements. *International Journal of Innovative Science and Research Technology, 5*(4), 544–550.

Soliman, H. M., & Leon-Garcia, A. (2014). Game-theoretic demand-side management with storage devices for the future smart grid. *IEEE Transactions on Smart Grid, 5*(3), 1475–1485.

Sriyono, E. (2020). Digitizing water management: Toward the innovative use of blockchain technologies to address sustainability. *Cogent Engineering, 7*(1), 1769366.

Sukhwani, H., Martínez, J. M., Chang, X., Trivedi, K. S., & Rindos, A. (2017). Performance modeling of PBFT consensus process for permissioned blockchain network (hyperledger fabric). Paper presented at the *2017 IEEE 36th symposium on reliable distributed systems (SRDS)*, Hong Kong.

Sundaresan, S., Kumar, K. S., Kumar, T. A., Ashok, V., & Julie, E. G. (2021). Blockchain architecture for intelligent water management system in smart cities. In *Blockchain for smart cities* (pp. 57–80): Elsevier.

Tian, F. (2016). An agri-food supply chain traceability system for China based on RFID & blockchain technology. Paper presented at the *2016 13th International Conference on Service Systems and Service Management (ICSSSM)*, Kunming, China.

Vranken, H. (2017). Sustainability of bitcoin and blockchains. *Current Opinion in Environmental Sustainability, 28*, 1–9.

Zhang, Y., Luo, W., & Yu, F. (2020). Construction of Chinese smart water conservancy platform based on the blockchain: Technology integration and innovation application. *Sustainability, 12*(20), 8306.

Zheng, Z., Xie, S., Dai, H.-N., Chen, X., & Wang, H. (2018). Blockchain challenges and opportunities: A survey. *International Journal of Web and Grid Services, 14*(4), 352–375.

2 Blockchain Technology
Tackling Global Challenges in Sustainable Development

Ritika Mahajan, Vaibhav Aggarwal, and Monika Agrawal

2.1 INTRODUCTION

2.1.1 Defining Sustainability in Environmental Systems

The concept of sustainability dates back to the 18th century when it was first used by Hans Carl von Carlowitz, a German forester, to recommend the method for long-term management of forests (Scoones, 2007). The term sustainability gained prominence only in the late 1980s as environmental activism during the late 1960s and 1970s linked environmental concerns to economic development leading to debates on the boundaries of development (Furze et al., 2022). The Brundtland Commission in the mid-1980s provided global attention and direction to these debates. Consequently, Brundtland's 1987 report defined the idea of sustainable development as an economic development activity "that meets the needs of the present without compromising the ability of future generations to meet their own needs" (WCED, 1987 p. 8).[1] The idea of sustainability is also promoted through sustainable ecosystems and circular economy which have been used interchangeably in the academic literature.

At the heart of sustainability lies the earth's biophysical environment (Farley and Smith, 2020). Until the 1990s,the sustainability agenda was generally focused on the use of natural resources for economic development needs without harming human, plant, and animal health (Geissdoerfer et al., 2017; Giovannoni and Fabietti, 2013). Sustainability in present times encompasses three parts or elements or pillars, namely environment, social equity, and economics referred to as the 3Es of sustainability or triple bottom line or 3Ps indicating people, planet, and profit (Dao et al., 2011; WCED, 1987). The equity dimension calls for the right of all communities to participate in social, political, and economic development. The economic perspective requires organizations to not only sustain with respect to profitability and productivity but also in managing the environmental and social aspects of the capital. These dimensions describe sustainability as a broad concept and convey that sustainability can be achieved when economic growth, environmental protection, and equity progress simultaneously. The sustainable development goals (SDGs) set up in 2012 in the United Nations conference were 17 goals that included all three aforementioned aspects at a global scale.

2.1.2 Addressing Sustainability Challenges: The Role of Technology

The challenges of meeting sustainable development goals are identified worldwide. The environmental challenges of sustainability include population growth, urbanization, energy consumption, global warming, water shortage, and waste management (Subramanian, 2007). The slow speed of social and political processes limits the achievement of SDGs within the time frame (Jansen, 2003). Fitting the goals to the local context is another barrier to a sustainable future. The integrated approach to sustainability presents additional challenges. The dimensions of sustainability sometimes contradict each other. Research reports contradictory results in terms of relationships between the three sustainability dimensions. For instance, some studies show a direct link between the financial and

DOI: 10.1201/9781003599593-3

social performance of an organization while others report an inverse relationship between the two (Boyd et al., 2009; Orlitzky et al., 2003). The same is true for environmental and financial dimensions (Giovannoni and Fabietti, 2013).

Technology offers solutions to the barriers to SDGs by leveraging innovations and systems renewal (Jansen, 2003). Harnessing the power of technological developments enables governments, organizations, and communities to execute strategies to promote sustainable practices. This may include the use of renewable energy sources (Adenle, 2020), eco-friendly manufacturing processes (Kim et al., 2014), smart infrastructure (Selim et al., 2018), and digital tools (Popkova et al., 2022) for managing resources in efficiency and effectiveness.

In today's digital age "digitization and sustainability" are of special interest among policy makers and practitioners. The latest technologies include blockchain, artificial intelligence (AI), internet of things, big data and so on. In the following paragraphs, the authors discuss blockchain technology and its potential for a sustainable future.

2.1.3 The Evolution of Blockchain Technology

Blockchain is an emerging technology –a distributed ledger that stores all transactions in a secured, verifiable, and transparent manner. As the technology is still under development, a universally accepted definition is missing. Various scholars create an understanding of blockchain in different ways; Table 2.1 provides a summary of the blockchain definitions in scholarly articles.

The concept of secured blockchain is not new and was initially proposed in 1991 by Stuart Haber et al. (Schneier, 1996). Based on the distinct features and usage, there are four types of blockchains, namely public blockchains, private blockchains, consortium blockchains, and hybrid blockchains. Public blockchains are open for anybody to join and engage in network management. Contrary to this, private blockchains are specific to a single organization and only invited people can join and manage the network. The consortium blockchain, also known as "Federated Blockchain," is a special case of private blockchain where people from more than one organization may be invited to join

TABLE 2.1
Blockchain Definitions

Study	Definition
Holotescu et al. (2018)	"Blockchain technology enables the creation of a decentralized environment, where the cryptographically validated transactions and data are not under the control of any third-party organization."
Upadhyay et al. (2021)	"Blockchain is a peer-to-peer system that eschews the role and power of centralized control over transactions, their record-keeping, and access to these records."
Bhutta et al. (2021)	"Blockchain is defined as the chain of digital blocks connected and associated with each other as an open distributed ledger."
Ghiro et al. (2021)	"The definition identifies three elements that, only when combined together, give to blockchains their specific features of openness, decentralization, security, and ability to preserve user privacy. These three elements are: relying on a strong distributed consensus protocol, which makes the blockchain immutable, hence secure from tampering attacks, and further frees the system from centralized trusted authorities (e.g., banks); maintaining a full and public history of transactions, which permits their distributed and completely transparent validation, and being open to anonymous users, thus allowing blockchains to preserve users privacy."
NIST (2022)	"Blockchains are tamper-evident and tamper-resistant cryptographically linked blocks of data (which create digital ledgers) implemented in a distributed fashion (i.e., without a central repository) and usually without a central authority (i.e., a bank, company, or government)."

and manage. The hybrid blockchain combines the privacy and permission mode of private blockchain with the flexibility, transparency, and simplicity of public blockchains.

The first generation of the blockchain (Blockchain 1.0) was based on the concept of distributed ledger technology (DLT) and proof of work (PoW) for the consensus process. DLT allows consensual sharing of data among the participants, thus eliminating double spending due to the presence of public witnesses. Its first application dates back to 2008 with its use in "Bitcoin" cryptocurrency for decentralized storing and tracking of transactions (Bhutta et al., 2021).

The limitations of poor scalability and wasteful mining prompted the development of the second-generation blockchain (Blockchain 2.0) by Buterin (2014). Blockchain 2.0 uses the principle of smart contracts along with the PoW mechanism. Smart contracts are computer programs that are automatically managed by the parties in an exchange relationship based on predefined provisions. From providing security and transparency to cryptocurrency transactions, the technology has evolved rapidly to be employed in areas other than digital payments. Blockchain 2.0 provides solutions to a range of platforms such as Ethereum, IBM Blockchain, Hyperledger Fabric, Hyperledger Sawtooth, and Quorum.

Both Blockchains 1.0 and 2.0 are not scalable as they are based on PoW, where transactions take a long time for confirmation. Thus, the current generation blockchain or the third generation blockchain (Blockchain 3.0) uses smart contracts along with decentralized apps (dApps); dApps are digital programs running on a blockchain network of computers, not just a single computer. Therefore, they are outside the view of any central authority, making interchain transfer possible. Blockchain 3.0 uses proof of stake and proof of authority for consent mechanisms to provide enhanced speed and power of computing to smart contracts. Blockchain 3.0 aims to enhance the scalability, privacy, interoperability, and security of transactions. It uses the Fast, Feeless, and Minerless (FFM) concept as it is not dependent upon miners to confirm transactions. Blockchain 3.0 led to the development of several platforms like ICON, Cardano, and IOTA to name a few. However, Blockchain 3.0 also suffers from limitations such as debugging, updating information constantly, and complicated consensus methods.

In the future, a promising development is the fourth-generation blockchain (Blockchain 4.0). It aims to bring blockchain technology into a business-usable platform by integrating blockchain with other technologies such as AI. Some of the utilities of Blockchain 4.0 are Unibright and SEELE platform. Blockchain 4.0 facilitates the rise of continuous assimilation of different digital platforms to work in coherence to meet business and industry demands. Figure 2.1 depicts different generations of blockchain and their underlying principle with some examples.

2.1.4 The Intersection between Blockchain and Sustainability

Blockchain technology opens new doors for management to address environmental and social challenges. There are numerous ways in which the two intersect. Blockchain creates transparency and traceability and reduces waste in supply chains; tracks carbon footprints, emissions, and reporting; allows decentralization in energy trading;advances social justice by ensuring minimum wages and working conditions, and generates smart contracts for environmental protection. The potential of blockchain is significant but its implementation is in its early stages.

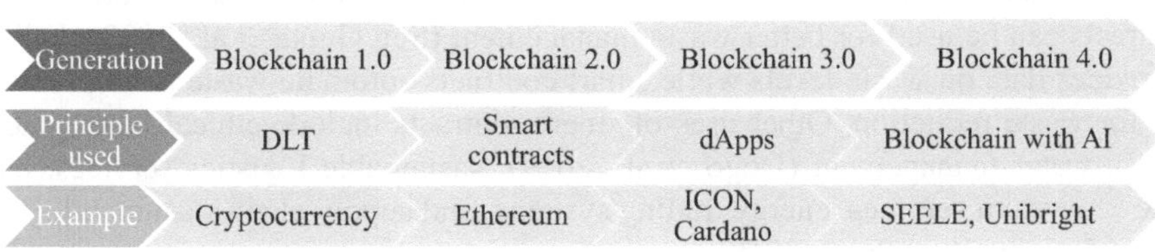

FIGURE 2.1 Blockchain evolution.

2.2 THE ROLE OF BLOCKCHAIN IN ADVANCING SUSTAINABILITY

The sustainability agenda puts government and public pressure on organizations to advance their sustainability practices (Jeppsson and Olsson, 2017). Sustainability awareness motivates stakeholders to take an interest in firms' sustainability practices and consumers' demand for sustainable products.

2.2.1 Transparency and Traceability: Blockchain's Impact

Traceability provides product information by tracking the products (e.g., components, originality) in the production and distribution process (Sodhi and Tang, 2018). Customers demand complete information on products, but traceability of the supply chain is a challenge. Blockchain technology enables distinctive, real-time system tracking status updates timely and automatically, thus allowing organizations to make effective and efficient decisions. Blockchain offers time efficiency by saving organizationshuge amounts oftime byharmonizing data and bypassing the verification authenticity due to the presence of verified actors in the distributed ledger.

Transparency is the extent to which information can be accessed by parties in exchange relationships as well asoutsiders. The product travels through various actors (e.g., producers, retailers, distributors, storage) before reaching the final consumers. Ensuring compliance with environmental, social, and economic sustainability requires access to accurate data by various stakeholders (Centobelli et al., 2022). With blockchain technology, the distributed tamper-proof information gets stored in realtime in a private-permissioned ledger. This provides information transparency and security, operational and time efficiency, and system automation.

2.2.2 Decentralization for Resilient and Sustainable Systems

Blockchain technology works on the principle of distributed data at multiple nodes rather than collected data at a centralized point (Mukherjee et al., 2021). This feature of decentralization eliminates the need for intermediaries and adds simplicity and time efficiency in social and financial transactions. Decentralization reduces settlement lead times and makes external verification redundant. The elimination of intermediaries also reduces the cost of transactions and enhances trust between the stakeholders, which builds system integrity. Supply chain real-time information and coordination by blockchain technology aid manufacturers and retailers to advance their accuracy in forecasting demand and inventory replenishment capacities (Cui and Idota, 2018). The capabilities of Blockchain 3.0 have been found in the Decentralized Autonomous Organization (DAO). DAO enables system robustness and builds resilience against disruptions such as war, geopolitical tensions, and natural emergencies (e.g. COVID-19) (Leng et al., 2023).

2.2.3 Smart Contracts: Automating Sustainability Processes

Blockchain technology enables transactions and documentation between parties in exchange relationships using digital automated contracts known as smart contracts. A smart contract is implemented based on agreed-upon conditions (Mukherjee et al., 2021). Smart contracts facilitate adherence to carbon emission agreements (Sadawi et al., 2021). The contracts remain in tamper-proof blockchain and provide extra security to carbon trading as compared to traditional contracts. Smart contracts can be used for better waste management (Sen Gupta et al., 2022). IoT along with sensors provides data on waste levels while smart contracts automate waste pickup, recycling, and incentives for waste reduction. Other uses of smart contracts include ethical labor practices (Tate et al., 2017), water management (Dogo et al., 2019), sustainable logistics services, payment for sustainable ecosystem services, energy trading systems, and supply chain sustainability (Salmerón-Manzano & Manzano-Agugliaro, 2019).

2.2.4 Enhancing Collaboration through Blockchain

The characteristic features of transparency, traceability, decentralization, being tamper-proof, and data privacy in blockchain technology enhance the possibility of collaboration in various industries. Blockchain streamlines information-sharing systems, supports decision and reward methods, and strengthens communication between parties in exchange relationships (Rejeb et al., 2021). Further, blockchain adds future capacities in exchange relationships by smoothing audibility, refining accountability, facilitating transparency of data and information, and enhancing trust in business relationships. The technology possesses the potential to enhance collaboration and manage vulnerabilities of moral risk and loopholes detected in legacy technologies.

Presently blockchain networks used by organizations are developed depending on their need. These networks were developed using different architectures and protocols. Thus, blockchain projects are highly fragmented with negligible or no interoperability between blockchain projects. The following section discusses some case studies and applications of the technology, followed by challenges, opportunities, and future trends.

2.3 CASE STUDIES AND APPLICATIONS

2.3.1 Blockchain for Carbon Credits and Emissions Trading

Kyoto Protocol was adopted in 1997 and came into force in 2005. It has provided a market-based mechanism to reduce carbon and greenhouse gas (GHG) emissions. It consisted of a carbon emission trading system (CETS), a clean development mechanism (CDM), and a joint implementation system. Carbon emission is curbed mainly through two methods: carbon tax and carbon trading. The carbon tax sets up a quota on the emission of carbon or GHG on a unit basis; exceeding units is mechanized to be punished by applying some taxes. However, carbon trading works on a cap-and-trade model (Fang et al., 2020). Carbon credits are generated, monitored, and auctioned through a legitimate agency.

Homogeneity, transparency, and trust are significant issues in the carbon trading system (Sovacool, 2011), causing leakages in the system (Martin et al., 2014). Therefore, public blockchain is a potential solution for these problems (Franke et al., 2020) by enhancing the monitoring and tracking of actual emissions (Al Sadawi et al., 2021) by averting the double accounting method (Schneider et al., 2019). Thus, the issues in the current ETS may be solved by integrating the inherent features of the public blockchain, like data transparency, distributed modeling, and data tracking, with the CETS systems.

Despite the advantages and government awareness about the blockchain, it is still not substantially utilized for the ETS. This is because of specific challenges in blockchain technology, like large energy consumption, scalability, system maturity, data privacy issues, and garbage data handling (Braden, 2019). However, with the updates in technology and growing research focusing on streamlining the system, blockchain is an essential technological solution that can enhance the efficiency of the CETS.

2.3.2 Sustainable Supply Chain Management with Blockchain

With the world becoming more interconnected, supply chains are also becoming more complex and distributed; hence, managing the same is a herculean task (Lambert and Enz, 2017). Different and incoherent regulatory policies, diverse cultures, and human behavior make it almost impossible to process information and manage risks in the supply chain network (Ivanov and Dolgui, 2019). This leads to inefficient transactions, trust issues, and a lack of transparency. For example, in the case of organic products from the field to the customer's plate, there are trust issues, and the market is still nascent in India.

Thus, blockchain technology also provides a promising solution in this scenario. Being distributed in nature, blockchain is a disruptive technology for the design, arrangements, operations, and management of supply chains. It offers reliability, traceability, and homogeneous information that enhance trustworthiness. This way, it doesn't only impact the supply chain but also product and financial management among the network nodes. Disintermediation of the financial mediators (payment network, exchanges, dealers, etc.) thus becomes one of the critical advantages of blockchain technology in the supply chain. This can also potentially reduce further inefficiencies related to operations and financial management in the supply chain, saving millions of dollars (Fanning and Centers, 2016).

Blockchain also provides hope for the way forward to environmental supply chain sustainability through identifying and tracking substandard products accurately, providing support for detailed analysis of connections for improvement in products, reducing wastage, enhancing reach and trust for green products, and so on.

2.3.3 Eco-Friendly Cryptocurrencies and Blockchain Initiatives

Cryptocurrency is one application of blockchain that has shown its potential to bring changes around the financial sector. As per Statista (2023),[2] the cryptocurrency market size is expected to reach around USD 162.84 billion by 2027. However, this growth comes at the cost of the environment through electricity consumption (FACT SHEET, 2022), and therefore, the requirement for green blockchains has gained momentum.

Green blockchain technology refers to implementing and utilizing this technology in an environmentally sustainable manner, with priority given to carbon reduction and a decline in electricity consumption (Oudani et al., 2023). This has found potential application in various fields, from agriculture to smart cities, through interconnected nodes to the IoT, thereby helping combat climate change and offset carbon emissions. Alzoubi and Mishra (2023) identified 23 green blockchain initiatives that are useful for sustainable development. Some of the blockchain initiatives in the different segments are listed as follows:

Renewable energy: solar coin, renewable energy certificate, blockchain for climate foundation, green digital finance, treelion, etc.

Energy efficiency: EFFORCE, chimpzee, decentralized energy exchange, green technology Asia, MIRIS, tomorrow, etc.

Carbon neutrality: greentrust, ecoterra, the blockchain climate institute, energy web foundation, etc.

2.3.4 Blockchain in Renewable Energy Trading

Sustainability in energy is another major challenge that can be solved through distributed energy production and supplies. Decentralized energy independence has commercial, social, and environmental benefits (Jiayi et al., 2008; Perera et al., 2019). This method has the potential to increase the efficiency in energy utilization (Basak et al., 2012), fulfilling the distributed scenario-based energy demands in local areas (Mengelkamp et al., 2018) and also has the potential to link to the global grid system for energy transition (Cao, 2019). Figure 2.2 presents a distributed model of energy production.

In this scenario, blockchain emerges as the foremost technological solution for integrating distributed energy networks from microgrids to the global grid system (Cao, 2019). As the cryptocurrency technology that introduces peer-to-peer financial transactions without needing a third-party trust system, the blockchain in the renewable energy segment can help create the energy internet with peer-to-peer energy transactions based on the crypto trust system. This system, therefore, presents

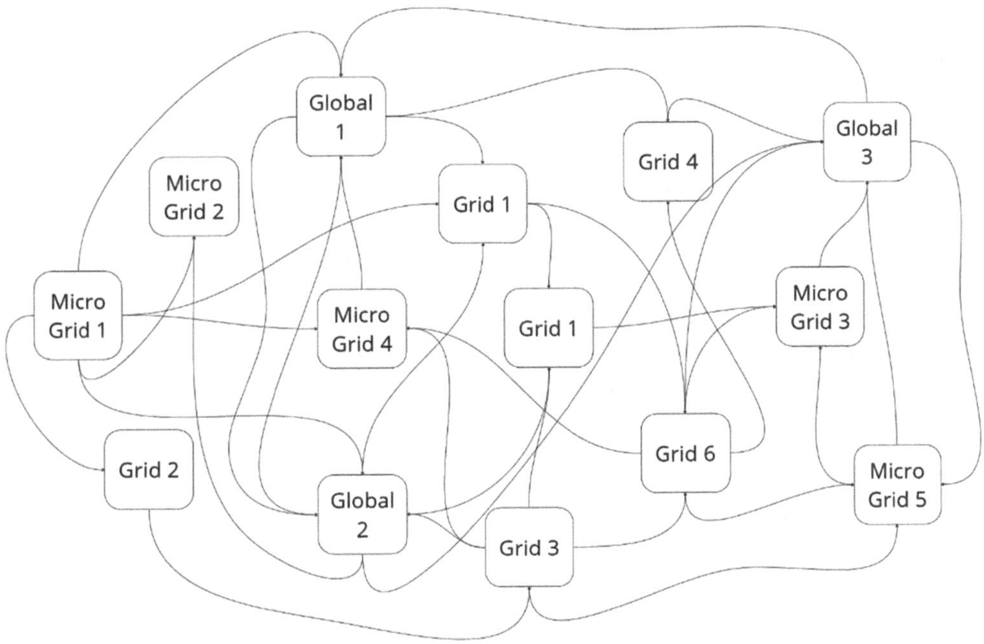

FIGURE 2.2 Energy internet (distributed model of energy production).

the following benefits: first, the use of big data and the energy internet with all refined data sources can help in achieving the just transition toward renewable energy with energy utilization efficiency (Ahl et al., 2019; Khan and Salah, 2018). Second, it can help give due place to the new entrants by disturbing the monopolies with reasonable pricing mechanisms (Queiroz and Fosso Wamba, 2019). Third, every participant and node in this system is equally essential (Mukherjee et al., 2021).

Therefore, blockchain technology in the space of renewable energy can support the just transition, facilitating a distributed model of energy production and utilization, enhanced transparency, a decision support system, and various other social benefits.

2.4 CHALLENGES AND ISSUES

2.4.1 Energy Consumption

One of the major challenges associated with using blockchain technology is the high consumption of energy involved in the implementation of projects based on this technology (Hastig and Sodhi, 2020). According to the UNCTAD[3] report, at some point, the total amount of energy required to sustain bitcoins was equal to that of the entire country of Switzerland. Thus, the fundamental requirement of this technology exacerbates environmental challenges like global warming, even if it is being used to address SDGs. As the need and demand for scaling up the technology increases, the energy challenge becomes more complex. The technology usage also involves security concerns; dealing with those concerns also makes the energy challenge more difficult to address. While decentralization is a vital feature of this technology, balancing decentralization and energy efficiency is also critical. The issues become complicated as the scale of adoption of technology becomes wider. However, this challenge can be mitigated by utilizing opportunities for solving environmental and social challenges arising out of the use of blockchain technology, as discussed in the next section on opportunities.

2.4.2 Regulatory and Compliance Issues

Decentralization, as a key feature of blockchain technology, can pose a significant challenge during its execution. The possibility of criminal activities increases due to this feature. Access is also limited to a few hands. Thus, those with the privilege of this access are more likely to engage in illegal or illicit activities, creating compliance issues related to this technology. Many scams have taken place, for instance, concerning cryptocurrencies. This is a threat to the attainment of sustainability as it will create loopholes related to compliance. Lack of clarity in regulations makes the matter more challenging (Govindan, 2022). Since the technology continuously evolves, the laws are also in a work-in-progress mode. The lack of clear and consistent national and international regulations can be problematic. Since blockchain operates across borders, such inconsistencies can also create complexities. Regulatory issues could be related to taxation, data privacy, ownership, access, consumer protection, and environmental conservation from a sustainability viewpoint. Interoperability is also a concern. Navigating these concerns is necessary to avail benefits of depending upon blockchain for sustainable development.

2.4.3 Social and Cultural Factors

From a social angle, technology that contributes to sustainable development must be inclusive and accessible to all (Kshetri, 2021). In the absence of digital infrastructure, or what we may refer to as the digital divide, blockchains will not be accessible to all sections of society. Similarly, even if there is access, there may be resistance among communities and sections to adopting new technology due to issues related to privacy, security, or lack of a regulatory framework. Diversity with respect to literacy rates or lack of training facilities can also pose a challenge to the adoption of such new technology. If the new technology creates unemployment and associated social issues, that can also be a hindrance. Community engagement, capacity building, and narrowing of the digital divide are critical elements to be focused upon to promote blockchain technology and sustainability integration.

2.4.4 Ethical Considerations

Blockchain's promise of transparency can create a conflict and ethical dilemma for privacy concerns; balancing both while achieving sustainability goals is also a challenge. The distribution of benefits of blockchain technology to different sections of society may also be problematic and may lead to equity concerns (Friedman and Ormiston, 2022). Sometimes, errors in technology design or implementation may cause potential harm, generating ethical issues. Consider a smart contract governing a significant financial transaction; any mistakes in its execution shall lead to financial losses to the parties involved in it.

Automation achieved through the adoption of this technology shall lead to changes in employment rates and the need to develop different skill sets. If there is no facilitation provided in this regard, some ethical concerns may arise in this case, too. Further, inequalities and ownership conflict, as discussed previously, may also need consideration from an ethical lens. Eventually, a long-term assessment of this technology is to be made to find out how it enables sustainable development and whether the pros outweigh the cons. Figure 2.3 presents the key challenges and issues related to blockchain and sustainability.

2.4.5 Lessons Learned from Challenges

As evident from the challenges explained in the preceding sections, some key considerations significantly impact the role of blockchain technology in meeting SDGs. The main lessons that emerge out of the explained challenges are as follows. One, energy is the first and foremost concern; high energy consumption hits the main pillar of the 3Ps approach and hence needs to be

FIGURE 2.3 Blockchain and sustainability: Key challenges and issues.

addressed to leverage technology successfully for sustainability concerns. Two, compliance is a concern, and lack of clear and consistent regulations is a bigger issue. In the absence of clarity on rules and regulations, technology cannot lead to sustainable development. Three, inclusivity and accessibility are the main challenges from a social standpoint, and the digital divide needs to be addressed. Four, while transparency is described as a benefit of leveraging blockchain, it leads to privacy and security issues. Five, using technology to deal with social and environmental issues, creating an ecosystem for equitable distribution of benefits, and long-term ethical assessment are some emerging areas from an ethical viewpoint to be taken into account while using this technology for sustainable development.

2.5 OPPORTUNITIES AND TRENDS

If we analyze the challenges, these can become the pathway to opportunities and future trends in using blockchain technology for sustainability issues and concerns (Andoni et al., 2019; Upadhyay, 2020). First, considering the high energy consumption associated with blockchains, there are opportunities to explore energy-efficient mechanisms and incentivize green practices, including using renewable resources for generating energy.

Second, regulators must proactively engage in a dialogue with stakeholders and ascertain the global best practices to develop clear, concise, and consistent regulations that align with the global context. The regulator shall need the industry's support and collaboration to develop legal frameworks and compliance, leading to sustainable growth of the industry and the economy. Concerns related to privacy, security, and ownership can only be addressed through stakeholder engagement, cooperation, and collaboration. Some level of standardization will also facilitate the adoption of blockchain technology and its effective functioning across countries.

Third, to mitigate social and cultural challenges, there is an emerging opportunity to promote inclusive education and training. A community-centric strategy will be required to facilitate the effective implementation of blockchain technology, especially when it comes to sustainability challenges that encompass economic, social, and environmental dimensions. Thus, community engagement will be helpful in this regard. This could be done through training programs for building skill sets, awareness generation programs, and so on.

Fourth, ethical challenges can be dealt with through strategic and responsible practices by all the stakeholders. Laws shall be designed to facilitate equitable distribution of benefits, and the industry shall ensure compliance. At an industry level, rigorous planning, testing, and auditing are prescribed.

Overall, there is a scope for long-term assessment for evaluating the role of blockchain in sustainable development. The role of the blockchain in achieving the SDGs was discussed and asserted extensively by Hughes et al. (2019). It would involve continuous monitoring and evaluation through establishing mechanisms and adaptive strategies.

In light of the challenges and strategies explained in this chapter, the probable future trends may include (a) development of a national-level policy framework, aligned with global priorities and regulations, that takes into account environmental and social concerns about using blockchains for sustainable development; (b) impetus to start-ups and innovation hubs engaged in greening the blockchain technology to combat the environmental challenges arising out of its usage; (c) an increase in awareness generation and capacity building programs to sensitize and develop skills in individuals to leverage this technology for sustainable development; (d) development of organizational standing operating procedures, guidelines, and principles in line with the national framework; and (e) more focus on stakeholder engagement including international collaborations to build support for leveraging the relationship between blockchain and sustainable development.

2.6 CONCLUSION

This chapter traces the roots of sustainability to the 18th century and presents a brief overview of the contemporary triple-bottom-line approach that explains the underpinnings of sustainability. The interconnectedness between three pillars, that is economic, social, and environmental, is acknowledged as crucial to achieving sustainable development. This multidimensional nature is also demonstrated through the UN SDGs adopted by member states in 2015. In the preceding sections, we explain the role of technology, notably blockchain, for sustainable development with its unique features in terms of transparency, traceability, and the role of automation. We describe the intersection of blockchain and sustainability through examples like waste reduction in supply chains, carbon footprint tracking, and promotion of ethical business practices and case studies about carbon credits, supply chain management, and renewable energy trading. Limitations of blockchain technology are also discussed in terms of energy consumption as well as ethical, social, cultural, and legal problems, followed by opportunities emerging from ideas aimed at mitigating these challenges.

While we attempt to provide a comprehensive overview of the relationship between blockchain and sustainability, the study is restricted to secondary data. Future research may delve deeper into the cases, challenges, and opportunities to present an in-depth exploration of the uses and problems of blockchain for sustainability using primary data. A deeper insight into stakeholder valuation may also be useful to deeply understand the socio-cultural factors that will enable the adoption of blockchain technology. Future research may focus on the issue of designing and using blockchain, considering energy consumption, and addressing the associated environmental concerns. Researchers may explore regional variations in the adoption of blockchain technology to take into account the role of "context" in technology adoption. Other avenues for future research could be related to the evolution of regulatory frameworks, addressing ethical dilemmas, stakeholder engagement strategies, innovative financial models, and so forth.

In conclusion, this chapter offers a foundational understanding of the intersections between blockchain and sustainability; future research may provide a more nuanced understanding of the barriers, enablers, and strategies for sustainable development through blockchain technology.

NOTES

1 World Commission on Environment and Development.
2 https://www.statista.com/outlook/fmo/digital-assets/cryptocurrencies/worldwide.
3 https://unctad.org/publication/world-investment-report-2021.

REFERENCES

Adenle, A. A. (2020). Assessment of solar energy technologies in Africa-opportunities and challenges in meeting the 2030 agenda and sustainable development goals. *Energy Policy*, 137, 111180.

Ahl, A., Yarime, M., Tanaka, K. and Sagawa, D. (2019). Review of blockchain-based distributed energy: Implications for institutional development. *Renewable and Sustainable Energy Reviews* 107, 200–211. https://doi.org/10.1016/j.rser.2019.03.002

Al Sadawi, A., Madani, B., Saboor, S., Ndiaye, M., and Abu-Lebdeh, G. (2021). A comprehensive hierarchical blockchain system for carbon emission trading utilizing blockchain of things and smart contract. *Technological Forecasting and Social Change*, 173(August), 121124. https://doi.org/10.1016/j.techfore.2021.121124

Alzoubi, Y.I. and Mishra, A., (2023). Green blockchain: A move towards sustainability. *Journal of Cleaner Production*430, 139541. https://doi.org/10.1016/j.jclepro.2023.139541

Andoni, M., Robu, V., Flynn, D., Abram, S., Geach, D., Jenkins, D., McCallum, P., and Peacock, A. (2019). Blockchain technology in the energy sector: A systematic review of challenges and opportunities. *Renewable and Sustainable Energy Reviews*, 100, 143–174

Basak, P., Chowdhury, S., Halder Nee Dey, S., and Chowdhury, S.P. (2012). A literature review on integration of distributed energy resources in the perspective of control, protection and stability of microgrid. *Renewable and Sustainable Energy Reviews*16, 5545–5556. https://doi.org/10.1016/j.rser.2012.05.043

Bhutta, M. N. M., Khwaja, A. A., Nadeem, A., Ahmad, H. F., Khan, M. K., Hanif, M. A., Song, H., Alshamari, M., and Cao, Y. (2021). A survey on blockchain technology: Evolution, architecture and security. *IEEE Access*, 9, 61048–61073. https://doi.org/10.1109/ACCESS.2021.3072849

Boyd, B., Henning, N., Reyna, E., Wang, D., Welch, M. and Hoffman, A. J.(2009). *Hybrid organizations: New business models for environmental leadership*. Routledge, 184

Braden, S. (2019). *Blockchain for Mexican climate instruments: Emissions trading and MRV systems*, Deutsche Gesellschaft für Internationale Zusammenarbeit.

Buterin, V. (2014). A next-generation smart contract and decentralized application platform. *EthereumWhite Paper*53. https://ethereum.org/whitepaper/

Cao, Y. (2019). Energy Internet blockchain technology, in: Su, W., Huang, A.Q. (Eds.), *The Energy Internet*. Woodhead Publishing, pp. 45–64. https://doi.org/10.1016/B978-0-08-102207-8.00003-5

Centobelli, P., Cerchione, R., Vecchio, P. Del, Oropallo, E., and Secundo, G. (2022). Blockchain technology for bridging trust, traceability and transparency in circular supply chain. *Information and Management*, 59(7). https://doi.org/10.1016/j.im.2021.103508

Cui, Y., and Idota, H. (2018). Improving supply chain resilience with establishing a decentralized information sharing mechanism. *ACM International Conference Proceeding Series*, 1–7. https://doi.org/10.1145/3227696.3227723

Dao, V., Langella, I., and Carbo, J. (2011). From green to sustainability: Information Technology and an integrated sustainability framework. *Journal of Strategic Information Systems*, 20(1), 63–79. https://doi.org/10.1016/j.jsis.2011.01.002

Dogo, E. M., Salami, A. F., Nwulu, N. I., and Aigbavboa, C. O. (2019). Blockchain and internet of things-based technologies for intelligent water management system. In *Artificial Intelligence in IoT, Transactions on Computational Science and Computational Intelligence*. Springer, 129–150. https://doi.org/10.1007/978-3-030-04110-6_7

FACT SHEET: Climate and Energy Implications of Crypto-Assets in the United States|OSTP [WWW Document](2022). The White House. https://bidenwhitehouse.archives.gov/ostp/news-updates/2022/09/08/fact-sheet-climate-and-energy-implications-of-crypto-assets-in-the-united-states/ (accessed 09.10.25).

Fang, D., Wang, D., and Kleit, A.N. (2020). How a choice between emission trading and tax schemes affects an environment-biased Environmental Protection Bureau's discretion? *Science of the Total Environment* 736, 139141. https://doi.org/10.1016/j.scitotenv.2020.139141

Fanning, K., and Centers, D.P. (2016). Blockchain and its coming impact on financial services. *Journal of Corporate Accounting Finance*27, 53–57. https://doi.org/10.1002/jcaf.22179

Farley, H. M., and Smith, Z. A. (2020). *Sustainability: If it's everything, is it nothing?* Routledge.

Franke, L., Schletz, M., and Salomo, S. (2020). Designing a blockchain model for the paris agreement's carbon market mechanism. *Sustainability*12, 1068. https://doi.org/10.3390/su12031068

Friedman, N., and Ormiston, J. (2022). Blockchain as a sustainability-oriented innovation?: Opportunities for and resistance to Blockchain technology as a driver of sustainability in global food supply chains. *Technological Forecasting and Social Change*, 175, 121403.

Furze, J. N., Eslamian, S., Raafat, S., and Swing, K., (2022). *Mathematical Advances Towards Earth Systems Protection and Sustainability* Vol. 1, Springer, Switzerland, 337.

Geissdoerfer, M., Savaget, P., Bocken, N. M. P., and Hultink, E. J. (2017). The circular economy: A new sustainability paradigm? *Journal of Cleaner Production*, 143, 757–768. https://doi.org/10.1016/j.jclepro.2016.12.048

Ghiro, L., Restuccia, F., D'Oro, S., Basagni, S., Melodia, T., Maccari, L., and Cigno, R. Lo. (2021). What is a blockchain? A definition to clarify the role of the blockchain in the Internet of Things. https://arxiv.org/abs/2102.03750

Giovannoni, E., and Fabietti, G. (2013). What is sustainability? A review of the concept and its applications. In C. Busco, M. Frigo, A. Riccaboni, &P. Quattrone (Eds.), *Integrated Reporting: Concepts and Cases that Redefine Corporate Accountability* (Issue 2010, pp. 1–350). Springer, Cham. https://doi.org/10.1007/978-3-319-02168-3

Govindan, K. (2022). Tunneling the barriers of blockchain technology in remanufacturing for achieving sustainable development goals: A circular manufacturing perspective. *Business Strategy and the Environment*, 31(8), 3769–3785.

Hastig, G. M., and Sodhi, M. S. (2020). Blockchain for supply chain traceability: Business requirements and critical success factors. *Production and Operations Management*, 29(4), 935–954.

Holotescu, C., Holotescu, V., and Holotescu, T. (2018). Understanding Blockchain Technology and How to Get Involved. *The 14th International Scientific Conference ELearning and Software for Education*, 2018 (April), 1–9. https://doi.org/10.13140/RG.2.2.25185.33126/1

Hughes, L., Dwivedi, Y.K., Misra, S.K., Rana, N.P., Raghavan, V., and Akella, V., (2019). Blockchain research, practice and policy: Applications, benefits, limitations, emerging research themes and research agenda. *International Journal of Information Management*, 49, 114–129.

Ivanov, D., and Dolgui, A. (2019). Low-Certainty-Need (LCN) supply chains: A new perspective in managing disruption risks and resilience. *International Journal of Production Research*,57, 5119–5136. https://doi.org/10.1080/00207543.2018.1521025

Jansen, L. (2003). The challenge of sustainable development. *Journal of Cleaner Production*, 11(3), 231–245. https://doi.org/10.1016/S0959-6526(02)00073-2

Jeppsson, O., and Olsson, O. (2017). Blockchains as a solution for traceability and transparency [Lund University]. https://lup.lub.lu.se/student-papers/search/publication/8919957

Jiayi, H., Chuanwen, J., and Rong, X. (2008). A review on distributed energy resources and MicroGrid. *Renewable and Sustainable Energy Reviews*,12, 2472–2483. https://doi.org/10.1016/j.rser.2007.06.004

Khan, M.A., and Salah, K. (2018). IoT security: Review, blockchain solutions, and open challenges. *Future Generation Computer Systems*,82, 395–411. https://doi.org/10.1016/j.future.2017.11.022

Kim, J. Y., Jeong, S. J., Cho, Y. J., and Kim, K. S. (2014). Eco-friendly manufacturing strategies for simultaneous consideration between productivity and environmental performances: A case study on a printed circuit board manufacturing. *Journal of Cleaner Production*, 67, 249–257.

Kshetri, N. (2021). Blockchain and sustainable supply chain management in developing countries. *International Journal of Information Management*, 60, 102376.

Lambert, D.M., and Enz, M.G. (2017). Issues in supply chain management: Progress and potential. *Industrial Marketing Management*,62, 1–16.

Leng, J., Zhong, Y., Lin, Z., Xu, K., Mourtzis, D., Zhou, X., Zheng, P., Liu, Q., Zhao, J. L., and Shen, W. (2023). Towards resilience in Industry 5.0: A decentralized autonomous manufacturing paradigm. *Journal of Manufacturing Systems*, 71, 95–114.

Martin, R., Muûls, M., de Preux, L.B., and Wagner, U.J. (2014). On the empirical content of carbon leakage criteria in the EU Emissions Trading Scheme. *Ecological Economics*,105, 78–88. https://doi.org/10.1016/j.ecolecon.2014.05.010

Mengelkamp, E., Gärttner, J., Rock, K., Kessler, S., Orsini, L., and Weinhardt, C. (2018). Designing microgrid energy markets: A case study: The Brooklyn Microgrid. *Applied Energy*210, 870–880. https://doi.org/10.1016/j.apenergy.2017.06.054

Mukherjee, A. A., Singh, R. K., Mishra, R., and Bag, S. (2021). Application of blockchain technology for sustainability development in agricultural supply chain: justification framework. *Operations Management Research*, 0123456789. https://doi.org/10.1007/s12063-021-00180-5

NIST. (2022). Blockchain for Access Control Systems. https://doi.org/10.6028/NIST.IR.8202

Orlitzky, M., Schmidt, F., and Rynes, S. (2003). Corporate social and financial performance: a meta_analysis. *Organization Studies*, 24(3), 403–441.

Oudani, M., Sebbar, A., Zkik, K., El Harraki, I., and Belhadi, A. (2023). Green Blockchain based IoT for secured supply chain of hazardous materials. *Computers & Industrial Engineering*175, 108814. https://doi.org/10.1016/j.cie.2022.108814

Perera, A.T.D., Nik, V.M., Wickramasinghe, P.U., and Scartezzini, J.L. (2019). Redefining energy system flexibility for distributed energy system design. *Applied Energy*253, 113572. https://doi.org/10.1016/j.apenergy.2019.113572

Popkova, E. G., De Bernardi, P., Tyurina, Y. G., and Sergi, B. S. (2022). A theory of digital technology advancement to address the grand challenges of sustainable development. *Technology in Society*, 68(December 2021), 101831. https://doi.org/10.1016/j.techsoc.2021.101831

Queiroz, M.M., Fosso Wamba, S., (2019). Blockchain adoption challenges in supply chain: An empirical investigation of the main drivers in India and the USA. *International Journal of Information Management* 46, 70–82. https://doi.org/10.1016/j.ijinfomgt.2018.11.021

Rejeb, A., Keogh, J. G., Simske, S. J., Stafford, T., and Treiblmaier, H. (2021). Potentials of blockchain technologies for supply chain collaboration: A conceptual framework. *International Journal of Logistics Management*, 32(3), 973–994. https://doi.org/10.1108/IJLM-02-2020-0098

Salmerón-Manzano, E., and Manzano-Agugliaro, F. (2019). The role of smart contracts in sustainability: Worldwide research trends. *Sustainability (Switzerland)*, 11(11). https://doi.org/10.3390/su11113049

Schneider, L., Duan, M., Stavins, R., Kizzier, K., Broekhoff, D., Jotzo, F., Winkler, H., Lazarus, M., Howard, A., and Hood, C. (2019). Double counting and the Paris agreement rulebook. *Science*366, 180–183. https://doi.org/10.1126/science.aay8750

Schneier, B. (1996). *Applied Cryptography, Protocols, Algorithms and Source Code in C*(2nd ed.). Wiley.

Scoones, I.(2007). Sustainability. *Development in Practice*, 17, 589–596. https://doi.org/10.1080/09614520701469609

Selim, A. M., Yousef, P. H., and Hagag, M. R. (2018). Smart infrastructure by (PPPs) within the concept of smart cities to achieve sustainable development. *International Journal of Critical Infrastructures*, 14(2), 182–198.

Sen Gupta, Y., Mukherjee, S., Dutta, R., and Bhattacharya, S. (2022). A blockchain-based approach using smart contracts to develop a smart waste management system. *International Journal of Environmental Science and Technology*, 19(8), 7833–7856. https://doi.org/10.1007/s13762-021-03507-8

Sodhi, M. M. S., and Tang, C. S. (2018). Corporate social sustainability in supply chains: A thematic analysis of the literature. *International Journal of Production Research*, 56(1–2), 882–901. https://doi.org/10.1080/00207543.2017.1388934

Sovacool, B.K. (2011). Four problems with global carbon markets: A critical review. *Energy and Environment*22, 681–694. https://doi.org/10.1260/0958-305X.22.6.681

Statista (2023). Accessed from https://www.statista.com/outlook/fmo/digital-assets/cryptocurrencies/worldwide

Subramanian, N. (2007). Sustainability: Challenges and solutions. *The Indian Concrete Journal*, 8(12), 39–50.

Tate, M., Johnstone, D., and Fielt, E. (2017). Ethical issues around crowdwork: How can blockchain technology help? *28th Australasian Conference on Information Systems*, University of Tasmania, Hobart, Australia, 5–6 December, 1–11.

Upadhyay, A., Mukhuty, S., Kumar, V., and Kazancoglu, Y. (2021). Blockchain technology and the circular economy: Implications for sustainability and social responsibility. *Journal of Cleaner Production*, 126–130. https://doi.org/10.1016/j.jclepro.2021.126130

Upadhyay, N. (2020). Demystifying blockchain: A critical analysis of challenges, applications and opportunities. *International Journal of Information Management*, 54, 102120

WCED. (1987). *Our Common Future: Report of the World Commission on Environment and Development*, World Commission on Environment and Development, United Nations.

Part II

Blockchain Evolution and Applications

3 Blockchain Evolution
From Cryptographic Beginnings to Societal Applications

Ana Rodriguez and João Casqueira Cardoso

3.1 INTRODUCTION

Blockchain technology has been at the forefront of a revolution regarding populations' behavior with online tools – relevant for both market and nonmarket economics. The exchange of products in daily life, where blockchain systems can be applied, is being modified in what concerns to goods and values. Nakamoto was the precursor of blockchain technology with the creation of Bitcoin software, in 2009, and the introduction of a cryptocurrency in the financial market for the first time (Bitcoin, 2023). Throughout the years, other versions, updates and frameworks were developed, popularizing the use of blockchain technology beyond economic transactions.

Blockchain is a technology that allows keeping track of a set of transactions in a decentralized, secure and transparent way, in the form of blocks. Tequi et al. (2019) established an analogy between blockchain technology and a train, in which each 'wagon' (containing data on transactions) is attached to the train guaranteeing the solidity of the whole chain. At the same time, only authorized individuals could add a block and have access to all the stored information, making blockchain a trustworthy and reliable option.

The last two decades have profoundly reinforced the importance of blockchain technology. Around 2007, with the economic crisis that affected the United States and the world, consumers and citizens looked for an alternative to the classical operation, organization and functioning of the financial institutions (Eslamian and Eslamian, 2022). This circumstance motivated the renowned work of Nakamoto with cryptocurrencies (Bitcoin, 2023), among other contributions. Already in 2000, Berners-Lee, who was the pioneer of the World Wide Web, seems to have predicted that, in the future, there would be a technology capable of analyzing the information available on the web and inter-related data, as well as transactions involving individuals and computer devices (Berners-Lee, 2000, cit. in Sfetcu, 2019). The underlying idea was, in fact, to create a true web community capable of extracting the enormous potential of the internet. This system would ensure security and accuracy, apart from improving efficiency in the contacts between citizens and the various sectors of activity – the economic market, the administration and other key services or public goods (Berners-Lee, 2000, cit. in Sfetcu, 2019).

Several other authors have innovated on theories, computations and methods in order to allow what currently integrate the system of blockchain technology. This chapter provides an historical perspective on blockchain technology, from the 80s till today, emphasizing, chronologically, influential conceptual accounts since Merkle's theory on cryptography until Nakamoto's contribution with Bitcoin, almost thirty years later. Sections 3.2 and 3.3, consisting of the most extensive, encompass the mentioned perspectives thoroughly described, whereas other relevant developments on this field led by Haber and Stornetta (1991, 1997), Back (Hashcash.org, n.d.), Konst (2000), Szabo (1994, 1997, 2005) and Buterin with Ethereum (2023a) are concisely explained. Sections 3.4

DOI: 10.1201/9781003599593-5

and 3.5 focuses on the role that blockchain systems have assumed in contemporary societies and a range of scopes where this technology can be applied.

3.2 MERKLE'S THEORY ON CRYPTOSYSTEMS AND THE BACKGROUND OF BLOCKCHAIN TECHNOLOGY

Blockchain technology has a recent history in what concerns the use of cryptocurrency software and the applications of blockchain technology in diverse areas – financial market, governments, and smart contracts, among others. However, the conceptual frameworks behind this technology could be dated from several decades ago, when scientists started to conceive theories using cryptographic systems and the core compounds of blockchain were developed (Figure 3.1). These methods continue to be applied in contemporary blockchain programs, as will be further detailed in this chapter. Although they represent crucial contributions in this field, a vast proportion of the papers were not published in scientific journals, but rather shared within peer networks or available on cryptography or blockchain-related websites.

One of the most important theoretical approaches on cryptosystems took place around the beginning of the 80s – Merkle's cryptography for public key account (Merkle, 1980). At this time, key distribution verification and digital signing with timestamp were already being used, however with limitations concerning breaches in security and malicious intrusions in the network. The author considered the main weaknesses of these processes proposing a series of strategies to improve security, based on the available technology, resources and the structure where cryptosystems operate. According to Merkle (1980), communications using public key cryptosystems between two parties enabled keys to be deciphered by those trying to invade the network, eventually posing risks to the introduction of false contracts and signatures. In this communication, a key distribution center was thus required to share a key with traditional encryption, among both network users. In case the central keys were corrupted, the network functioning could not be operated normally.

For this reason, relying on more than one center would enhance security, as the possible loss of one center key would not compromise the remaining centers. An additional measure was related to encryption of centralized master keys, which should be safely stored with backup options to reinstate the keys, whenever deemed necessary. The author suggested ways of public key distribution, in a non-complex form, which had the drawback of being reachable by unwanted network users, beyond the two interested parties. The keys would be randomly generated by computer software and the exchange of information encrypted through traditional systems. As soon as the communication would end, the key would be eliminated, even if the users resume the conversation afterwards, for which the same procedure would have to start anew (Merkle, 1980).

In this approach, no interference could be made prior to the communication between the two parties or after the conversation was finished, since the keys would be self-destroyed. Whereas this measure was itself a safety guarantee, it also brought a wider exposure of keys to external users, eventually connected to the network, who in the worst-case scenario could actively introduce a key in the message sent between the two interested parties. Other users of the network could verify public key distribution when keys were shared in public networks, following an agreement involving the two parties. These keys were then applied in the encryption of subsequent communications between them (Merkle, 1980).

At this point, both parties would hold the same authenticated key, only known by them. To avoid losing the keys, a physical backup of these private keys could be kept. There was a disadvantage, however, regarding the modification of keys added to the files shared publicly. To address this limitation, public key distribution was able to be performed with certificates and 'tree authentication' (Merkle, 1980). The former followed the described process of authentication, although requiring a 'central authority' (CA) to verify and sign the keys added to the public network, thus producing the

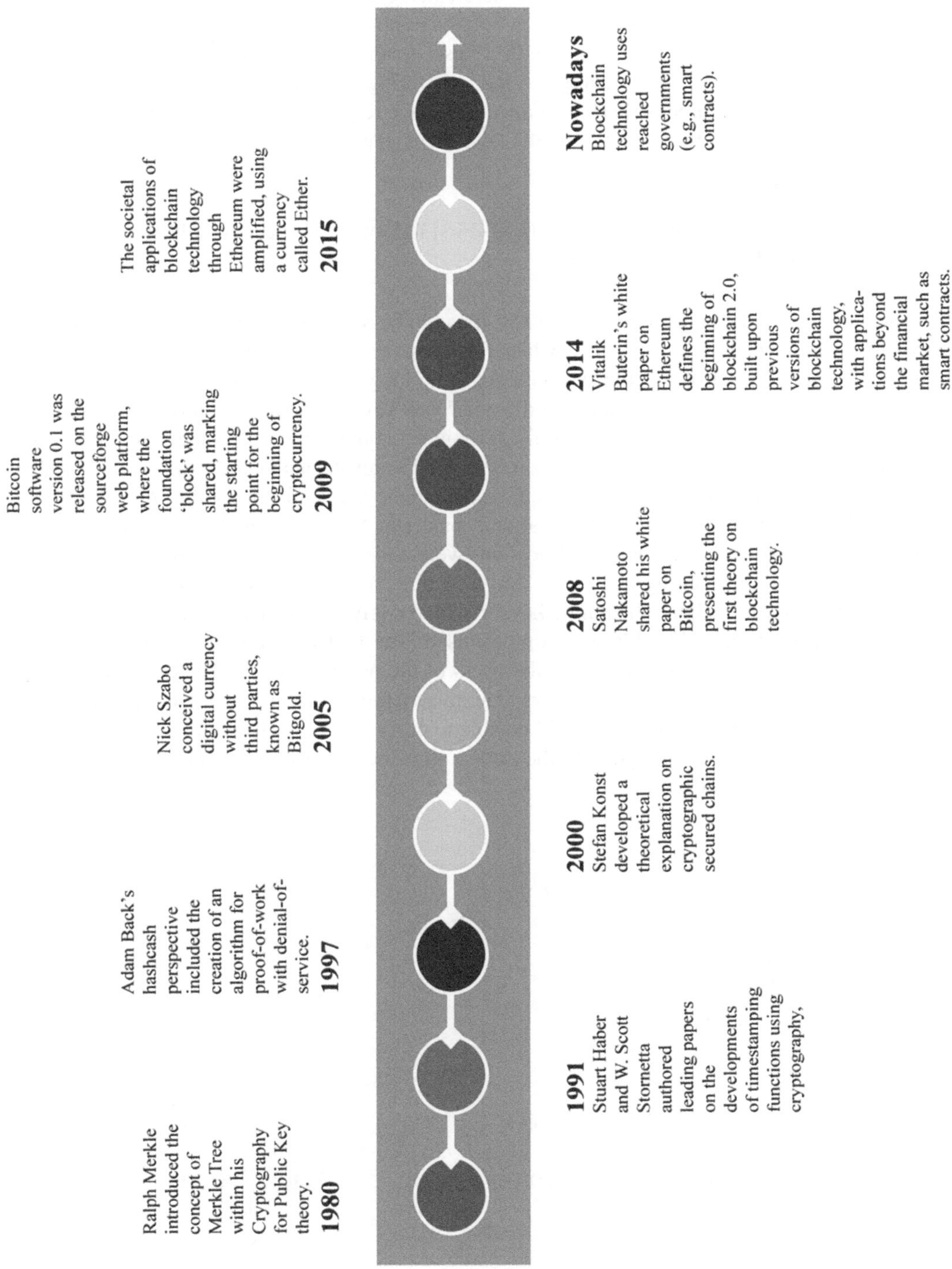

FIGURE 3.1 Chronological timeline of blockchain technology.

certificates. When the two parties checked the certificates from one another, both could be sure that they had the correct keys, avoiding the possibility of receiving corrupted enciphering keys (Merkle, 1980).

Even if with this improved security method, the created certificates were falsified, a model still used nowadays and known as the Merkle Tree would be applied. The 'tree authentication' enabled the verification of public keys components with a hash system for the complete public file. This procedure is, particularly, relevant for using a system of storing information keeping the foundation hash without the remaining blocks' hash to reduce the space requirements. In the author's paper, a conversation between A and B, where the public keys were shared, would define:

> "Y = Public File = $Y_1, Y_2, \ldots Y_n$", while
> "H (Public File) = H (Y)", being "H (Y) = F (H (First half of Y), H (Second half of Y)" and F unidirectional (Merkle, 1980, p. 126).

For the user A to verify B's public enciphering key, it would be necessary to find the Y_B, which could be known by the initial part of the public file, when divided in two. Additionally, the user A would have to figure out the second part of this file (H) with a size no longer than 100 bits. Dividing the public entries allowed users to obtain information on the other user key relying in half of the data shared in the public file. In this way, if the whole public file is divided in quarters, the calculation to identify Y_B remains the same. Considering that we compute this method again and again, the user A would be able to determine Y_B public key by verifying r and log2 n data in the 'authentication path' (Merkle, 1980). The Figure 3.2 highlights this way of authentication considering Y_5, preserving the original content of the author's graphic explanation, where modifications were only performed in the layout.

The main threats to the security of this system were three and could eventually occur when the users A or B inaccurately decipher the public keys; r was wrongly verified by either of the two users; or if the information in the public file was compromised. The first vulnerability described was, in fact, the most relevant, depending solely on the users of the network, while the remaining threats could be overcome because with an incorrect r value, this method would not work and no other user attempting to interfere with the process would be able to proceed. Similarly, data shared in the public file was unchangeable and accessible to the public, thus the integrity of the public file could be easily assessed (Merkle, 1980).

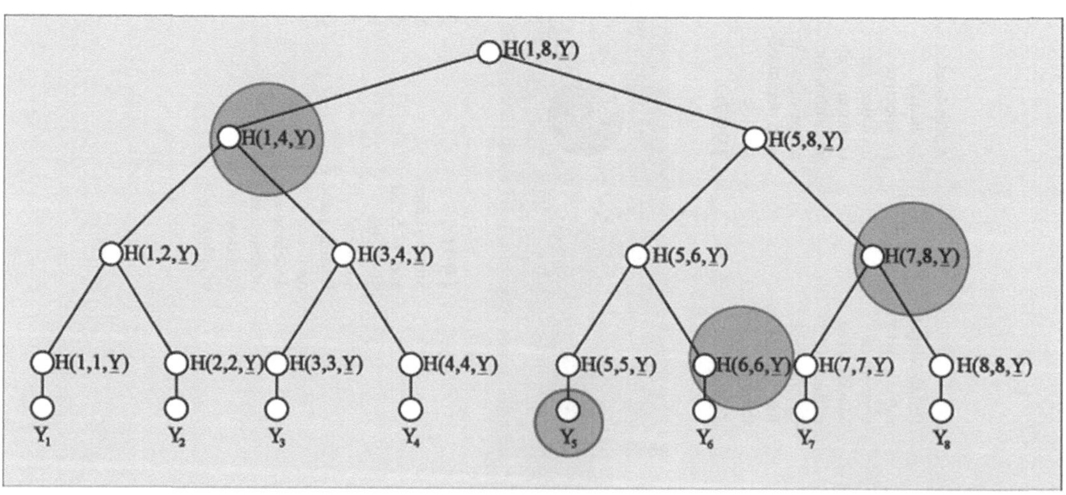

FIGURE 3.2 Merkle's tree authentication method for public file. Adapted from Merkle (1980) keeping the original content with a slightly different layout.

Users of a specific network were also allowed to check their entries to confirm its accuracy. Digital signatures were proposed in Merkle's model not necessarily using key cryptosystems. A signature could be confirmed by computations, such as "m = E_A (D_A (m))" (Merkle, 1980, p. 128), where the message (m) includes data that could only be sent by the user A, including enciphering and deciphering keys. This information comprised sequential bits, which could be interpreted correctly in a unique form, eliminating the possibility of accepting signatures using wrong messages with random bits. Apart from the user A who was responsible to sign the message, and the user B who had to verify this signature, digital signatures required a system of external validation.

The traditional way was based on the already detailed concept of CA, charging this trusted-party with the authentication and mediation between the users A and B. The disadvantages of this method were also described earlier. Alternatively, Merkle explored the 'basic digital signature protocol', which was a similar procedure to the computation "m = E_A (D_A (m))" (Merkle, 1980, p. 128), where the user B or a third party verified the digital signature. The weaknesses of this method with recommended options to overcome them were addressed earlier in this chapter, encompassing the possibility of inappropriate manipulation of the public file and the public availability of deciphering key A.

Timestamp was contemplated in Merkle's theory as a step within the authentication after digital keys are signed, establishing a time point and accounting for the possible compromise of A deciphering key. 'Timekeepers' were accountable for timestamp data, which was further confirmed by a CA. In case there were notifications of deciphering keys missing, this incident would be assessed by a CA. In practical terms, when the user A would sign in with A deciphering key and this information would reach the user B, the latter would be the 'timekeeper' of the message and would have to then timestamp it, requesting also authentication from a CA. If the loss of A deciphering key would occur, B could not approve the signature (Merkle, 1980).

The requirements to validate a digital signature included the use of A enciphering key, shared publicly, along with the timestamp of data previous to the possible incidence that A deciphering key was missing. In this way, when all users involved in the network followed the recommended procedure, the authentication process entailed the safety that the main vulnerabilities of other methods could have been properly addressed. In this procedure, few criticisms could be drawn, particularly in comparison with the 'basic digital signature protocol'. As the user B would perform timestamp and verification assessments of data, simultaneously, the requirement of a network supposes higher costs, while enabling breaches in the reliability of the procedure (Merkle, 1980).

Another way to carry out this model used an offline operation, allowing for the validation to take place in the subsequent days following a transaction date. While this strategy would save money and enhance reliability, it did not account for the messages that could have been lost, preventing the user B to be aware of eventual notifications from A, such as the missing of A deciphering key. The author added a witness of the digital signature, considering situations of transactions involving a relevant amount of money or value. The role of the witness was to physically testify that the user A had real intentions of proceeding with the transaction, confirming the signature of the message. The selection of a witness would be consensual between users A and B, although this extra step on security introduced the inconvenience of needing a witness available to physically testify the transaction (Merkle, 1980).

Until this point, the majority of the mentioned methods were referred as to digital signature procedures where two parties were potentially in dispute. However, Merkle (1980) also conceived a digital signature system whose network users would belong to the same party, either by representing various of its elements or one element conducting validity-checks at diverse time-points. This situation could be applied when a given software was operated by several users with specific nodes allocated for each person. In this context, authentication procedures could avoid the misuse of software components or validate the adequate transmission of different updates within the network. Regarding operating systems, it would be feasible to use the 'privilege mode' comprising of previously signed codes by the programmer or supervisors, in order to prevent codes not included in this mode from being executed.

Several other scientists contributed to explore the mechanisms underlying blockchain operations and functioning. The team of Haber and Stornetta (Bayer et al., 1993, Haber and Stornetta, 1991, 1997), for example, detailed thorough procedures of timestamping, inspiring subsequent theoretical frameworks and practical applications, such as Nakamoto's Bitcoin technology account (Bitcoin, 2023). Since the early 90s, these authors (Haber and Stornetta, 1991, 1997) were concerned with the technological evolution that would lead to an era where all documents would be stored in a digital format. The issues raised with physical records were already challenging, as the authors had contact with a case where a scientific publication of a renowned researcher was modified. In this situation, a deeper analysis had, fortunately, demonstrated that the content of the original book had been changed, since the ink found in the forged work was not the same as the one used in the remaining documents written by the author.

However, when digital records are kept in files, these documents are highly prone to manipulation, thus requiring a system to guarantee the integrity of these records (Haber and Stornetta, 1991, 1997). The researchers suggested the attachment of a hash to each document, cryptographically. Irrespective of the size of the record, when a single bit of the original record would be altered, the hash would unequivocally be modified. Moreover, this team worked with timestamp operations applied to the hash, instead of using the whole document, and then proceeding with the validation steps. Haber and Stornetta (1991, 1997) further detailed 'tree authentication' methods to minimize disk space usage and computer resources, as well as alternatives to increase the length of cryptographic functions, such as signatures (Bayer et al., 1993). In this way, a timestamped signature would serve as the proof for its integrity, even if later on the documented signature would be compromised or corrupted (Bayer et al., 1993).

Similarly, Back provided a crucial development on this scope, proposing a cryptographic method to prevent email spam and denial-of-service attempts. Although initially published in 1997, the author released a more comprehensive version of hashcash functioning in 2002 – *Hashcash – A denial of service countermeasure* (Hashcash.org, n.d.). This paper defined a proof-of-work (PoW) algorithm requiring an insignificant power cost, expended through cryptographic challenges, to add a hash stamp to email headers, enabling the determination of the computation needed to generate emails. This strategy, while serving as the PoW, would, in turn, minimize the production of spam emails. The proof could be reliably confirmed by checking the time spent by the email sender and the quantity of central processing unit (CPU) power used in this procedure, which for spammers, would be unusually common to verify a similar amount of computer resources to honest senders.

Whereas the average user expenses in sending an email would be minimal, for a person trying to send thousands or millions of emails, the computer functioning needed would refrain them from these attacks. In Back's algorithm function, the sender had to guess a random number starting with a predefined amount of zeros. Once the computer had deciphered a sequence with those variables, a hash stamp would be embedded to the email as PoW and sent to the recipients, who could verify the header, deciding whether to accept or reject the email as genuine. The hashcash function was intended to calculate partial hash collisions, considered at a pre-established number of zeros, at the beginning of the random number, where k represented the number of bits in the sequence 0^k, exponentially increasing the difficulty of the challenge the higher k would be. Hashcash used the hash SHA-1, which generally needed 20 zeros out of 160 hash bits (Hashcash.org, n.d.).

Meanwhile, Konst worked on a method for enhancing security in digital chains, published in an academic dissertation in 2000 (Konst, 2000). The research question being explored regarded the way authenticity could be preserved in digital chains where log files could be expanded. Konst based his theory on cryptographic models, conceiving a specific cryptographic pattern of chain structures. If the vertices and edges of a chain would be altered by intruders, these modifications would be detectable. Even though chains could be enlarged in vertices and edges, the integrity and original sequence of a log file were able to be assessed for potential modifications. Konst's pattern was built with symmetric and public key operations by means of cryptosystems, exemplifying diverse possible presentations for cryptographic chain structures when a log file is added (Konst, 2000).

Preceding Bitcoin, Szabo invented Bitgold – a digital currency with no interference of third-parties, sharing several elements of blockchain models. Although the author published this work in 2005, it was long before this date that he started developing these concepts (Szabo, 2005). Szabo's approach comprised timestamp procedures to prevent double-spending, across different points of a network, along with the definition of value for Bitgold, measured in computational expenses, by a mathematical calculation expressed in bit units. Regardless of being an influential theoretical account, Bitgold was not applied for financial use. Another concomitant contribution of this author included the idea of linking cryptographic systems with legal forms and electronic commerce, as the example of smart contracts (Szabo, 1994, 1997).

3.3 ORIGINS OF BITCOIN AND NAKAMOTO'S BITCOIN FRAMEWORK

Conversely to Bitgold cryptocurrency, in 2009, Bitcoin emerged in the real economic market, using what is known today as Blockchain systems. The global financial crisis that took place at that time with significant losses and a decrease in citizens' trust in banks and governments prompted the development of alternative pathways to centralized banking. Facing this panorama, the creator of Bitcoin's theory – who concealed their identity, remaining anonymous with the pseudonym Satoshi Nakamoto – seems to have intended that money, rather than being controlled by the banks, could actually be managed by individuals who owned it –a movement known as decentralized banking. The core purpose was embedded in the ideology of financial freedom for the society and individuals instead of being regulated by governments and large corporations, thus avoiding the profits that banks were receiving in money exchange (e.g., transfers with different coins) and transaction fees.

In the white paper written by Nakamoto, titled *Bitcoin: A peer-to-peer electronic cash system*, retrieved from Bitcoin (2023), the author detailed the basis and operations underlying his theoretical approach on cryptocurrency systems, including blockchain technology. According to the information available and corroborated on the website Bitcoin.com (2023), Nakamoto's white paper was first shared on 31st October 2008 on a cryptography mail list. This document explained the idea of an electronic system to perform transactions, using a peer group network, where a series of nodes would validate all transactions, hence improving the security of the procedure against attackers. The previously covered mechanisms of timestamp, chain of hashes, and PoW were incorporated in Nakamoto's model, elaborating on PoW measures to prevent data from being altered.

As opposed to Szabo, Nakamoto conceived that the value of Bitcoin depended on the market price, which was independent from the computational resources used. The theory developed would also solve the need of trusted third parties (e.g., financial institutions) to avoid simultaneous transactions, reducing the costs inherent to these procedures. As such, electronic payments, instead of relying on trusted entities, would be secured by unchangeable records capable of confirming these operations. The issues of double-spending would, in turn, be addressed with a sequential chain of hashes encompassing PoW verification within a network of timestamp transactions. The paper included ten main components, from which Nakamoto developed the different parts of their approach on digital transactions, such as: 'transactions', 'timestamp server', 'PoW', 'network', 'incentive', 'reclaiming disk space', 'simplified payment verification', 'combining and splitting value', 'privacy', and 'calculations' (Bitcoin, 2023).

For Nakamoto (Bitcoin, 2023), the mechanism of transactions was operated through diverse steps, involving the need to verify and sign, by the owners, with a hash code linked by blocks – nowadays referred to as blockchain. Briefly explained, this concept was conceived as defining a link between each block, which was made from the first 'block' created by the system, to which the following agglomerates of data were connected, subsequently one by one. These blocks were coded with hashes, individually and collectively. Regardless of attributing one unique hash to the foundation 'block', the next 'block' would have a distinctive hash that was conceived upon the previous, allowing the inter-connection of 'blocks' within this digital network of data. Financial transactions were, therefore, available without the involvement of third-party approval, which was considered less expensive and more feasible than traditional alternatives.

Nakamoto's model comprised a signature that enabled the coin to be transferred from one owner to the following, containing a hash and a 'public key'. In this chain, double-spending was complex to be assessed, and past owners could have signed previous transactions without the possibility of the payee to confirm these spendings. The creation of a reliable chain of transactions, in a chronological path, would be necessary by sharing publicly the payments among network users who would actively provide a confirmation. To overcome the requirement of other entities reviewing the transaction for double-spending, the author highlighted the efficiency of adding a timestamp server. In this operation, all transactions were proven, creating a chain and a PoW method where the different nodes could verify the validity of the exchange, thus guaranteeing a safe operationalization (Figure 3.3).

Each hash would be timestamped and would be made available for all network users, providing reliable records of the information involving the transaction at each time-point. The foundation of hashes would follow a time sequence where the most recent hash would be linked to the older, previously introduced to that chain, continuing to grow inherently connected to the earlier hashes. This procedure is similar to the methods proposed by Massias and colleagues (Massias et al., 1999), and Haber and Stornetta's team (Bayer et al., 1993; Haber and Stornetta, 1991, 1997). Additionally, Nakamoto built upon Back's hashcash algorithm (Hashcash.org, n.d.) to enhance robustness in PoW. In the context of Bitcoin currency, the number scanned was the hash SHA-256 and could be confirmed calculating one hash. The Bitcoin method for PoW added on Back's hashcash algorithm an extra security measure using a peer-to-peer network where all CPUs could look for the PoW (Bitcoin, 2023).

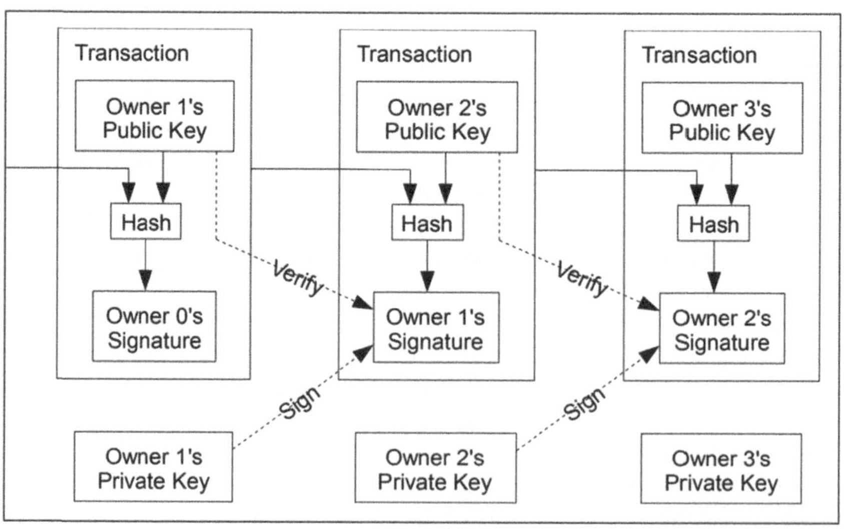

FIGURE 3.3 Transactions within Nakamoto's peer-to-peer electronic cash system. Reprinted from Bitcoin (2023).

The timestamp comprised a 'nonce' that was incorporated to the block, while network users attempted to discover the value of the hash with the needed zero bits. The first node that solved the function would be able to send it to the remaining network users, who could verify the value, adding the block to the chain with the timestamp and the transactions. As the PoW relied on the computer power used, if an intruder attempted to corrupt the system, a block could only be modified by altering all the blocks that were added afterwards. This type of attack would be less likely to happen considering the amount of work that would have to be performed again. The difficulty would increment exponentially with higher number of blocks to be changed. In this method, honest users of the network would be rewarded with a faster pace of increasing the chain, when compared to other users trying to outsmart the system (Bitcoin, 2023).

The author further explored the way in which the network of nodes would work, including how transactions were broadcasted to all points, compiled into blocks, and the search for a strong PoW for the block. Subsequently, the procedure required the 'block' to be validated prior to the connection with the rest of the chain. The initial transaction in a block would create a coin, which would belong to the founder of the block. With this incentive, the remaining users of the network would be motivated to expand the distribution of this currency. Incentives were also added to transactions, in the form of fees, based on the difference between the output and the input of the transaction. Different incentives for individuals integrate the peer-to-peer network, including the idea of "mining" new coins, or the encouragement of users to be honest to the network instead of behaving in a way as to corrupt it, considering that the benefits of mining would be higher than the manipulation of the system. Nodes usually choose to extend the largest chain, assuming this one to be the most valid option (Bitcoin, 2023).

Another aspect of this electronic transaction framework refers to the blockchain usage of insignificant disk space using the theory of Merkle Tree (Merkle, 1980), where transactions are eventually compacted together, reducing the size of each block to 80 bytes. Nakamoto used SHA-256 hashes in an algorithm that would allow to control for the creation of six blocks in an hour, corresponding to an approximate disk space of 4.2 Mb per year (Bitcoin, 2023). Regarding the authentication of payment, the author dismantled this step, proposing an easier method of verification without the need of a thorough review of the whole chain. This process would not be compromised, providing the network nodes had a fair conduct, however the odds of suffering external attacks could be incremented. In this context, a potential threat may be related to fake transactions that would eventually be accepted by the network, if the system was subjugated to malicious nodes.

The paper also defined the way the value of transactions could be divided or combined from its parcels during the exchange between parties. Due to the importance of privacy, Nakamoto explained how the details related to the owners could be kept anonymous, while the transaction was made available to all nodes of the network, in order to be reviewed and validated. In addition, the author addressed the obstacles faced by attackers who might find it nearly impossible to bypass the number of blocks needed to affect a chain, as the possibilities were significantly reduced the larger the number of blocks in the chain (Bitcoin, 2023).

Overall, the theoretical development presented by Nakamoto on Bitcoin transactions served as the basis for other cryptocurrencies frameworks and, of most importance, for the creation of Bitcoin software. Nakamoto seems to have worked with other collaborators improving Bitcoin software until 2009, when Bitcoin software version 1.0 was made available to all internet users on the sourceforge website (Satoshi Nakamoto Institute, 2009). This was the first time the genesis block was released, giving place to the introduction of other blocks and the growth of cryptocurrencies.

3.4 THE RISE OF CRYPTOCURRENCIES AND EVOLUTION OF BLOCKCHAIN APPLICATIONS

To understand the importance of cryptocurrency today, it is worth mentioning that, in the early stages, Bitcoin value was seen as so irrelevant that it was given in exchange for pizza. In fact, this was the first commercial use of this currency, in 2010, when an individual purchased two pizzas for 10,000 coins. Compared to recent years, the acquisition of these goods, in 2023, would be worth more than 250 million dollars (Hankin, 2023).

Following Nakamoto's original version of Bitcoin software, other versions were released with updates on codes and functions amplifying its use. In the original software, there were already mining operations and PoW computations with rewards offered to the users conducting these authentications. With Blockchain 2.0, the applications of cryptocurrencies were innovated, incorporating smart contracts operations. Ethereum was the precursor of this functionality in blockchain frameworks, being positioned among the most impactful uses of blockchain technology, along with Bitcoin. The name refers to a currency – Ether, which was the second big cryptocurrency that was materialized in the market and is also applied to smart contracts. The initial creator was Buterin, in 2013, with further developments until 2015 (Ethereum, 2023a).

However, digital smart contracts have existed for decades, with the aim of establishing an agreement of conditions and terms to process transactions and other protocols, with no third-parties (Szabo, 1994). The risks underlying this technology pertain to corruption of the system and ensuring confidentiality. In contemporary uses of smart contracts, efforts have been performed to improve these vulnerabilities, such as within Ethereum's technology (Ethereum, 2023c). Smart contracts usually claim to avoid the operation of unexpected functions or codes that are not built into the software (Ethereum, 2023c). Even though this feature brings higher safety for its users than other methods, intruders who tend to have a command of the software functions often access and attempt to manipulate the content and functioning. The same applies to the information recorded that, although it might not be modified once stored, the records were changeable prior to that step. Another weakness is evidenced by the extent of the data accuracy, which is limited to the context where it was shared.

Having considered the flaws in the system, with a smart contract, the time required for the operations is considerably minimal, while the protocols and functions are comprehensive. This technology is also restrictive in a way that it will not carry out a certain function, if the agreed conditions are not satisfied, thus not allowing any alterations that could occur in normal circumstances, usually solved when humans assess and perform these procedures. Being a public blockchain, Ethereum has smart contracts–related functions accessible for scrutiny, where privacy is enhanced using cryptographic codes for identity, instead of recognizable data linked to transactions and other operations. The applications of blockchain technology are expected to increase in the foreseeable future, encompassing, currently, currencies, data storage, tokens, and decentralized finance, among others (Ethereum, 2023a). Pragmatic uses of cryptocurrencies and their impact in our societies are highlighted in Table 3.1, unraveling some of the phenomena originated by the use of these coins in the financial market.

In a more recent update, Blockchain 3.0 has other uses for companies intending to exploit domains beyond those linked to financial gains or operations. This is also a decentralized functionality, where industries can take advantage of a system that ensures them with security and low-cost alternatives to get the results they pre-conceived. Software such as Avalanche or Solana improved on Bitcoin technology options with a higher number of transactions available (thousands of them), as opposed to Bitcoin, which limited the transactions to seven in the same timeframe (Originstamp, 2023). To comply with increasing demands on the ecological use of cryptocurrencies (Zhang et al., 2023), Ethereum recently updated its protocols using proof-of-stake to replace the traditional PoW function, already envisioned in 2014 by Buterin (Ethereum, 2023a).

TABLE 3.1
Cryptocurrencies Behavior in Real-Life

Since the launch of Bitcoin, countless companies have adopted the blockchain system to offer financial products by means of cryptocurrency.
In spite of providing individuals with autonomy to manage their own assets, the lack of governmental or financial regulations is among the main factors that make these currencies highly volatile to investments that may originate a cascade of trades expanding the economic potential of that coin, as it occurred with Whales. Another example might be considered in this perspective.
Elon Musk, the owner of X (formerly known as Twitter), Tesla, and SpaceX, could have influenced media followers with his tweets (Musk, 2022), considering the changes verified in the economic market of cryptocurrencies. In one of these occurrences, the entrepreneur announced on Twitter – not yet owned by themselves at that time, however having millions of followers – that the purchase of Tesla merch could be done with Dogecoin. After this event, a tsunami of small trades was observed with a considerable increment on the price and market cap of Dogecoin, possibly returning in their favor considerable benefits.
The movement of decentralized banking was paramount for the development of blockchain technology, where cryptocurrencies attracted crowds motivated for rapid investment profits, however, the volatility of these currencies can also be a threat to financial losses. The use of social influence leading to an inflated or deflated value for each currency reveals an exogenous permeability that could contribute to a market breach of insecurity in the economic behavior. As cryptocurrencies do not tend to be controlled by any governmental entity, the financial product poses both potential benefits and threats considering the fluctuations in its value within the market.

Token systems within Ethereum (2023a) consist of subcurrencies, all sharing similar characteristics. However, there are also tokens, each one of which with a distinctive representation. This is the case of non-fungible tokens (NFTs), considerably useful in the attribution of a unique token value per product – being it artistry, special collections, or even properties (Ethereum, 2023b). In digital items, for example, it is possible to acquire files (e.g., music records) in a decentralized way, contrary to the common use of platforms, such as Spotify or Apple Music. With NFTs, the products are owned by the owner, thus not being possible to be removed by the application provider, keeping a market value that allows items to be sold. The NFTs are based on smart contract computations of Ethereum blockchain using proof-of-stake norms to ensure the inviolability of the products (Ethereum, 2023b).

3.5 BLOCKCHAIN TECHNOLOGY IN THE SOCIETY: SPECTRUM OF USES

Given the security guaranteed by the blockchain technology, the methods used to trace operations, and its network philosophy have been regarded as a potential tool for several social applications, including the cooperative field. PricewaterhouseCoopers ([PwC], 2023) referred to areas such as healthcare, where data related to clinical conditions and health status could be encrypted and kept confidentially in the network to be accessed by its providers.

Another important use of blockchain technology is in humanitarian activities, considering that the need for funding in social and humanitarian interventions is growing at a pace never seen before. These organizations and sponsors are thus looking for new ways of financially supporting projects and have started using blockchain. Even though this alternative is not yet generalized among the nongovernmental organizations, major humanitarian structures resort to this way of raising funds. UNICEF has opened the precedent with the Game Chaingers action, between February and March, in 2018, within the cause of the Syrian crisis. During this initiative, gamers were invited to offer donations simply by using their computers. Since then, the program 'UNICEF Ventures' began to use systematically blockchain technology to help children and adolescents worldwide by making informative content available for them, and promoting better options and possibilities to grow (UNICEF, 2023).

According to UNICEF (2023), the benefits derived from blockchain may include offering different ways of financing to allocate resources adequately, ensuring improved methods for internal procedures that can be assessed for reliability by its users, while fostering a wider availability of digital products for the whole population in open-source services. In the area of ecology or sustainable development, the applications of blockchain are ingenious, considering that the definition of 'value' or currency can be attributed to residues (e.g., plastic waste). A recycling company in less developed countries may be able to use blockchain in activities that promote climate change and recycling awareness. Each contributor in a participating country who recycles waste might obtain compensation in digital tokens that could be exchanged for other goods (Uzsoki and Guerdat, 2019).

As it was previously detailed, blockchain technology may have many advantages, including the reduction of costs for organizations. However, it could also bring several potential problems. Firstly, it is necessary to acknowledge the fact that individuals who are not familiarized with this method might be discouraged of using these services. Additionally, the legal aspects could be a serious concern in certain territories. Indeed, more and more states have decided to regulate the use of cryptocurrencies by introducing, in some situations, legal restrictions in the market materialization of cryptocurrencies. China, for instance, is among the almost twenty countries that have banned these coins. Regions as Argentina, Colombia, Iran, and Taiwan implemented partial bans, mainly focused on avoiding the exchange of cryptocurrencies. In several cases, even if citizens were able to mine cryptocurrencies or digital goods, the financial institutions would not be allowed to perform transactions with these coins (Techopedia, 2023). Other deterrents refer to the complexity of the technology used, operationalization barriers, and the existence of competing platforms (PwC, 2023).

3.6 CONCLUSIONS

From the origins of theoretical models and software development related to blockchain, a long path has been drawn marked by key achievements and novelties in societal scopes. Blockchain developments have provided its users with easier methods and additional safety measures in data sharing and storage, where information reliability should also be assessed considering the available data and timeframe when it was added to the system. The use of cryptocurrencies, being a central purpose in this field, has revealed encouraging opportunities for investment, even though individuals have to ponder the concomitant constraints related to the unsteady value of these coins.

Apart from being a framework that has been applied foremost in the financial area, blockchain technology can also facilitate bureaucratic procedures, reducing costs and simplifying common tasks in contemporary societies. The versatility and historical progress of blockchain technology have created a paradigm where this system is becoming more widely known and used. Projections suggest that its full potential is far from being reached, highlighting areas where blockchain technology could be expanded and implemented by individuals and societal structures.

ACKNOWLEDGMENTS

All efforts were made to contact the authors of the graphical materials presented. However, either the email addresses were not working, or no reply was received. The images are shown in the original format or properly acknowledged the alterations performed, whenever applicable.

REFERENCES

Bayer, D., Haber, S. and Stornetta, W. S. (1993) Improving the efficiency and reliability of digital time-stamping. In R. Capocelli, A. De Santis & U. Vaccaro (Eds.), *Sequences II* (329–334 pp.). New York: Springer. https://doi.org/10.1007/978-1-4613-9323-8_24

Berners-Lee, T. (2000) *Weaving the Web: The Original Design and Ultimate Destiny of the World Wide Web.* (226 p.). New York USA Harper Collins.

Bitcoin (2023) Bitcoin: A peer-to-peer electronic cash system. Available online: https://bitcoin.org/en/bitcoin-paper

Bitcoin.com (2023) Bitcoin P2P e-cash paper. Available online: https://www.bitcoin.com/satoshi-archive/emails/cryptography/1/

Eslamian, S. and Eslamian, F. (2022) *Disaster risk reduction for resilience: Disaster economic vulnerability and recovery programs*, Cham: Springer Nature Switzerland, 339 p.

Ethereum (2023a) Ethereum whitepaper. Available online: https://ethereum.org/en/whitepaper/

Ethereum (2023b) Non-fungible tokens. Available online: https://ethereum.org/en/nft/

Ethereum (2023c) Smart contracts. Available online: https://ethereum.org/en/smart-contracts/

Haber, S. and Stornetta, W. S. (1991) How to time-stamp a digital document. *Journal of Cryptology*, *3*, 99–111. https://doi.org/10.1007/BF00196791

Haber, S. and Stornetta, W. S. (1997) Secure names for bit-strings. In: *4th ACM Conference on Computer and Communications Security*. (28–35 pp.). New York: Association for Computing Machinery. https://doi.org/10.1145/266420.266430

Hankin, A. (2023) Bitcoin pizza day: Celebrating the $250 million pizza order. Available online: https://www.investopedia.com/news/bitcoin-pizza-day-celebrating-20-million-pizza-order/

Hashcash.org (n.d.) Hashcash - A denial of service counter-measure. Available online: https://www.hashcash.org/papers/hashcash.pdf

Konst, S. (2000) *Secure logfiles on base of cryptographic chained blocks (English).* Braunschweig: Technische Universitat Braunschweig. 10.24355/dbbs.084-201707191306

Massias, H., Avila, X. S. and Quisquater, J.-J. (1999). Design of a secure timestamping service with minimal trust requirements. Available online: https://nakamotoinstitute.org/literature/secure-timestamping-service/

Merkle, R. C. (1980). Protocols for public key cryptosystems. In: *IEEE Symposium on Security and Privacy*, 122–134 pp, Oakland, USA: IEEE. https://doi.org/10.1109/SP.1980.10006

Musk, E. [@elonmusk]. (2022) Tesla merch can be bought with Doge, soon SpaceX merch too [Tweet]. Twitter. Available online: https://twitter.com/elonmusk/status/1530209049261658112

Norman, A. (2007) *Blockchain technology explained: The ultimate beginner's guide about blockchain wallet, mining, Bitcoin, Ethereum, Litecoin, Zcash, Monero, Ripple, Dash, IOTA, and smart contracts*, 26 p, South Carolina: CreateSpace Independent Publishing Platform.

Originstamp (2023) Blockchain 1.0 vs. 2.0 vs. 3.0 – What's the difference? Available online: https://originstamp.com/blog/blockchain-1-vs-2-vs-3-whats-the-difference/

PwC (2023) Making sense of Bitcoin, cryptocurrency and blockchain. Available online: https://www.pwc.com/us/en/industries/financial-services/fintech/bitcoin-blockchain-cryptocurrency.html

Satoshi Nakamoto Institute (2009) Cryptography mailing list. Bitcoin v0.1 released. Available online: https://satoshi.nakamotoinstitute.org/emails/cryptography/16/

Sfetcu, N. (2019) *La philosophie de la technologie Blockchain*. Bucharest: Ontologies, SetThings, Multimedia Publishing. https://doi.org/10.13140/RG.2.2.22996.14724

Szabo, N. (1994) Smart contracts. Available online: https://nakamotoinstitute.org/smart-contracts/

Szabo, N. (1997). The idea of smart contracts. Available online: https://nakamotoinstitute.org/the-idea-of-smart-contracts/

Szabo, N. (2005) Bit gold. Available online: https://nakamotoinstitute.org/bit-gold/

Techopedia (2023) Cryptocurrency bans explained: Which countries have restricted crypto and why? Available online: https://www.techopedia.com/cryptocurrency-bans-explained-which-countries-have-restricted-crypto

Tequi, C., Hiault, F. and Della Chiesa, M. (2019). *Blockchain: Vers de nouvelles chaines de valeur.* (320 p.). Paris, France: Eyrolles.

UNICEF (2023) Blockchain – Exploring blockchain applications to accelerate impact. Available online: https://www.unicef.org/innovation/blockchain

Uzsoki, D. and Guerdat, P. (2019) Ten case studies demonstrating the benefits of impact tokens. In: *Impact tokens: A blockchain-based solution for impact investing*, 13–28 pp., Winnipeg, Canada: International Institute for Sustainable Development.

Zhang, D., Chen, X. H., Lau, C. K. M. and Xu, B. (2023) Implications of cryptocurrency energy usage on climate change. *Technological Forecasting & Social Change*, *187*. https://doi.org/10.1016/j.techfore.2022.122219

4 Balancing Blockchain Sustainability

Analyzing Consensus Mechanisms and Environmental Impact

Tanuj Surve and Amit Kumar Tyagi

4.1 INTRODUCTION

Blockchain technology has emerged as a transformative force, revolutionizing how we conceptualize and perform digital transactions (Mwandosya & Luhanga, 2021). Blockchain is founded on the ideas of decentralization, transparency, and security (Puthal et al., 2018). This chapter presents an overview of blockchain and distributed ledger technology, emphasizing the vital role of consensus algorithms in the operation of blockchain networks. Now the impact of blockchain is not limited to the cryptocurrencies but has also been shifted to the big picture, such as impact of blockchain and consensus algorithm on environment.

4.1.1 Background on Blockchain and Distributed Ledger Technology

Blockchain, which was originally envisioned as the underpinning technology for cryptocurrencies such as Bitcoin, has evolved into a versatile tool with uses in a variety of industries (Morabito, 2017). A blockchain is essentially a decentralized and distributed ledger that records transactions over a network of computers (Sarmah, 2018). Because of this decentralized nature, no single entity has control, fostering trust and transparency.

The overarching concept that covers blockchain is known as distributed ledger technology (DLT). A decentralized database that is shared, copied, and synchronized among several participants is referred to as a DLT (Aviv, 2021). Blockchain is a type of DLT that organizes data into blocks that are linked together using cryptographic hashes, resulting in an immutable and visible chain of transactions (Laroiya et al., 2020).

4.1.2 Importance of Consensus Algorithms in Blockchain Networks

Consensus algorithms are at the heart of blockchain networks, allowing for transaction agreement and validation in a decentralized setting (Yao et al., 2021). A central authority validates and records transactions in traditional centralized systems. However, the absence of a central authority in a decentralized blockchain network needs a mechanism to maintain agreement among participants on the state of the ledger (Pereira et al., 2019).

The importance of consensus algorithms can be paralleled to how a group of people make a collective decision. In the absence of a central decision-maker, consensus algorithms serve as the protocol that directs how participants agree on transaction legality and ledger state. This decentralized consensus protects the blockchain's integrity and security.

 DOI: 10.1201/9781003599593-6

4.1.3 Impact of Consensus Algorithms and Blockchain on Environment

Consensus algorithms are fundamental procedures that guarantee agreement among network nodes on the ledger's state, forming the heart of blockchain functionality. These algorithms govern the consensus among nodes and the validation of transactions. Prevalent consensus algorithms comprise proof of work (PoW), proof of stake (PoS), and delegated proof of stake (DPoS). The environmental impact of blockchain, particularly its carbon footprint, is frequently linked to PoW consensus algorithms because of their energy-intensive characteristics. The resolution of intricate riddles in PofW necessitates considerable computational capacity, leading to large electricity usage and corresponding carbon emissions. Initiatives are in progress to implement more sustainable consensus algorithms, such as PoS and proof of authority (PoA), to reduce the environmental impact while maintaining the integrity of blockchain networks.

4.1.4 Overview of Consensus Algorithms Covered

This chapter digs into several consensus algorithms, each of which is meant to cope with different challenges and cater to specific use cases in blockchain networks. Understanding these algorithms is necessary for understanding the complex landscape of blockchain technology. This chapter includes the following algorithms:

Proof of Work (PoW): PoW, which originated with Bitcoin, requires participants (miners) to solve challenging mathematical puzzles in order to validate transactions and create new blocks. Although it's secure, PoW consumes a lot of energy (Lasla et al., 2022).

Proof of Stake (PoS): Validators are chosen based on the amount of cryptocurrency they own and are willing to "stake" as collateral. This energy-efficient alternative tries to address the environmental issues presented by PoW (Gola & Sedlmeir, 2022).

Delegated Proof of Stake (DPoS): In DPoS, coin holders elect a small number of delegates to validate transactions and generate blocks. This method improves scalability and efficiency (Saad & Radzi, 2020).

Practical Byzantine Fault Tolerance (PBFT): PBFT focuses on reaching consensus in a network with a specific number of malicious or faulty nodes (Meshcheryakov et al., 2021).

Directed Acyclic Graphs (DAGs): Directed Acyclic Graphs (DAGs) propose a new paradigm for reaching agreement. DAGs employ a graph structure where each transaction is linked to numerous prior transactions, generating a directed acyclic graph, rather than traditional blocks and chains (He et al., 2021).

Hybrid Consensus Models: Hybrid consensus models incorporate aspects of multiple consensus algorithms to capitalize on their particular strengths (Aggarwal et al., 2021). These approaches strive for a balance between security, scalability, and decentralization.

As we explore each consensus algorithm in detail, we will uncover the strengths, weaknesses, and real-world applications that make them integral components of the blockchain ecosystem.

4.2 PROOF OF WORK

Proof of Work is a pioneering consensus mechanism in the blockchain realm, gaining prominence with the introduction of Bitcoin (Romano & Schmid, 2021). In this section, we will delve into the mechanics of PoW, investigating its functioning, its important components, and its ramifications for blockchain networks.

4.2.1 How Proof of Work Functions

Proof of Work is, at its core, a mechanism for achieving consensus in a decentralized network by compelling users, known as miners, to solve challenging mathematical puzzles (Kumar et al., 2019). The major goal is to display computational effort, demonstrating that the participant has contributed significant resources, particularly processing power and energy.

A transaction is broadcast to the network to start the process. Miners then compete to solve the transaction's cryptographic puzzle. This puzzle, also known as a hash puzzle, challenges miners to find a certain value (once) that, when paired with transaction data, gives a hash that fits specific criteria. The solution is broadcasted to the network by the first miner who correctly solves the problem, and other players verify it.

The proposed block is added to the blockchain after verification, and the miner is paid with newly minted cryptocurrency and transaction fees. This competitive and resource-intensive procedure secures the network by making it economically impossible for any single party to control the bulk of computing power on the network.

4.2.2 Mining, Hash Puzzles, and Block Difficulty

Mining: Mining is the process by which participants compete to solve cryptographic problems and add new blocks to the blockchain in the context of PoW (Han et al., 2019). Miners execute the necessary computations using specialized hardware known as mining rigs. As more miners join the network, the competition heats up, making solving problems and validating transactions more challenging.

Hash Puzzles: Hash puzzles are mathematical problems that require miners to find a specific input value (nonce) that results in a hash output with specific characteristics, such as a certain number of leading zeros (Mardiansyah & Sari, 2022). The puzzle's difficulty can be adjusted, ensuring that blocks are added at a somewhat steady rate; in the case of Bitcoin, it's every 10 minutes.

Block Difficulty: The block difficulty is a parameter that adjusts the complexity of the cryptographic puzzle based on the total computational power of the network. The difficulty of the problem grows as the network's hash rate increases to maintain a consistent block production time (Dahiya et al., 2022).

4.2.3 Pros and Cons of Proof of Work

<u>Pros</u>

Security: PoW is well known for its security, as the computational effort required to solve puzzles makes it economically unfeasible for malicious actors to manipulate the blockchain.

Decentralization: PoW encourages decentralization by allowing anyone with the required hardware to participate in the consensus process.

<u>Cons</u>

Energy Consumption: One of the most significant criticisms of PoW is its high energy consumption. The process of solving complicated riddles necessitates a significant amount of processing power, raising worries about the environmental impact.

Scalability: PoW has had scalability issues, as increasing participation might cause network congestion and slower transaction processing times.

4.2.4 Examples: Bitcoin and Ethereum (PoS Transition)

Bitcoin: The first cryptocurrency, Bitcoin, uses the PoW consensus algorithm (Bach et al., 2018). PoW's success and endurance over the years have cemented its image as a reliable security technique.

Ethereum (PoS Transition): Ethereum, the second-largest cryptocurrency by market capitalization, began with PoW. However, due to concerns regarding energy consumption and scalability, Ethereum 2.0 is undertaking a shift to Proof of Stake. PoW's environmental effects and scalability challenges are addressed via PoS.

Proof of Work has shaped the blockchain environment by providing a safe and decentralized mechanism for reaching consensus. The limits of PoW are becoming increasingly evident as the market evolves, prompting the investigation of alternative consensus algorithms such as PoS. The ongoing shift of Ethereum to PoS underlines the industry's dedication to sustainability and scalability, highlighting blockchain technology's evolving and adaptable nature.

4.2.5 Consensus Algorithm and Environment

PoW consensus algorithms are frequently linked to their environmental impact, notably the carbon footprint, because of their energy-intensive nature. The study (Ali & Jaber, 2023) acknowledges the ongoing endeavours to implement more environmentally friendly alternatives, such as PoS and PoA, in order to preserve the integrity of the blockchain network and reduce the environmental impact.

4.3 PROOF OF STAKE

Proof of Stake emerged as an appealing alternative to Proof of Work, resolving concerns about energy consumption and scalability while upholding decentralization and security principles.

4.3.1 How Proof of Stake Functions

Unlike PoW, which selects validators based on the amount of bitcoin they are willing to "stake" as collateral, PoS picks validators based on the amount of cryptocurrency they are willing to "stake" as collateral (Saleh, 2021). The bigger the stakes, the more likely it is that you will be chosen to create a block. In a decentralized network, this presents a more energy-efficient and cost-effective method for attaining consensus.

Validator Nodes: Validator nodes replace miners in a PoS system. These nodes are in charge of generating and validating new blocks. Validators are chosen depending on how much cryptocurrency they lock up as collateral, indicating their dedication to the network.

Staking: Staking involves participants locking up a certain amount of cryptocurrency in a wallet to participate in the consensus process. The staked coins act as collateral, aligning validators' interests with the overall health and security of the network.

Slashing: Slashing is a mechanism used to penalize validators for malicious or negligent behavior. If a validator attempts to validate a fraudulent transaction, their staked coins may be forfeited in part or entirely. Slashing acts as a disincentive, maintaining the PoS system's integrity.

Unlocking: Staked coins are usually locked for a period during which they cannot be accessed. Once their participation in the consensus process is complete, the staked coins are returned to the validator as part of the unlocking process.

Governance: PoS often involves a governance structure that allows coin holders to participate in protocol development and parameter decision-making processes. This democratic method strengthens PoS's decentralization.

4.3.2 Pros and Cons of Proof of Stake

Pros

Energy Efficiency: PoS is significantly more energy-efficient than PoW since it doesn't rely on the intense computational power needed for solving complex puzzles.

Decentralization: PoS maintains decentralization by allowing anyone with a stake to participate in the consensus process, promoting a more inclusive network.

Scalability: PoS is frequently more scalable than PoW since validator selection is less resource-intensive, resulting in faster transaction processing times.

Cons

Wealth Concentration: Some critics say that PoS can lead to wealth concentration because members with greater holdings have more power in the network. This concentration has the potential to undermine the decentralization premise.

Security Concerns: Some opponents question PoS's security, particularly in cases where a significant number of coins are held by a single party.

4.3.3 Examples: Ethereum (Beacon Chain), Solana, Cardano

Ethereum (Beacon Chain): With its Ethereum 2.0 upgrade, Ethereum, a pioneer in blockchain technology, has been at the vanguard of the transition from Proof of Work to Proof of Stake (Leonardos et al., 2020). The Beacon Chain is a critical stage in this shift. The Beacon Chain, launched as a distinct PoS blockchain, operates in parallel with the existing Ethereum network, building the groundwork for a more scalable and sustainable ecosystem.

The switch to PoS overcomes the environmental concerns raised by PoW by drastically reducing the energy required for consensus (Platt et al., 2021). This change demonstrates Ethereum's dedication to adaptation and scalability in the ever-changing blockchain world.

Solana: Solana stands out in the blockchain realm for its novel approach to PoS, which incorporates an innovative technique known as Proof of History (PoH). PoH timestamps transactions before adding them to a block, producing a historical record that improves the network's overall efficiency and scalability.

In the Solana ecosystem, validator nodes play an important part in maintaining consensus. With the addition of PoH, Solana became one of the quickest blockchain platforms in terms of performance, with faster transaction confirmation times and increased throughput.

Solana's use of PoS in conjunction with PoH demonstrates a dedication to addressing the scalability issues that are sometimes linked with blockchain networks. The platform's ability to execute a large number of transactions in a short period of time has made it a feasible choice for decentralized applications with high throughput requirements.

Cardano: Cardano uses a PoS consensus mechanism to accomplish decentralization, security, and scalability. It is well known for its rigorous academic approach to blockchain development. Cardano is distinguished by its layered architecture, which separates the settlement layer from the computation layer.

Staking is an important part of the Cardano ecosystem. Participants can delegate their ADA (Cardano's native cryptocurrency) to a pool of their choosing, and the success of the chosen pool has a direct impact on the incentives obtained by the participants. This method fosters decentralization by allowing a broad range of participants to engage in the staking process.

The emphasis on formal verification and peer-reviewed research publications demonstrates Cardano's dedication to research-driven development. This thorough approach aims to maintain the platform's security and durability, positioning Cardano as a significant player in the PoS environment.

These examples demonstrate not just technological advances in attaining consensus, but also the variety of tactics used by various blockchain projects in implementing PoS. As the industry evolves, these platforms help to pave the way for more sustainable, scalable, and secure decentralized networks.

4.4 DELEGATED PROOF OF STAKE

Delegated Proof of Stake is a consensus method that expands on the principles of PoS, introducing a more streamlined and efficient technique for obtaining consensus in blockchain networks.

4.4.1 How Delegated Proof of Stake Functions

By allowing coin holders to vote on a limited number of delegates or representatives who will act as validators, Delegated Proof of Stake adds a democratic aspect to the process of block validation (Calcaterra & Kaal, 2018). DPoS's primary features are as follows:

Validator Nodes: The consensus process in DPoS is carried out by a predetermined number of elected validators, also known as delegates or witnesses. These validators alternate between creating blocks and validating transactions.

Voting and Electing Validators: In a DPoS system, coin holders can vote for their preferred delegates (Li et al., 2023). A participant's ability to vote is usually proportionate to the number of cryptocurrencies they own or are willing to stake. The delegates with the most votes are elected validators, who are in charge of safeguarding and validating blockchain transactions.

Block Production: Unlike regular PoS, which allows all validators to construct a block, DPoS uses a rotational method (Liu et al., 2021). Elected validators generate blocks in a preset order. This approach improves efficiency and minimizes the risk of blockchain forks.

Slashing Mechanism: To dissuade malicious behavior, DPoS systems frequently include a slashing mechanism. If a validator is caught lying or attempting to influence the network, they may face penalties such as having a percentage of their staked coins seized.

4.4.2 Voting and Electing Validators, Governance

Voting and Electing Validators: Voting is an important part of the consensus process in a DPoS system. Coin holders participate in network governance by voting for their chosen validators. The process is dynamic, allowing participants to change their votes based on the performance of a validator or other relevant criteria.

The selection of validators is a continual process, with voting taking place regularly. Validators with the most votes join the active set, assisting in the development and validation of new blocks. This democratic approach improves network decentralization by allowing coin holders to actively alter the composition of the validator set.

Governance: DPoS frequently incorporates a governance structure that allows currency holders to participate in decision-making procedures for protocol updates, parameter tweaks, and other network improvements. Proposals, voting periods, and the execution of agreed changes are common features of DPoS governance.

Participants can make suggestions or vote on existing ones, affecting the blockchain protocol's direction. This participatory model adheres to the ideals of decentralization by allowing the community to have a role in the network's progress.

4.4.3 Pros and Cons of Delegated Proof of Stake

Pros

Efficiency: DPoS is noted for its efficiency, as having a fixed number of elected validators decreases decision-making time and speeds up block generation. When compared to typical PoW or PoS systems, this results in speedier transaction confirmation times.

Scalability: DPoS is extremely scalable because of the fixed number of validators, which reduces the possibility of network congestion. Rotational block production improves scalability even more.

Decentralization (to some degree): While not as decentralized as some other consensus systems, DPoS offers some decentralization by allowing coin holders to vote on validators.

Cons

Centralization Concerns: Critics argue that DPoS can lead to centralization, as the election process may be influenced by the concentration of wealth. Validators with substantial support may wield excessive power in the network.

Vote Buying Risk: There is a risk of vote buying in DPoS, where validators may give incentives to coin holders in return for their votes. This practice has the potential to undermine the system's democratic essence.

Security Concerns: The fixed number of validators in DPoS may pose a security risk if a substantial number of them become compromised or collude. This is a potential flaw that critics point out.

4.4.4 Examples: EOS, TRON, LISK

EOS: EOS, a blockchain platform for decentralized applications (DApps), uses a DPoS consensus method. EOS token holders can vote for block producers, who are in charge of validating transactions and creating new blocks. The platform has gained popularity because of its scalability and capacity to handle a high volume of transactions per second.

Token holders can propose and vote on network enhancements under the EOS governance model, giving them some say over the platform's development. EOS is an example of how DPoS may be used to build a scalable and efficient blockchain infrastructure.

Tron: Tron is another type of blockchain technology that uses DPoS to reach consensus. TRX holders can vote for a group of Super Representatives who validate transactions and generate blocks. Tron's DPoS approach prioritizes speed and scalability, making it appropriate for high-volume transaction applications.

Super Representatives propose and vote on protocol updates as part of Tron's governance process. The platform's goal is to encourage community participation in decision-making processes, which aligns with the DPoS principles.

Lisk: Lisk uses a variant of DPoS known as Delegated Proof of Stake with Sidechains (DPoS+SC). Lisk allows developers to build their sidechains that are linked to the main blockchain. Delegates, who are chosen via DPoS, protect the main chain and validate transactions.

The dynamic fee scheme in Lisk's DPoS architecture allows users to vote for delegates without locking their coins. This adaptability improves user participation and engagement in network governance.

These platforms show how DPoS may be adapted to meet unique needs, with a focus on efficiency, scalability, and community participation. The success of DPoS, like any other consensus mechanism, depends on its implementation and the balance struck between centralization and decentralization.

4.5 BYZANTINE FAULT TOLERANCE

Byzantine fault tolerance (BFT) is an important feature of consensus algorithms that aims to address the issues posed by the Byzantine Generals Problem in distributed systems.

4.5.1 Byzantine Generals Problem

The Byzantine Generals Problem is a classic quandary in fault-tolerant and distributed computer systems. It depicts a situation in which a group of generals, each commanding a section of the Byzantine army, must coordinate an attack or retreat (Dimitrios, 2022). The problem is that certain generals may be traitors, transmitting conflicting commands to produce chaos or misinformation.

Despite the possibility of disloyal generals and the risk of unreliable communication methods, the generals must reach an agreement on whether to fight or retreat. In distributed systems, where nodes must agree on a common state despite the risk of malicious or malfunctioning components, solving this problem is critical.

4.5.2 BFT Algorithms: PBFT, RBFT, SBFT

Practical Byzantine Fault Tolerance (PBFT): Castro and Liskov developed the foundational BFT algorithm PBFT. A consensus is formed in PBFT through a multiphase procedure that includes the following steps:

Request: A client sends a request to the network.

Preparation: The primary node proposes a block and transmits it to the other nodes for approval.

Prepare: Nodes validate the preapproved block and broadcast their agreement to the network.

Commit: Nodes accept the validity of the block, finalizing the consensus.

In the presence of malfunctioning nodes, PBFT provides consistency and safety, but it requires a two-thirds majority of honest nodes to function properly.

Redundant Byzantine Fault Tolerance (RBFT): An enhancement of PBFT, RBFT improves the consensus algorithm's security and fault tolerance (Qin et al., 2022). RBFT adds redundancy to the consensus process by allowing multiple nodes to function as leaders. This redundancy reduces the chance of a single point of failure and improves the overall resilience of the system.

RBFT strengthens the resilience of the consensus process against Byzantine faults by allowing multiple leaders to propose blocks at the same time.

Simplified Byzantine Fault Tolerance (SBFT): SBFT is a simplified variant of BFT that is intended to obtain consensus in a more lightweight manner. While adhering to the fundamental concepts of BFT, SBFT lowers the computational and communication costs associated with classic BFT algorithms (Wang & Nixon, 2023).

SBFT is particularly well-suited to circumstances requiring high efficiency and low latency. Despite its simplicity, SBFT preserves Byzantine fault tolerance, maintaining the consensus process's integrity.

4.5.3 Pros and Cons of BFT Algorithms

Pros

Fault Tolerance: By design, BFT algorithms are resistant to Byzantine faults, allowing the system to establish consensus even in the presence of malevolent or malfunctioning nodes.

Finality: BFT algorithms provide finality, meaning once consensus is reached, it is irreversible. This property is of paramount importance in applications where assurance is essential.

Security: Since BFT algorithms are well suited for situations in which security is a key priority, they are appropriate for financial transactions as well as critical systems.

Cons

Scalability: As the number of nodes increases, several BFT algorithms may encounter scalability issues. The amount of communication overhead and the quantity of messages exchanged can affect performance.

Centralization Tendencies: Depending on how they are implemented, BFT algorithms may exhibit centralization tendencies, as a few trusted nodes frequently play a crucial role in the consensus process.

4.5.4 EXAMPLES: HYPERLEDGER FABRIC, STELLAR

Hyperledger Fabric: Hyperledger Fabric is a permissioned blockchain platform with a modular and scalable design. To ensure that all participating nodes reach a consistent picture of the ledger, it implements a PBFT consensus process.

Nodes in Hyperledger Fabric form a network of endorsing peers, validating peers, and ordering nodes. The ordering service, which employs PBFT, plays an important role in reaching an agreement on the order of transactions. This technique improves the Hyperledger Fabric blockchain's security and stability, making it appropriate for enterprise applications.

Stellar: Stellar is a decentralized platform designed to facilitate fast and low-cost cross-border transactions. The Stellar Consensus Protocol (SCP), a federated Byzantine agreement technique, is used by Stellar. SCP combines features of BFT and federated consensus to achieve network-wide agreement.

Stellar organizes nodes, or validators, into quorums, and a consensus process establishes transaction legitimacy. SCP enables Stellar to preserve decentralization while swiftly attaining agreement, making it an appealing solution for financial institutions and payment networks.

BFT algorithms, from the classic Byzantine Generals Problem to practical implementations such as PBFT, RBFT, and SBFT, provide a solid foundation for achieving consensus in a trustless and distributed context. Real-world examples such as Hyperledger Fabric and Stellar show how BFT can be used in a variety of scenarios.

4.6 DIRECTED ACYCLIC GRAPHS (DAGs)

Directed acyclic graphs (DAGs) represent a unique approach to achieving consensus in decentralized systems. DAGs, as opposed to typical blockchain topologies with linear blocks, use a graph-based model that increases scalability and efficiency (Wang et al., 2020).

4.6.1 How DAGs Work Without Blocks

Directed acyclic graphs operate on a fundamentally different principle compared to traditional blockchains, as they eliminate the concept of blocks and introduce a more dynamic and scalable structure.

Individual Transactions as Nodes: Individual transactions are represented as nodes in a DAG–based system. Each node contains data about a single transaction, such as the sender, receiver, amount, and unique identifier (Yang et al., 2020). Unlike traditional blockchains, which arrange transactions into blocks, nodes in DAGs are not bound by a specified block structure.

Transaction Interlinking: DAGs use a web-like structure in which transactions are directly related based on references. Transactions in a DAG refer to other transactions directly rather than waiting for a block to be created and added to a chain. This results in a web of interconnected transactions that serve as the DAG's foundation.

Tips as Unreferenced Transactions: A key element in DAGs is the concept of "tips." Tips are DAG transactions that have no future transactions that reference them (Pervez et al., 2018). They are the most recent transactions that have not yet been incorporated into the following blocks. When nodes in the network create new transactions, they often use tips as references, allowing for a continuous and developing structure without the linear limits of blockchains.

Timestamps for Temporal Order: While DAGs abandon the traditional block structure, establishing a temporal order of transactions is still crucial. Timestamps are essential in identifying transaction orders in a DAG. Nodes use timestamps to reach a consensus on the chronological sequence of events, allowing for a coherent understanding of the transaction history.

Asynchronous Confirmation Mechanism: DAGs introduce an asynchronous confirmation mechanism, allowing for transactions to be confirmed without waiting for the next block to be created. When a new transaction is issued, it refers to a set of DAG tips. Other network nodes authenticate and verify this transaction by examining its references and ensuring that the referenced transactions are valid. When compared to typical blockchains, this asynchronous confirmation approach helps to lower latency and faster transaction processing.

Consensus Mechanisms in DAGs: In DAGs, consensus is achieved through the approval of the following transactions. To be declared genuine, a new transaction must be approved by a specified number of prior transactions, usually tips. This approval method ensures that the following transactions validate each transaction, resulting in a decentralized and distributed consensus model that eliminates the need for centralized entities such as miners.

4.6.2 Pros and Cons of DAGs

Pros

Scalability: DAGs's intrinsic scalability is one of their most significant features. In contrast to traditional blockchains, which need the entire network to validate a single block, DAGs allow transactions to occur concurrently, boosting throughput.

Reduced Latency: The absence of a block creation cycle in DAGs reduces latency. Since there is no need to wait for the next block to be constructed, transactions can be confirmed more quickly.

Low Transaction Fees: Due to its more efficient structure and the absence of miners, DAG–based systems often incur cheaper transaction fees. This makes them attractive for applications that require frequent and low-cost transactions.

Adaptability to IoT: DAGs are particularly well suited for applications in the Internet of Things (IoT). IOTA, for example, is specifically designed for IoT devices, enabling them to exchange data and value seamlessly.

Cons

Security Concerns: The security of DAGs relies on the assumption that the majority of participants in the network are honest. If an attacker gains control of a major piece of the network, the DAG's integrity may be jeopardized.

Complexity of Consensus: Achieving consensus in a DAG can be more complex than in traditional blockchains. Creating a single legitimate transaction history without the usage of blocks necessitates the implementation of robust consensus mechanisms.

Limited Adoption: While gaining popularity, DAGs are still relatively new compared to traditional blockchains. The lack of acceptance has an impact on the availability of tools, resources, and developer communities.

4.6.3 EXAMPLES: IOTA, NANO

IOTA: IOTA is a pioneering example of DAG technology. IOTA is a DAG–based framework that was designed specifically for the Internet of Things. The Tangle replaces the traditional blockchain, and transactions are confirmed by directly referencing and approving two previous transactions. As a result of its structure, IOTA is well suited for microtransactions and IoT applications.

IOTA's unique consensus mechanism, known as the Coordinator (though the goal is to remove it eventually), assists in achieving network consensus. IOTA's DAG design enables feeless transactions and efficient data transfer, demonstrating the technology's unique potential.

Nano: Formerly known as RaiBlocks, Nano is another cryptocurrency that utilizes a DAG–based structure. Nano's architecture employs a block-lattice structure, where each account has its own blockchain. Transactions are added to the associated account chain, removing the need for global network consensus.

Nano efficiently achieves transaction agreement by utilizing a dPoS consensus mechanism. With its lightweight structure and quick confirmation times, Nano seeks to provide a scalable and efficient alternative for peer-to-peer transactions.

Directed acyclic graphs (DAGs) are a paradigm shift in decentralized systems, providing unparalleled scalability and efficiency. DAGs, by foregoing the standard block-based structure, provide a platform for low-latency, feeless transactions, making them particularly appealing for IoT applications. Real-world uses of DAG technology, such as IOTA and Nano, demonstrate its adaptability and potential in the wide landscape of decentralized technologies.

4.7 HYBRID MODELS IN BLOCKCHAIN CONSENSUS

Hybrid consensus models are a dynamic strategy that combines the strengths of different consensus algorithms to achieve a balance of security, scalability, and decentralization.

4.7.1 Overview of Hybrid Consensus Models

Hybrid consensus models seek to capitalize on the benefits of several consensus algorithms by integrating them in a synergistic manner. The goal is to overcome the constraints of individual algorithms to build a more resilient and flexible system (Pieroni et al., 2020).

4.7.2 Combining PoW and PoS Security Guarantees

Proof of Work: PoW, the original Bitcoin consensus mechanism, entails miners solving challenging mathematical puzzles to validate transactions and create new blocks. The computational effort necessary to complete these activities provides the security guarantee of PoW. The greater a participant's computational capacity, the more secure the network becomes against attacks.

Proof of Stake: PoS selects validators based on the amount of cryptocurrency they own and are prepared to "stake" as collateral. This energy-efficient alternative tries to address the environmental issues raised by PoW. The economic incentive for validators to operate honestly, given they have a financial stake at risk, provides the security guarantee in PoS.

4.7.3 Pros and Cons of Hybrid Consensus Models

Pros

Enhanced Security: Hybrid models use the security guarantees of both algorithms by mixing PoW and PoS. PoW provides a strong foundation against Sybil attacks and computational vulnerabilities, whereas PoS includes economic incentives that align validators' interests with network stability and security.

Scalability: By adding the efficiency of PoS, hybrid models can overcome scalability challenges. While PoW is well known for its security, its resource-intensive nature might provide scalability issues. Integrating PoS provides for a more efficient validation process, which contributes to increased scalability.

Decentralization: Including PoS components in hybrid architectures often enhances decentralization. PoS implements methods that allow stakeholders to engage in decision-making processes, resulting in a more inclusive and decentralized governance structure.

Flexibility and Adaptability: Hybrid models are adaptable to changing situations. If a specific consensus algorithm encounters unexpected obstacles or weaknesses, a hybrid system may be able to change its parameters or switch to a different consensus method without requiring a hard fork.

Cons

Complexity: Putting in place and maintaining hybrid consensus models can be difficult. Coordination of interactions between PoW and PoS components, as well as their seamless integration and avoidance of potential conflicts, necessitate careful design and ongoing development.

Centralization Risks: While hybrid models aim to enhance decentralization, there is a risk of unintentional centralization if the parameters are not well balanced. To avoid centralization, issues such as mining power distribution, stake concentration, and governance systems must be carefully taken into account.

Transition Challenges: If a blockchain network transitions from one consensus mechanism to a hybrid model, challenges may arise during the transition period. To ensure a smooth move without jeopardizing security or causing disruptions, careful preparation and community involvement are required.

Hybrid consensus models, particularly those that combine PoW and PoS, are an important step toward addressing the issues that individual consensus algorithms confront. These approaches provide increased security, scalability, and decentralization by using the security guarantees of both PoW and PoS.

4.7.4 Examples of Hybrid Models

(PoW/PoS) Decred: Decred is a cryptocurrency that uses a hybrid consensus model that combines PoW and PoS components. Miners on the Decred network continue to use the PoW procedure, solving cryptographic challenges to generate new blocks. The PoS component, on the other hand, adds a layer of governance by allowing stakeholders to influence decision-making and validate blocks.

PoS participants in Decred can vote on proposed network improvements, providing for a more decentralized and inclusive decision-making process. This hybrid strategy improves security by retaining PoW's robustness while incorporating PoS's economic incentives and governance features.

Casper (ETH): Casper is a consensus algorithm created by Ethereum as part of its PoW-to-PoS transition. To maintain security throughout the transition period, the Casper protocol employs a hybrid technique known as Casper FFG (Friendly Finality Gadget).

Casper FFG is a hybrid of the standard PoW technique with a PoS overlay. To participate in block validation, validators, also known as "bonded validators," must lock up a specific amount of cryptocurrency as collateral. This collateral acts as an economic incentive for validators to act honestly, thus enhancing network security.

4.8 ENVIRONMENT AND CONSENSUS ALGORITHM WITH BLOCKCHAIN

A consensus algorithm transitions from proof of work to proof of useful work in order to reduce the environmental impact of a blockchain. The concept, which is derived from RECAPTCHA, is to substitute the work performed by the nodes with work that is potentially beneficial. This will address the issue of emissions by opting for a proof of useful work mining process and granting green nodes the ability to perform block production.

The promotion of responsible practices within the blockchain community is also a component of sustainable blockchain. This encompasses the promotion of energy-efficient technologies among developers and consumers, the encouragement of sustainable blockchain solutions through research and development, and the promotion of awareness regarding the environmental implications of blockchain technology.

Sustainable blockchain technology has the potential to improve the security and scalability of blockchain networks, in addition to its environmental and economic potential. The network is more resilient to attacks and the risk of centralization is reduced by energy-efficient consensus mechanisms, such as DPoS and PoS, which require less computational capacity. In addition, these mechanisms facilitate quicker transaction processing and increased throughput, which mitigates some of the scalability challenges encountered by conventional blockchains.

4.9 CONCLUSION

In the preceding sections, we embarked on a journey through the diverse landscape of consensus algorithms that underpin blockchain networks. Each consensus approach provides its own set of advantages and challenges, ranging from the fundamental PoW to innovative approaches such as DAGs and hybrid models. As we conclude our investigation, let us review the important points raised, look into current challenges, envisage future paths, and emphasize the vital significance of consensus in the blockchain ecosystem.

4.9.1 Summary of Different Consensus Algorithms

Our journey began with Bitcoin's robust and widely recognized proof of work (PoW technology. PoW relies on miners solving computationally demanding puzzles to validate transactions and generate new blocks, maintaining the blockchain's security and immutability.

We then looked into proof of stake (PoS), a more energy-efficient alternative that selects validators based on the amount of cryptocurrency they hold and are willing to "stake" as collateral. PoS creates economic incentives for honest participation while also providing scalability benefits.

DAGs (directed acyclic graphs) represented a paradigm change, replacing the traditional block-based structure with a dynamic and scalable architecture. DAG transactions are closely interrelated, allowing for greater scalability and efficiency, making them particularly ideal for applications such as the IoT.

Delegated proof of stake (DPoS) established a governance layer in which coin holders elect a small number of delegates to validate transactions and generate blocks. While incorporating the community in decision-making, DPoS improves scalability and efficiency.

Byzantine fault tolerance (BFT) techniques, such as practical Byzantine fault tolerance (PBFT) and redundant Byzantine fault tolerance (RBFT), aim to achieve consensus in networks with potentially malevolent nodes, hence maintaining system integrity.

As demonstrated in projects like Decred (PoW/PoS) and Ethereum's Casper (PoW/PoS), hybrid models integrate parts of several consensus methods. These hybrids strive for a balance of security, scalability, and decentralization.

4.9.2 Current Challenges and Future Directions

Current Challenges

Scalability: As blockchain technology becomes more widely used, scalability remains a critical issue for many consensus algorithms. The increasing volume of transactions and network size can put certain algorithms to the test.

Energy Consumption: Concerns have been raised about the environmental impact of energy-intensive consensus algorithms such as PoW. The blockchain community faces a significant problem in addressing sustainability concerns while maintaining security.

Interoperability: Achieving interoperability across different blockchain networks and consensus methods is an essential requirement for decentralized system integration (Belchior et al., 2022). Cross-chain communication standards and protocols are still under development.

Security and Governance: Striking a balance between security and decentralization is an ongoing challenge. While enhancing decentralization, the implementation of governance mechanisms must be carefully structured to avoid unintentional centralization risks.

Future Directions

Layer 2 Scaling Solutions: The development and integration of Layer 2 scaling solutions, such as Bitcoin's Lightning Network and Ethereum's state channels, aims to relieve scalability difficulties by processing transactions off-chain and settling them on the main blockchain.

Sustainability Measures: To lower the carbon footprint of blockchain networks, the industry is investigating alternate consensus techniques and sustainability measures. A viable avenue is to transition to more energy-efficient algorithms or to incorporate environmental considerations into consensus design (Hasanzadeh Saray et al., 2022).

Cross-Chain Compatibility: Improving interoperability through standardized protocols and frameworks would allow for easy communication and collaboration among various blockchain networks, enabling a more integrated and diverse ecosystem.

Advanced Security Protocols: Continued research into advanced security protocols and cryptographic techniques will enhance the resilience of consensus algorithms against emerging threats, ensuring the ongoing security of blockchain networks.

4.9.3 Concluding Remarks on the Importance of Consensus

Consensus algorithms are the foundation of blockchain networks, providing the necessary trust, security, and decentralization that characterize the technology's essence. Consensus is important since it defines the integrity of the distributed ledger, transaction immutability, and the entire functionality of the blockchain ecosystem.

Consensus algorithms not only permit agreement on the state of the ledger but also allow blockchain networks to evolve and adapt to suit the different needs of users and applications. They are the key method for establishing trust in decentralized contexts, enabling innovation, collaboration, and the democratization of different industries.

As the blockchain landscape evolves, the search for optimal consensus methods continues. Today's issues are stepping stones toward a future in which blockchain technology is more scalable, sustainable, and inclusive. Consensus algorithms will continue to determine the trajectory of decentralized systems, laying the groundwork for a new era of trust and collaboration. The discovery of consensus mechanisms is a never-ending adventure in the ever-expanding cosmos of blockchain, and the advances made today pave the path for the decentralized landscapes of tomorrow.

REFERENCES

Aggarwal, A., Gaba, S. and Mittal, M. (2021). A Comparative Investigation of Consensus Algorithms in Collaboration with IoT and Blockchain. *Transforming Cybersecurity Solutions using Blockchain*, Springer, pp. 115–140.

Ali, M. and Jaber, T. (2023). Quantifying the Environmental Impact: A Comparative Analysis of Consensus Algorithms in Blockchain for Carbon Footprint Reduction and Mitigating Climate Change. *Qeios*. https://doi.org/10.32388/WB9U29

Aviv, I. (2021). The Distributed Ledger Technology as Development Platform for Distributed Information Systems. In: *Proceedings of the International Conference on Intelligent Vision and Computing (ICIVC 2021), Oman, by Harish Sharma, Vijay Kumar Vyas, Rajesh Kumar Pandey, Mukesh Prasad, Springer*, pp.344–355.

Bach, L.M., Mihaljevic, B. and Zagar, M. (2018). Comparative analysis of blockchain consensus algorithms, *IEEE, Opatija, Croatia*, pp. 1545–1550. https://doi.org/10.23919/MIPRO.2018.8400278

Belchior, R., Vasconcelos, A., Guerreiro, S. and Correia, M. (2022). A Survey on Blockchain Interoperability: Past, Present, and Future Trends. *ACM Computing Surveys*, 54(8), pp. 1–41.

Calcaterra, C. and Kaal, W.A. (2018). Secure Proof of Stake Protocol. *SSRN Electronic Journal*. https://doi.org/10.2139/ssrn.3125827

Dahiya, A., Gupta, B.B., Alhalabi, W. and Ulrichd, K. (2022). A Comprehensive Analysis of Blockchain and Its Applications in Intelligent Systems Based on IoT, Cloud and Social Media. *International Journal of Intelligent Systems*, 37(12). https://doi.org/10.1002/int.23032

Dimitrios, P. (2022). *Distributed Consensus Inference and Blockchain*, Master Thesis, Faculty of Sciences School of Informatics, Aristotle University of Thessaloniki, Greece

Gola, C. and Sedlmeir, J. (2022). Addressing the Sustainability of Distributed Ledger Technology. *SSRN Electronic Journal*. https://doi.org/10.2139/ssrn.4032837

Han, R., Foutris, N. and Kotselidis, C. (2019). Demystifying Crypto-Mining: Analysis and Optimizations of Memory-Hard PoW Algorithms. In: *2019 IEEE International Symposium on Performance Analysis of Systems and Software (ISPASS)*, 1–12.

Hasanzadeh Saray, M., Baubekova, A., Gohari, A., Eslamian, S. S., Klöve, B. and Torabi Haghighi, A. (2022). Optimization of Water-Energy-Food Nexus considering CO2 Emissions from Cropland: A Case Study in Northwest Iran. *Applied Energy*, 307, 118236.

He, J., Wang, G., Zhang, G. and Zhang, J. (2021). Consensus Mechanism Design based on Structured Directed Acyclic Graphs. *Blockchain: Research and Applications*, 2(1). https://doi.org/10.1016/j.bcra.2021.100011

Kumar, G., Saha, R., Rai, M.K., Thomas, R. and Kim, T.-H. (2019). Proof-of-Work Consensus Approach in Blockchain Technology for Cloud and Fog Computing Using Maximization-Factorization Statistics. *IEEE Internet of Things Journal*, 6(4), 6835–6842.

Laroiya, C., Saxena, D. and Komalavalli, C. (2020). Applications of Blockchain Technology. In: *Handbook of Research on Blockchain Technology*. pp. 213–243.

Lasla, N., Al-Sahan, L., Abdallah, M. and Younis, M. (2022). Green-PoW: An energy-efficient blockchain Proof-of-Work consensus algorithm. *Computer Networks*, 214, 109118.

Leonardos, S., Reijsbergen, D. and Piliouras, G. (2020). PREStO: A Systematic Framework for Blockchain Consensus Protocols. *IEEE Transactions on Engineering Management*, 67(4), 1028–1044.

Li, C., Xu, R. and Duan, L. (2023). Liquid Democracy in DPoS Blockchains. In: *BSCI'23: Proceedings of the 5th ACM International Symposium on Blockchain and Secure Critical Infrastructure*, pp. 25–33. https://doi.org/10.1145/3594556.3594606

Liu, Y., Wang, B., Ye, W., Ning, X. and Gu, B. (2021). Global Estimation Method Based on Spatial–Temporal Kalman Filter for DPOS. *IEEE Sensors Journal*, 21(3), 3748–3756. https://doi.org/10.1109/JSEN.2020.3027582

Mardiansyah, V. and Sari, R.F. (2022). SimBlock Simulator Enhancement with Difficulty Level Algorithm Based on Proof-of-Work Consensus for Lightweight Blockchain. *Sensors*, 22(23), 9057. https://doi.org/10.3390/s22239057

Meshcheryakov, Y., Melman, A., Evsutin, O. and Morozov, V. (2021). On Performance of PBFT Blockchain Consensus Algorithm for IoT-Applications With Constrained Devices. *IEEE Access*, 9, 80559–80570. https://doi.org/10.1109/ACCESS.2021.3085405

Morabito, V. (2017). *Business Innovation Through Blockchain*. https://doi.org/10.1007/978-3-319-48478-5

Mwandosya, M.J. and Luhanga, M.L. (2021). Blockchain: A Disruptive and Transformative Technology of the Fourth Industrial Revolution. *Business Management Review*, 23(2), pp. 16–31.

Pereira, J., Tavalaei, M.M. and Ozalp, H. (2019). Blockchain-Based Platforms: Decentralized Infrastructures and Its Boundary Conditions. *Technological Forecasting and Social Change*, 146, pp. 94–102.

Pervez, H., Muneeb, M., Irfan, M.U. and Haq, I.U. (2018). A Comparative Analysis of DAG-Based Blockchain Architectures. In: *2018 12th International Conference on Open Source Systems and Technologies (ICOSST)*, 19–21. https://doi.org/10.1109/ICOSST.2018.8632193

Pieroni, A., Scarpato, N. and Felli, L. (2020). Blockchain and IoT Convergence: A Systematic Survey on Technologies, Protocols and Security. *Applied Sciences*, 10(19). https://doi.org/10.3390/app10196749

Platt, M., Sedlmeir, J., Platt, D., Xu, J., Tasca, P. and Vadgama, N. (2021). The Energy Footprint of Blockchain Consensus Mechanisms Beyond Proof-of-Work. In: *2021 IEEE 21st International Conference on Software Quality, Reliability and Security Companion (QRS-C)*, pp. 1135–1144.

Puthal, D., Malik, N., Mohanty, S.P., Kougianos, E. and Yang, C. (2018). The Blockchain as a Decentralized Security Framework [Future Directions]. *IEEE Consumer Electronics Magazine*, 7(2), pp. 18–21.

Qin, H., Cheng, Y., Ma, X., Li, F. and Abawajy, J. (2022). Weighted Byzantine Fault Tolerance Consensus Algorithm for Enhancing Consortium Blockchain Efficiency and Security. *Journal of King Saud University - Computer and Information Sciences*, 34(10, Part A), pp. 8370–8379.

Romano, D. and Schmid, G. (2021). Beyond Bitcoin: Recent Trends and Perspectives in Distributed Ledger Technology. *Cryptography*, 5(4), 36. https://doi.org/10.3390/cryptography5040036

Saad, S.M.S.K.H. and Radzi, R.Z.R.M. (2020). Comparative Review of the Blockchain Consensus Algorithm Between Proof of Stake (POS) and Delegated Proof of Stake (DPOS). *International Journal of Innovative Computing*, 10(2). https://doi.org/10.11113/ijic.v10n2.272

Saleh, F. (2021). Blockchain Without Waste: Proof-of-Stake. *The Review of Financial Studies*, 34(4), pp. 1156–1190. https://doi.org/10.1093/rfs/hhaa075

Sarmah, S.S. (2018). Understanding Blockchain Technology. *Computer Science and Engineering*, 8(2), pp. 23–29.

Wang, G. and Nixon, M. (2023). X-assisted BFT Consensus Protocols. In *International Conference on Blockchain*, Honolulu, HI, USA, September 23–26, Proceedings, pp. 54–71.

Wang, Q., Yu, J., Chen, S. and Xiang, Y. (2020). Diving into DAG-Based Blockchain Systems. *Computer Science*. https://doi.org/10.48550/arXiv.2012.06128

Yang, W., Dai, X., Xiao, J. and Jin, H. (2020). LDV: A Lightweight DAG-Based Blockchain for Vehicular Social Networks. *IEEE Transactions on Vehicular Technology*, 69(6), 5749–5759. https://doi.org/10.1109/TVT.2020.2963906

Yao, W., Ye, J., Murimi, R. and Wang, G. (2021). A Survey on Consortium Blockchain Consensus Mechanisms. *IEEE Access*, 11. https://doi.org/10.48550/arXiv.2102.12058

Part III

Blockchain Technology and Smart Water Systems

5 Blockchain and IoT for Smart Water Management

Addressing Scarcity and Ensuring Sustainability

C. Beulah Christalin Latha, Sujni Paul, S.V. Evangeline Sonia, and Saeid Eslamian

5.1 INTRODUCTION

Though water covers 70% of our planet, freshwater that is used by us for drinking, cooking and various other purposes is incredibly scarce. Only 3% of the available water is freshwater. Two-thirds of the water is unavailable for use because it is hidden in frozen glaciers. About 1.1 billion people around the globe suffer from unavailability of water and 2.7 billion people are affected by scarcity of water at least for a month annually. Climate change, global warming, rise in global population, pollution, overwhelmed water systems, and inadequate management of water resources are considered to be primary factors contributing to such issues (Eslamian, 2014). Scarcity of water may lead to failure of sewage systems and surges the threat of contracting water-borne diseases such as cholera, typhoid and so on. Smart water management systems based on IoT, big data and the AI technologies can provide a means to measure, monitor and control water distribution networks and ensure that the quality of water is maintained. Controlling water wastage, regulation of water consumption and management of water quality can optimize the usage of water resources.

5.1.1 Smart Water Management

Smart water management is the process of planning, developing, distributing and managing water resources using IoT technologies to provide reasonable and more sustainable usage of water resources. IoT can find various applications in water resource management such as checking water consumption levels, checking the quality of water, detecting toxic leakages of plants near water bodies, and tracking pressure variations along pipes to avoid wastage of water. Various sensors such as ultrasonic, flow, temperature, salinity, conductivity, humidity, pressure, and luminosity sensors can be used for IoT–based water monitoring (Eslamian, 2014).

Smart water management systems provide various benefits. Smart irrigation systems and crop management systems can help in water conservation on a large scale in agriculture applications. IoT–enabled water management systems can help monitor and control water pollution thereby improving the water quality. For example, in the UK, Legionnaires' disease has been linked to *Legionella* bacteria, which develops in uncontrolled soil and water environments. The bacteria, when inhaled through contaminated water droplets, can cause an acute form of potentially fatal pneumonia.

DOI: 10.1201/9781003599593-8

5.1.2 Overview of Blockchain Technology

Emerging from the world of cryptocurrencies, blockchain technology has become a game-changer in many different sectors, far beyond its original use in Bitcoin. Increasing water shortages, changing weather patterns, and an increasing human population are reshaping the water management scene on a worldwide scale. Ineffectiveness, secrecy, and quality assurance are common problems with conventional water management systems. In this chapter, we will go over the basics of how smart water management systems can incorporate blockchain technology to tackle these problems and completely revamp the distribution, monitoring, and management of water resources.

5.1.3 The Origins of Blockchain

Some person or persons using the pseudonym of Satoshi Nakamoto published a groundbreaking treatise on Bitcoin after the financial crisis of 2008, introducing the idea of blockchain. This was the first step towards creating a distributed ledger system that may facilitate trustless transactions between individuals.

5.1.4 Rationale for the Integration of Blockchain in Water Management

Water, a precious and limited resource, is encountering more and more problems on a global scale. Ineffectiveness, secrecy, and quality assurance are common problems with conventional water management systems. Blockchain technology changes the game when it comes to water resource governance, and this chapter delves into the compelling reasons why water management systems should incorporate it.

5.2 SMART WATER: A NECESSITY IN MODERN TIMES

Rapid urbanization, burgeoning population growth, and escalating environmental concerns are the major factors that made smart water management to emerge as a critical necessity in modern times. Various challenges are faced by the current society, ranging from scarcity to pollution, demanding innovative solutions to ensure sustainable access to clean and potable water for all. The advent of smart water technologies, propelled by advancements in data analytics, sensor networks, and Internet of Things (IoT) capabilities, offers a promising avenue to address these challenges efficiently and effectively. This section sets the stage to delve deeper into the realm of smart water management, exploring its importance, underlying technologies, and the transformative potential it holds in shaping a more resilient and water-secure future.

5.2.1 Challenges in Traditional Water Management

5.2.1.1 Uneven Water Distribution

Disparities in the accessibility, reliability, and availability of water resources among communities and regions contribute to uneven water distribution, a persistent difficulty in traditional water management systems. Factors such as population growth, past patterns of resource distribution, and inadequate infrastructure all contribute to this problem. In order to find lasting solutions to this age-old problem, it is essential to comprehend the complexities of uneven water distribution.

a. **Infrastructure Disparities**:
   A lack of modern and adequate infrastructure is a major factor that leads to water distribution inequalities. The infrastructure for distributing water in many areas may have fallen behind the times, having been put in place decades before the current wave of urbanization

and population boom. As a result, water leakage, pressure changes, and an incapacity to meet the increased demand can occur in locations where pipes are inadequate or degrading (Figure 5.1).

b. **Population Dynamics**:
A problem made worse by an uneven distribution of people is an uneven distribution of water. Existing water infrastructure can be overwhelmed by rapid urbanization and population increase in certain places, leading to insufficient supply to fulfil the demands of dense populations. On the flip side, water access may be inadequate in less populous or rural areas due to a lack of investment in infrastructure.

c. **Geographical Disparities**:
Inequitable distribution of water is caused in part by geographical reasons. Building effective water distribution networks can be challenging in places with difficult topography,

FIGURE 5.1 Challenges faced in traditional water management.

such as those in isolated or hilly areas. So, compared to locations with flat terrain and easy access to water sources, these areas are more prone to experience water scarcity.

d. **Economic Disparities**:
Because more well-off communities often have more robust and well-maintained water systems, economic disparities exacerbate water scarcity. On the other hand, areas that are economically disadvantaged might not have the means to invest in water distribution system upgrades and maintenance, which would further widen the gap in water availability.

e. **Historical Resource Allocation**:
Patterns of water distribution are influenced by past choices and techniques for allocating resources. Persistent disparities in water availability may have resulted from prior policies and practices that favored particular regions over others. If we want a more fair distribution of water resources, we must fix these long-standing injustices.

f. **Social and Environmental Consequences**:
The environmental and social impacts of water distribution inequality are substantial. As a result of water constraint, some communities may engage in unsustainable practices, such as excessive groundwater extraction, which can cause aquifer depletion and environmental damage. Inadequate water availability disproportionately affects already-vulnerable populations, which can lead to societal problems.

g. **Mitigation Strategies**:
There must be a combination of strategies to fix the problem of unequal water distribution. Key approaches include investing in sustainable water supplies, developing demand control measures, and improving and extending infrastructure in underprivileged areas. Smart water meters and sensors, which can improve monitoring and management, can also help distribute water resources more efficiently and fairly.

5.2.1.2 Aging Infrastructure

The inefficiency, higher maintenance costs, and possible hazards to public health and the environment caused by aging infrastructure is a major problem with traditional water management systems. In this section, we'll take a closer look at the complex problems caused by water infrastructure that is getting worse in years, as well as the causes and effects of and solutions to these problems.

5.2.1.2.1 Causes of Aging Infrastructure

One of the biggest obstacles to achieving the goals of smart water management is aging infrastructure. Water distribution networks that are decades or even centuries old and plagued by leaks, inefficiency, and disruption susceptibility are a problem for many cities across the world. In addition to causing large water losses, these outdated systems make it more difficult to use contemporary technologies for real-time monitoring and control. Integrating smart sensors, meters, and predictive analytics into such infrastructure becomes an intricate task that frequently calls for large financial outlays and careful design. However, there is hope thanks to the incorporation of smart water technologies into deteriorating infrastructure. Through improved visibility, resilience, and operational efficiency, these technologies help utilities allocate resources more efficiently, find leaks quickly, and take proactive measures to address maintenance needs. All of these factors contribute to the longevity and performance of current water systems. In order to ensure sustainable water management practices for future generations, it is becoming more and more necessary for cities to integrate smart water solutions with traditional networks as they continue to face the problems faced by aging infrastructure. Let us discuss the causes of aging infrastructure in this section.

a. ***Limited Investment*:**
Underinvestment in infrastructure upkeep and upgrades in the past is a major contributor to its aging. As the world's population and urbanization have grown, so has the need for water – eventually beyond the capacity of our current systems.

b. ***Material Deterioration*:**
Iron pipes and asbestos cement are two examples of the materials used to construct ancient water infrastructure that can easily deteriorate. Leaks, cracks, and water quality problems can occur when chemical reactions, erosion, and corrosion weaken the infrastructure.

c. ***Lack of Regular Maintenance*:**
The aging process is hastened by neglecting regular maintenance. Ignoring little problems for too long causes them to grow, necessitating more significant and costly fixes.

5.2.1.2.2 Consequences of Aging Infrastructure

a. ***Water Loss and Leakage*:**
Significant water loss can occur due to leaks caused by ageing pipelines. Not only does this reduce the amount of water available to consumers, but it also makes water shortages much more of a problem.

b. ***Reduced Water Quality*:**
Water quality can be compromised by infrastructure that is deteriorating. Contaminants can infiltrate the water supply through damaged or corroded pipes, endangering public health and calling for expensive water treatment procedures.

c. ***Increased Operational Costs*:**
Higher operational costs are a result of increased energy consumption for pumping water via aging infrastructure and the necessity for frequent repairs and replacements. Water utilities and, by extension, consumers, will feel the financial pinch as a result of this.

d. ***Disruptions and Service Outages*:**
The likelihood of unanticipated breakdowns, which can interrupt or even completely shut down services, increases as infrastructure ages. For areas, companies, and rescue services that depend on a constant water supply, these disruptions can cause a domino effect.

5.3 LACK OF REAL-TIME MONITORING AND RESPONSE

A major obstacle to proactively addressing water quality issues, leaks, and other crucial events is the inability of existing water management systems to provide real-time monitoring and response capabilities. In this section, we will examine this topic from every angle, looking for its causes, effects, and ways to improve the responsiveness and agility of water management.

5.3.1 Causes of Lack of Real-time Monitoring

5.3.1.1 Outdated Monitoring Infrastructure

There are a lot of water management systems out there that use antiquated monitoring tools that can't keep up with the times. Delays in data gathering and response are caused by manual monitoring techniques and old sensors.

5.3.1.2 Limited Sensor Deployment

Data gaps occur as a result of insufficient sensor and monitoring device deployment across water distribution networks. It becomes difficult to gather real-time data on infrastructure health, water quality, and flow rates without a complete sensor network.

5.3.1.3 Inadequate Communication Systems

Delays in data transmission are caused by outdated communication methods between central control hubs and monitoring devices. Building a smooth, real-time monitoring network is made more difficult by inefficient communication infrastructure.

5.3.2 Consequences of Lack of Real-Time Monitoring

5.3.2.1 Delayed Detection of Water Quality Issues

The prompt identification of water quality concerns, including pollution or unusual chemical levels, is hindered in the absence of real-time monitoring. Widespread health and environmental consequences are more likely to occur as a result of this postponement.

5.3.2.2 Inefficient Leak Detection

It is essential to have real-time monitoring in order to identify and repair water distribution network leaks promptly. Lower operational costs, infrastructure damage, and water loss result from leak detection systems that are too sluggish.

5.3.2.3 Limited Ability to Respond to Emergencies

Responding quickly to emergencies, including pipeline breakdowns or contamination events, is made more difficult when water management agencies do not have access to real-time data. Incidents can become more dangerous and endanger the public if reaction times are too long.

5.3.3 Centralized Decision-Making and Governance

Traditional water management systems have centralized decision-making and governance, which brings bureaucratic red tape and a lack of stakeholder involvement. In this section, we will examine these problems in detail, looking at how they affect decision agility and why community-driven water management approaches need to be more inclusive.

5.3.3.1 Bureaucratic Decision Processes

Examination of Bureaucratic Hurdles:
Bureaucratic decision-making processes, with their complicated hierarchies, long approval procedures, and plethora of administrative steps, are common in centralized water governance. Delays in implementing crucial water management decisions are caused by these bureaucratic barriers, which impair the system's responsiveness to developing difficulties.
Impacts on Agility in Response:
The ability to quickly adapt to evolving water management demands is hindered by the inherently sluggish nature of bureaucratic decision-making procedures. Extreme weather, population development, and environmental changes are just a few of the ever-changing difficulties that water systems face that require swift solutions. Opportunities for preventative actions or timely interventions may be lost due to bureaucratic delays.

5.3.3.2 Lack of Stakeholder Engagement:

- *Challenges of Limited Stakeholder Involvement*:
  Local communities, water consumers, and other stakeholders with a stake in sustainable water management are frequently marginalized by centralized decision-making. When people aren't involved in making decisions, it can cause such judgements to ignore local preferences, requirements or environmental considerations.
- *Implications for Community-Driven Practices*:
  The possibility of community-driven and long-term water management strategies is diminished in the absence of stakeholder involvement. People on the ground have unique perspectives on issues like water conservation, cultural customs, and overall environmental health. Community support and contextual relevance could be lacking in the water management strategies that emerge from a decision-making process that excludes them.

5.3.3.3 Data Silos and Information Asymmetry

Disjointed data systems and restricted data accessibility are problems caused by information asymmetry and data silos in conventional water management systems. In this section, we'll delve into the intricacies of these problems and examine how they affect water resource planning, decision-making and collaboration.

5.3.3.3.1 Fragmented Data Systems

Large volumes of data have been gathered by companies in a variety of industries in the digital age, which has fueled insights, creativity, and decision-making. Fragmented data systems remain a persistent concern, though. Within an enterprise, data fragmentation is the dispersion of information among many applications, platforms, and storage silos. The organization's capacity to fully utilize its data assets is hampered by this fragmentation, which also affects data accessibility, interoperability, and coherence. The problem of fragmented data systems has grown more urgent as data continues to multiply at an unprecedented rate, which is made worse by the development of cloud computing, edge computing, and IoT devices.

- *Challenges of Fragmented Data Systems*:
  In the past, water management authorities frequently used data systems that were disjointed and uncoordinated. Some water management functions, such as quality monitoring, distribution or infrastructure repair, may necessitate the use of specialized databases maintained by other units or departments. Data silos arise from these systems' failure to communicate with one another, making it difficult to gain a holistic view of the water management ecosystem as a whole.
- *Impact on Holistic Decision-Making*:
  i. **Limited Data Accessibility**:
  Inability to take into account all pertinent elements is a major drawback of fragmented data systems when it comes to making holistic decisions. Water quality, consumption trends, and infrastructure health may not be fully and continuously displayed to decision-makers. Due to this restriction, less-than-ideal choices may be made without considering the interdependence of many aspects of water management.
- *Exploration of Limited Data Accessibility*:
  One cause of information asymmetry is the fact that important parties have a hard time getting their hands on and sharing data pertaining to water. Concerns about proprietary data ownership, institutional policies or technological limitations are all potential causes of access obstacles. Not all parties participating in water management have easy access to relevant data, which makes it difficult for them to draw useful conclusions.
- *Instances of Hindered Collaborative Efforts*:
  As a result, stakeholder collaboration suffers when data accessibility is limited. Organizations at the federal, state, and regional levels, as well as local communities, must work together to effectively manage water resources. Poor results in water resource management and planning may result from these parties' inability to make educated contributions due to a lack of access to vital data.
  ii. **Lack of Transparency and Accountability:**
  Traditional water management systems have serious flaws when it comes to accountability and transparency, which leads to problems like shady dealings with water usage records and other forms of corruption and fraud. The consequences of these difficulties for responsible water distribution and management are explored in this section, which digs into their complexity.

iii. **Opacity in Water Usage Records**:

- *Discussion on Lack of Transparency*:
  One major problem with conventional approaches to water management is the lack of transparency around data collection and reporting. No one, including stakeholders or regulatory agencies, may have easy access to water use records. We can't evaluate the efficacy of water distribution and use because of this opaque process, which makes us doubt the reliability of the data we do have.
- *Challenges in Accountability*:
  It is difficult to hold individuals accountable for water distribution and management when there are opaque records of water usage. To guarantee responsible water stewardship, transparency regarding the distribution and use of water resources is essential for detecting inefficiencies, resolving inequities, and enforcing rules.

iv. **Susceptibility to Fraud and Corruption**:

- *Examination of Vulnerabilities*:
  Corruption and fraud can easily infiltrate traditional water management systems. Manual recordkeeping, a lack of supervision, or insufficient controls in water resource allocation and distribution can lead to these risks. Every step of water management can be jeopardized if records of water usage are tampered with.
- *Real-World Cases and Consequences*:
  Malpractices in water management and their effects can be better understood by looking at actual cases. Unfair resource distribution, environmental degradation, and financial losses for water utilities can result from fraudulent actions such meter manipulation or unauthorized water withdrawals. Gaining insight into these instances highlights the critical need to resolve vulnerabilities in order to maintain the reliability of water management procedures.

5.4 NEED FOR EFFICIENT WATER DISTRIBUTION AND CONSERVATION

The repercussions of water management malpractices can be better understood by looking at actual instances. Water utilities lose money, resources are unequally distributed, and the environment suffers as a result of fraudulent operations like meter manipulation or illegal water withdrawals. It is critical to fix vulnerabilities immediately in order to maintain the integrity of water management operations, as these incidents demonstrate.

5.4.1 Global Water Challenges

Water, the essence of life, is a finite resource critical for sustaining ecosystems, supporting human health, and driving socio-economic development. But as the world's population grows and environmental stresses increase, humanity is faced with a growing number of challenging water issues. These issues, which include depleting freshwater supplies, water pollution, unequal access to water resources, and the effects of climate change, interact and intensify one another to pose serious threats to both current and future generations. This section provides background information for a thorough analysis of the complex global water issues, including their root causes, wide-ranging effects, and the pressing need for coordinated local, national, and international action to guarantee water security and sustainability for all. In order to protect the most valuable resource in the world, it is critical that we comprehend and handle the problems associated with water governance, conservation, and management.

- *Rising Demand and Population Growth*
  The need for water is increasing as the world's population keeps growing. There is an immediate need for efficient distribution of water due to the increased demand caused by rapid urbanization and industrialization.

- *Water Scarcity and Stress*
  When the demand for water exceeds the supply, many areas experience water scarcity and stress. The importance of efficient distribution in addressing issues caused by shortage is highlighted by the need to understand the spatial variations in water availability.

5.4.1.1 Environmental Impact of Water Mismanagement

- *Ecosystem Degradation*:
  Ecosystem deterioration is exacerbated by inefficient water delivery. The ecological repercussions of mismanagement of this valuable resource are highlighted by altered water flow patterns, habitat loss, and the depletion of aquatic ecosystems.
- *Energy Consumption and Carbon Footprint*:
  Significant energy is needed for water pumping, treatment, and distribution. A greater carbon footprint is a result of increased energy consumption caused by inefficient processes. Sustainable development objectives are congruent with the shift toward more efficient water distribution.

5.4.1.2 Socio-Economic Benefits of Water Conservation

- *Economic Efficiency*:
  Operating costs, resource allocation, and economic efficiency are all positively impacted by efficient water distribution. Water utilities and customers alike get the benefits of investments in cutting-edge procedures and technology in the long run.
- *Agricultural Productivity and Food Security*:
  Efficient water distribution can help agriculture, which uses a lot of water. Enhanced agricultural output and food security on a worldwide scale are outcomes of water conservation measures, smart farming methods, and precision irrigation.

5.4.1.3 Water as a Catalyst for Social Equity

- *Equitable Access*:
  An end to socioeconomic inequalities and the divide between urban and rural regions can be achieved through efficient water distribution, which guarantees equitable access. Optimized distribution networks can help communities who have experienced water scarcity in the past.
- *Health and Well-Being*:
  The availability of safe drinking water is a basic human right. Communities are better protected from waterborne diseases and enjoy higher quality of life when water is distributed efficiently.

5.5 TECHNOLOGICAL INNOVATIONS FOR EFFICIENT WATER MANAGEMENT

The need for effective water management has never been greater in light of the growing water scarcity, pollution, and climatic variability. Thankfully, a wave of technology advancements is transforming our understanding of water resource management and providing viable answers to these overwhelming problems. Advanced sensor networks, data analytics, artificial intelligence, and remote monitoring systems are just a few of the advances that enable water utilities, stakeholders, and regulators to optimize water usage, improve infrastructure resilience, and lessen the risks associated with water-related disasters. In an increasingly water-stressed world, this section lays the groundwork for a deeper examination of technical advancements for effective water management by stressing their revolutionary potential to promote sustainability, resilience, and equitable access to water resources. We are moving toward a future where water is handled wisely, sustainably, and for the good of all as we use technology to address water-related issues.

5.5.1 Smart Water Meters and Sensors

Implementing smart water meters and sensors allows for continuous monitoring and decision-making based on data. This system promotes responsible water consumption, finds leaks, and optimizes usage, all of which increase efficiency. With the help of the IoT, smart water networks can be built. IoT devices provide data interchange, which in turn enables automated controls, predictive maintenance, and more efficient water distribution systems.

Sustainable water management relies heavily on community engagement. A shared dedication to water conservation can be achieved through educating communities, promoting responsible practices, and including them in decision-making processes. Initiatives to educate the public about water conservation, efficiency, and the power of personal decision-making help foster an environment where water is used responsibly. On the road to more effective water distribution, educated residents are vital allies.

5.5.2 Emergence of Smart Water Management Solutions

Our perception, monitoring, and distribution of water resources have entered a new era with the advent of smart water management technology. A more effective, environmentally friendly, and technically sophisticated water management environment is the result of smart water management, which is examined in this chapter along with its development, advantages, and practical uses.

i. *Early Adoption of SCADA Systems*:
The first step toward automating water management was the implementation of Supervisory Control and Data Acquisition (SCADA) systems. These devices paved the way for future innovations by allowing for the remote control and monitoring of water infrastructure.
ii. *Integration of IoT in Water Networks*:
Modern water management practices have been radically altered by the use of IoT technologies. Decisions could be made based on data collected in real time thanks to smart meters, sensors, and other devices integrated into the water infrastructure.

5.5.2.1 Key Technologies Driving Smart Water Management

i. *IoT Sensors and Devices*:
The collection of real-time data on water usage, quality, and infrastructure health is greatly assisted by IoT devices. These gadgets make it possible to monitor everything at all times, which speeds up problem responses and allows for better water distribution.
ii. *Data Analytics and Machine Learning*:
Modern analytics and machine learning systems sift through mountains of data in search of trends, outliers, and patterns. Proactive water management is made possible by these technologies, which also boost operational efficiency and decision-making.

5.5.2.2 Benefits of Smart Water Management

i. *Improved Operational Efficiency*:
Distribution, maintenance, and resource allocation are all made more efficient with smart water management solutions. Reduced downtime and maximized system performance are the results of automation and real-time data that allow for rapid problem remedies.
ii. *Enhanced Resource Conservation*:
One fundamental advantage of smart water management is the efficient utilization of water resources. These solutions encourage conservation practices, decrease water losses, and help create sustainable water usage patterns by using data-driven insights.

5.5.2.3 Real-World Applications of Smart Water Management

i. *Smart Water Meters for Consumption Monitoring*:
The installation of smart water meters enables precise and immediate tracking of water usage. Both water utilities and consumers benefit from this data since it allows for better demand control techniques (Okoli & Kabaso, 2024). *Leak Detection and Prevention Systems*:

There are state-of-the-art leak detection devices that can detect leaks as they happen as part of smart water management solutions. Minimizing water losses, preventing infrastructure damage, and reducing repair costs are all achieved by the rapid handling of leaks by these systems.

5.5.3 Challenges and Considerations

Data Security and Privacy Concerns:
Data security and privacy are major issues with the integration of modern technology. For smart water management technologies to be widely adopted, it is critical to implement strong cybersecurity safeguards and address privacy concerns.
Initial Implementation Costs:
Smart water management technologies may save money in the long run, but they can be expensive to install. Important factors to consider for broad adoption include removing financial obstacles and creating ROI frameworks.

5.6 BLOCKCHAIN TECHNOLOGY: FOUNDATIONS AND FEATURES

Although blockchain technology was originally developed to facilitate cryptocurrency transactions, it has since developed into a robust and flexible system that finds use in many different sectors. This chapter explores blockchain's architecture, its main characteristics, and the methods that enable it to revolutionize data security, transparency, and decentralized governance.

Decentralized governance, made possible by blockchain technology, guarantees water distribution transparency. Stakeholders can see how their money is being used, which increases accountability and trust, thanks to an immutable ledger. Automating and enforcing water-related agreements, smart contracts streamline administrative processes and guarantee adherence. Improved efficiency, streamlined operations, and long-term water sustainability are all outcomes of this innovation. The immutability and integrity of data pertaining to water are guaranteed by blockchain technology, which creates a transparent and decentralized ledger. Water management practices that are more trustworthy, accountable, and transparent are better for all parties involved. Automating agreements in water management with smart contracts streamlines processes and guarantees adherence. The entire efficacy of water governance is improved by these self-executing contracts, which simplify procedures, increase efficiency, and cut costs.

5.6.1 Foundations of Blockchain Technology

Blockchain operates on the notion of decentralization. A distributed ledger, which is essential to blockchain technology, is not maintained by a single entity but rather by a network of computers called nodes. Data transactions are made more secure, resilient, and trustworthy by virtue of its decentralized structure. Data recorded on the blockchain cannot be edited or tampered with because of immutability. Blockchain technology is perfect for keeping open and immutable records because it uses cryptographic hash functions and consensus procedures to ensure the data remains intact.

5.6.1.1 Core Features of Blockchain

i. *Distributed Ledger*:
Every node in the blockchain network contributes to the distributed ledger, which is a shared database that is always up to date. Transparency is guaranteed and the need for a central authority to validate transactions is eliminated because all participants have access to the same ledger.

ii. *Consensus Mechanisms*:
Network nodes can reach an agreement on the legitimacy of transactions through consensus techniques, which are protocols. Tokenization and proof of stake are two well-known techniques. The reliability and safety of the blockchain are enhanced by these systems.

iii. *Cryptographic Hash Functions*:
Data integrity is ensured by means of cryptographic hash functions in blockchain technology. Hashes are stored in blocks, and if the content of a block is changed, a new hash is generated. With this cryptographic connection between blocks, the entire blockchain is guaranteed to be immutable.

iv. *Smart Contracts*:
Smart contracts are agreements that can be "smart" and "self-executed" according to certain criteria. They eliminate middlemen by automating and enforcing agreements directly. The use of smart contracts enhances the reliability, openness, and efficiency of blockchain transactions. Smart contracts, a fundamental component of blockchain technology, provide a programmable and self-executing method for automating agreements. This chapter thoroughly examines smart contracts, including what they are, how they work, and the revolutionary impact they have on automating procedures in several sectors.

5.6.1.2 Understanding Smart Contracts

A "smart contract" is an agreement whose terms are encoded and may be executed automatically. Without the need for middlemen, these contracts can execute, validate, or even help with the negotiation of contracts thanks to blockchain technology. A smart contract's terms and conditions are controlled by the code itself, according to the "code as law" notion. After deployment, the code executes in a predictable and rule-based fashion, doing away with room for interpretation and guaranteeing an open and trustworthy execution of agreements.

When certain circumstances are met, smart contracts are activated. The completion of a delivery or the fulfilment of a milestone are examples of external data inputs that could occur as events. Other examples include a specified date and time. The smart contract will carry out the predetermined actions without human intervention once it is activated. There is an unchangeable record of the transaction thanks to the automated enforcement of the contract terms recorded on the blockchain.

5.6.1.3 Key Features of Smart Contracts

i. *Trustless Execution*:
Participants in a trustless environment, where smart contracts function, are not required to have faith in one another. The rules are defined by the code, and execution is deterministic, so everyone is on an equal footing.

ii. *Transparency and Immutability*:
The immutability and transparency of smart contract results and transactions are recorded on the blockchain. Every party can be certain that the terms agreed upon have been fulfilled by checking the status of the contract's execution.

iii. *Automation of Transactions*:
By automating transactions, standardizing agreements, and improving overall efficiency, smart contracts—a crucial part of blockchain technology—revolutionize water management. Smart contracts are examined in this chapter as they pertain to water management. The automation of water transactions and agreements, optimization of billing and payments, and use cases in water conservation programs are the main areas of attention.

Automated and trouble-free water exchange between parties is now possible thanks to smart contracts. The efficient and open transfer of water rights can be achieved by the use of predetermined regulations in water transactions between farmers, businesses, and cities. In water management, smart contracts make it possible to make agreements with conditions. Automated responses to changing circumstances can be achieved, for instance, by setting agreements to trigger in response to certain weather conditions, water quality standards, or specified consumption restrictions.

In water management, smart contracts automate the billing process. These contracts automatically create invoices, compute usage based on real-time data from smart meters, and apply predefined rates. This shortens billing cycles, decreases the likelihood of human error, and lessens the need for manual intervention. Automating the delivery of funds upon the fulfillment of established conditions, smart contracts enable secure and timely payments. This makes sure that water providers get paid quickly and that customers have an easy and clear way to pay.

5.6.1.4 Use Cases of Smart Contracts in Water Conservation Programs

i. *Incentivizing Water Conservation*:
In order to encourage water conservation programs to use water efficiently, smart contracts can be used. A culture of responsible water usage and conservation can be fostered through contracts that offer incentives to users who keep below predetermined consumption criteria.

ii. *Dynamic Water Pricing*:
Water prices can be adjusted in real-time according to supply and demand thanks to smart contracts. Water pricing can be set to automatically adjust during periods of scarcity to encourage conservation, while during seasons of abundance, rates can be made more reasonable to promote optimal resource utilization.

5.6.1.5 Challenges and Considerations in Implementing Smart Contracts in Water Management

i. *Data Accuracy and Reliability*:
In order to make decisions, smart contracts need trustworthy data. It is critical to guarantee the accuracy of sensor and meter data to avoid errors that can affect contract execution.

ii. *Regulatory Compliance*:
Existing regulatory frameworks should be considered when smart contract usage in water management is considered. In order to apply it widely, it is essential to ensure compliance with legal requirements and resolve any regulatory issues.

iii. *Code Vulnerabilities*:
Code mistakes and security holes can compromise smart contracts. Reducing the likelihood of code vulnerabilities requires thorough testing, regular security audits, and the use of industry best practices.

iv. *Legal and Regulatory Challenges*:
Smart contracts' regulatory and legal environment is always changing. For wider acceptance, it is vital to address difficulties linked to being enforceable, resolving disputes, and complying with current laws.

5.6.1.6 Future Developments and Opportunities

i. *Integration with IoT Sensors*:
The capabilities of smart contracts are enhanced through their continued integration with sensors connected to the IoT. More adaptive and real-time water management systems are possible thanks to sensor data that can initiate smart contract executions.

ii. *Decentralized Autonomous Organizations (DAOs)*:
For community-driven water management, the idea of decentralised Autonomous Organizations (DAOs) supported by smart contracts shows potential. DAOs eliminate the

need for a single, authoritative body to handle matters such as decision-making, resource allocation, and governance.

5.6.2 The Intersection of Blockchain and Smart Water Management

Combining smart water management with blockchain technology offers a revolutionary way to fix major problems with conventional water systems. Effective water management relies on maintaining water quality. In this chapter, we'll look at how blockchain technology can work in tandem with water management systems, bringing attention to the unique problems that blockchain technology can solve as well as the opportunities it presents to improve openness, effectiveness, and longevity.

5.6.2.1 Key Challenges in Water Management Addressed by Blockchain

i. *Lack of Transparency and Accountability*:
By generating an immutable and auditable record, blockchain brings accountability and openness. Water usage records and management decisions are made public, immutable, and traceable by virtue of blockchain's decentralized nature.

ii. *Data Silos and Limited Accessibility*:
By creating a distributed ledger that anybody can access, blockchain technology eliminates data silos. A more unified and thorough understanding of water management data is fostered, and stakeholder collaboration is improved. Data exchange is also made easier.

iii. *Standardization of Data Formats*:
Data format standardization is necessary to provide compatibility across various data sources. Coordinated efforts to standardize data exchange protocols can make blockchain integration a breeze.

iv. *Sensor Accuracy and Reliability*:
For water quality monitoring to be effective, the data collected by IoT devices must be accurate and reliable. Before deploying sensors, make sure they are calibrated, will be maintained regularly, and adhere to quality standards.

v. *Lack of Real-time Monitoring and Response*:
With the help of IoT devices and blockchain technology, water usage, quality, and infrastructure health can be tracked in real time. Because of this, new problems can be addressed quickly, shortening the lag time between water management issues being identified and their resolution (Nokoli and Kabaso, 2024).

vi. *Centralized Decision-Making and Governance*:
Traditional water management decision-making systems are being challenged by the decentralized governance paradigm of blockchain technology. Blockchain technology allows for more transparent stakeholder engagement and the distribution of decision nodes, leading to more inclusive and agile government.

vii. *Technological Adoption Barriers*:
Resistance to technological change, lack of expertise, and the need for significant infrastructure upgrades present challenges to the widespread adoption of blockchain in smart water management. Strategies for overcoming these barriers include education, training, and collaborative implementation efforts.

viii. *Interoperability and Integration*:
Ensuring seamless integration of blockchain with existing water management systems and promoting interoperability among diverse technologies remain significant challenges (Namasudra and Akkaya, 2023). Standards and protocols for data exchange need to be established to facilitate integration across different platforms.

ix. *Scalability Issues*:
As the volume of water-related data continues to grow, scalability becomes a concern. Blockchain networks must evolve to handle increasing transaction volumes and data storage requirements without compromising efficiency or decentralization (Smart Water Management).

x. *Compliance with Existing Regulations*:
Adhering to existing regulations while integrating blockchain in water management is crucial. Legal frameworks may need to be adapted to accommodate blockchain technologies, ensuring compliance with water rights, data protection, and environmental regulations.

xi. *Data Privacy and Security*:
Blockchain's transparent nature must align with data privacy regulations. Striking the right balance between transparency and protecting sensitive information is a key consideration, necessitating robust privacy-enhancing technologies.

5.6.2.2 Potential Benefits of Integrating Blockchain Technology

Integrating data from decentralized sources, such as IoT sensors, labs, and monitoring stations, is made possible by blockchain technology. By bringing all of these systems together, we can build a robust system for monitoring pollutants and pinpointing their origins. Blockchain technology improves traceability by creating an immutable record of all transactions pertaining to water quality. The blockchain records every step of data collection, allowing us to see where the data came from and how it was handled.

i. *Enhanced Data Security and Integrity*:
Ensuring secrecy and preventing unauthorized manipulation, blockchain's cryptographic security features safeguard data related to water. A further safeguard against data manipulation is the immutability of blockchain records.

ii. *Automation of Water-Related Agreements*:
The use of blockchain technology allows for the automation of water-related agreements, which streamlines several operations including payments, billing, and the transfer of water rights. This guarantees the efficient execution of agreements while reducing administrative overhead and minimizing errors.

iii. *Increased Trust and Accountability*:
Stakeholders are more likely to trust blockchain due to its immutable and transparent ledger. It promotes a more responsible and accountable approach for managing water by making everyone pay for their actions. Water quality data may be recorded transparently and simultaneously across numerous nodes using blockchain technology because of its decentralized nature. Stakeholders are able to independently confirm the data's accuracy and integrity, which cultivates confidence.

iv. *Decentralized Decision-Making*:
Decisions may be made with more participation thanks to blockchain technology's decentralized governance architecture. Decisions about water management can be better informed and tailored to local circumstances when stakeholders like communities and environmental groups are involved.

v. *Tamper-Proof Data Logging*:
Once data on water quality is recorded via blockchain, it cannot be changed or tampered with because of the immutable ledger. The result is an open and trustworthy record of water quality over time, which is useful for meeting regulatory requirements and making informed decisions.

vi. *Detection of Water Quality Issues*
Instantaneous data updates are made possible through the integration of blockchain technology with real-time monitoring systems. The ability to construct smart contracts to send out warnings or notifications when certain water quality measures reach certain levels allows for quick reactions to new problems as they arise. Rapid decision-making for water quality restoration is made possible by the decentralized governance paradigm of blockchain technology. Stakeholders can work together to identify and evaluate problems, distribute resources, and take corrective action without bureaucratic red tape.

5.6.3 How Blockchain Enhances Water Data Security

Protecting the authenticity and trustworthiness of data that is vital to efficient water management is of the utmost importance when dealing with water. In this chapter, we will look at how blockchain technology improves the security of water data. Specifically, we will see how it safeguards data on water quantity and quality, prevents data manipulation, and reduces cyber risks, all of which contribute to making water management systems more trustworthy. Due to the requirement for technical infrastructure and knowledge, blockchain implementation in water management may encounter opposition. It will take teamwork and educational campaigns to overcome these adoption hurdles. Integrating blockchain technology into water management requires careful navigation of the regulatory environment. One important thing to think about is making sure that blockchain technologies are legal and that current regulations are followed (Soori et al., 2023).

Once data on water amount and quality is recorded on blockchain, it cannot be erased or altered because of the immutable ledger. This feature ensures that crucial information is recorded securely and cannot be altered, providing a trustworthy historical viewpoint that can be used for decision-making. Water data is protected by blockchain technology, which uses cryptographic methods. Cryptography, digital signatures, and public/private key pairs all work together to protect water-related data from prying eyes and ensure its legitimacy and integrity.

Because it is distributed and not centralized, blockchain cannot be manipulated. A secure and interconnected series of transactions is created when each block in the chain includes a cryptographic hash of the block before it. Data manipulation would be extremely difficult if not impossible because changing one block would need modifying all blocks that followed it. Consensus mechanisms, such as roof of work (PoW) or proof of stake (PoS), ensure that transactions are validated by the majority of nodes in the network. This consensus-building process adds an extra layer of security, preventing malicious actors from manipulating data without the agreement of the network (Soori et al., 2023). Blockchain's decentralized architecture minimizes the vulnerability to cyber threats. Unlike centralized databases, where a single breach can compromise all data, blockchain's distributed nature disperses data across multiple nodes, reducing the impact of potential cyberattacks.

All nodes in the network can see all transactions involving water since blockchain is transparent. Stakeholders are able to independently confirm the data's correctness and integrity thanks to this openness, which promotes a culture of responsibility and confidence. The necessity for a governing body to supervise water management is rendered obsolete by blockchain's decentralized governance paradigm. Without members intrinsically trusting one another, blockchain creates a trustless ecosystem that secures data and executes transactions.

There are a couple of challenges also faced by blockchain while it is implemented in water data security. The following are the challenges:

i. *Integration with Existing Systems*:
Possible stumbling blocks include integrating blockchain with preexisting water management systems. It takes meticulous preparation and teamwork to resolve interoperability issues, ensure compatibility, and integrate seamlessly.

ii. *Data Privacy and Compliance*:
Sensitive information is frequently included in water statistics. Careful consideration of how to balance security and privacy is necessary to ensure compliance with data privacy requirements while making use of blockchain technology's benefits.

5.6.4 Decentralization for Efficient Water Distribution

The decentralized nature of blockchain technology presents a significant departure from the conventional wisdom on water distribution. More efficient and fair distribution of water can be achieved through decentralization, which is made possible in this chapter by blockchain technology.

Decentralization allows for peer-to-peer transactions, monitors water infrastructure, and lowers reliance on centralized agencies.

The smart contracts built within the blockchain make it possible for water distribution to be conducted directly between users, cutting out middlemen. Users of water can streamline operations and reduce costs by engaging in direct, automated, and transparent exchanges. Efficient water trading among users is made possible by decentralized platforms built on blockchain. Water transactions can be directly negotiated and executed by farmers, companies, and municipalities according to their demands, allowing for transparent and real-time optimization of resource distribution.

Distributed and real-time water infrastructure monitoring is made possible by the integration of blockchain technology with IoT devices. Optimal resource allocation, predictive maintenance, and early problem identification are made possible by the continuous data provided by sensors and devices installed throughout the water network. The immutable ledger of blockchain technology guarantees the security and integrity of data pertaining to water infrastructure, including records of maintenance and quality evaluations. More faith in the health and efficiency of water assets is fostered by this openness (Figure 5.2).

FIGURE 5.2 IoT–based water quality management system.

With blockchain technology, local people may take part in water management choices because of its decentralized governance approach. Communities can develop a sense of agency and responsibility when they are less reliant on top-down decision-making and can instead make decisions based on their own unique circumstances. The decision-making procedures are made transparent by decentralization. Everything from transactions to consumption trends and governance choices pertaining to water is transparent to all parties. Promoting an egalitarian and inclusive water management system, this transparency lessens the likelihood of favoritism, corruption, and unfair distribution.

A few of the challenges that may be faced for implementation are as follows: unfamiliarity with blockchain technology and decentralized governance could lead to resistance to the implementation of decentralized water distribution systems. Efforts to raise awareness and educate stakeholders are vital in creating an environment where everyone can work together. There are technological hurdles to overcome when integrating decentralized systems with the current water infrastructure. Thorough preparation and teamwork are essential to guarantee compatibility, smooth integration, and minimum interference with current systems.

5.6.4.1 Future Trends and Opportunities

i. *Community-Driven Water Markets*:
 The possibility of community-driven water markets is made possible by decentralized water distribution. Communities can take part in water transactions according to their own demands if transparent pricing mechanisms and localized trade platforms come into existence.
ii. *Machine Learning Integration*:
 Potential for predictive analytics exists in water quality monitoring with the integration of blockchain and machine learning. Water quality problems can be better detected and prevented in their early stages when machine learning algorithms examine blockchain data for trends and patterns (Thinakaran et al., 2023).
iii. *Citizen Science Participation*:
 Blockchain's distributed ledger technology and immutability make it an ideal tool for public water quality monitoring projects. By allowing users to contribute data using blockchain-based decentralized applications (DApps), the monitoring network can grow and community involvement can flourish (Verma et al., 2024).

5.7 CASE STUDIES AND SUCCESS STORIES

To fully grasp the potential uses and results of blockchain technology in water management, it is essential to have access to relevant case studies and success stories. This chapter explores real-world cases of fruitful blockchain projects, shedding light on the successes, failures, and lessons learned from each. It also covers global instances where blockchain solutions have revolutionized water governance.

5.7.1 Implementation of Blockchain in Real-World Water Management Projects

- *IBM's Water Management Platform*:
 IBM has integrated data from multiple sources, including IoT devices, into a water management platform that is blockchain-based. In order to support sustainable water management practices and enable optimal allocation of resources, the platform offers a full perspective of water usage.
- *WaterChain Project in California*:
 Using blockchain technology, the California WaterChain project is establishing a decentralized water market. An improved and more adaptable method for allocating water is

possible thanks to this project, which facilitates the transparent and efficient trading of water rights among farmers and other users of the resource.

- *Singapore's Smart Water Grid*:
  The Smart Water Grid in Singapore uses blockchain technology to improve water management. The system makes use of IoT sensors to track water usage in real time, and blockchain technology guarantees the safe and transparent storage of data pertaining to water, which helps to make better use of available resources.
- *Western Australia's Water Allocation Platform*:
  For the purpose of managing water rights, Western Australia has implemented a platform that is based on blockchain technology. The platform ensures compliance with legal requirements while facilitating transparent and auditable transactions, which allows users to efficiently trade water allotment.
- *Dutch Pilot Project for Transparent Water Quality Monitoring*
  To improve the traceability and transparency of water quality monitoring, the Netherlands—a country renowned for its intricate water management systems—launched a pilot program. The necessity for real-time monitoring and the problems caused by disjointed data sources were the driving forces behind the project. Integrating data from numerous water quality sensors placed across the nation, a blockchain-based platform was created for implementation. Recorded on the blockchain in real time was every single data point, including contamination alarms and chemical levels. The public, water utilities, and government organizations all had access to an unchangeable and transparent record of water quality data. More efficient tracing of pollutant sources allowed for more targeted cleanup. Regional water authorities were able to work together more effectively due to decentralized data access.
- *Australian Blockchain Platform for Water Trading*
  The Australian government used blockchain technology to improve the efficiency of water trading in the face of water constraints. Complexities, such as delays and a lack of transparency, plagued the old water trading system. The implementation of a blockchain platform was made possible to enable farmers, industries, and municipalities to engage in peer-to-peer water trade. Water transactions were mechanized through the use of smart contracts, which automate the negotiation, execution, and settlement processes. The elimination of middlemen in blockchain-enabled peer-to-peer transactions improved the efficiency and transparency of water trading. Integration with IofT sensors supplied stakeholders with real-time data on water availability, allowing them to make educated decisions. Water trading market participants saved money due to the elimination of middlemen and the automation of transactions.
- *Chinese City's Blockchain-Based Water Billing System*
  An ever-expanding Chinese metropolis encountered difficulties in effectively overseeing its water billing procedures. Problems with inaccurate billing, disputes, and delays led to the investigation of blockchain technology. To improve transparency and automate billing operations, a water billing system based on blockchain technology was put into place. Using real-time data from smart meters, smart contracts were utilized to calculate water usage, and payments were completed automatically. The following are the benefits:
  - More Accurate and Transparent Water Usage Calculations: Thanks to automation, billing processes are now more precise and error free.
  - Automated and Timely Payments: Smart contracts improved water utilities' cash flow by facilitating automated and timely payments.
  - A decrease in billing disputes occurred as a result of the immutable and transparent data kept on the blockchain, which improved confidence between the water company and its customers.

Case studies like this highlight the versatility and potential good impact of blockchain technology by illustrating its various uses in solving particular problems in different water management situations.

5.8 CONCLUSION

Combining blockchain with advanced data analytics, including machine learning and artificial intelligence, opens avenues for more sophisticated water management insights. Predictive analytics and pattern recognition can enhance early detection and decision-making. Research on cross-chain interoperability aims to enable seamless communication and transactions between different blockchain networks. In the context of smart water management, this innovation can enhance collaboration among diverse stakeholders using different blockchain platforms.

Global collaboration is essential for developing common standards and protocols in blockchain for water management. International organizations, governments, and technology providers can work together to establish guidelines for interoperability and data exchange. Collaborative research initiatives and pilot projects can facilitate the testing and validation of blockchain solutions in diverse water management contexts. These projects provide valuable insights and contribute to the development of best practices.

REFERENCES

Eslamian, S. (ed.), (2014). *Handbook of Engineering Hydrology, Vol. 3: Environmental Hydrology and Water Management*, Taylor & Francis Group, CRC Group, USA.

Namasudra, S., Akkaya, K., (2023). *Blockchain and its Applications in Industry 4.0*, Singapore: Springer.

Nokoli, N. J., Kabaso, B., (2024). Building a smart water city: IoT smart water technologies, *Applications, and Future Directions, Water, 16*(4), 557.

Okoli, N. J., & Kabaso, B. (2024). Building a smart water city: IoT smart water technologies, applications, and future directions. *Water, 16*(4), Article 557. https://doi.org/10.3390/w16040557

Smart Water Management, https://www.saft.com/energizing-iot/smart-water-management-5-innovative-solutions-water-scarcity-offered-iot#:~:text=Improving%20the%20efficiency%20of%20water,widely%20deployed%20on%20infrastructures%20nowadays

Soori, M., Dastres, R., Arezoo. B., (2023). AI-powered blockchain technology in Industry 4.0, a review, *Journal of Economy and Technology*, 1, 222–241.

Thinakaran, J., Paul, S., Christalin Latha, Ch. B., Jacob. G., (2023). Blockchain in Big Data for Agriculture Supply Chain. In Soori, M., Dastres, R., Arezoo, B. (Eds.), *Blockchain and Its Applications in Industry 4.0*, Singapore: Springer.

Verma, V. K., Tiwari, S., Mazumdar, B. D., Rajnath, Y. K. (2024). Role of Blockchain Technology in Protecting a Cloud Architecture. In Lakshmi, D., Tyagi, A. (Eds.), *Emerging Technologies and Security in Cloud Computing*, pp. 76–104, IGI Global.

6 Blockchain in Water Trading Rights

Enhancing Transparency, Efficiency, and Sustainable Management

Saeid Eslamian and Mousa Maleki

6.1 INTRODUCTION

Water resource management has undergone a significant shift since the 1970s, moving away from traditional supply augmentation strategies toward demand management. This shift is driven by increasing urbanization, agricultural changes, and heightened environmental concerns. Governments are increasingly implementing measures like voluntary water transfers to address growing water demands. However, current water rights systems based on queuing principles often lead to inefficient resource allocation, potentially resulting in land overuse and underutilization of conservation technologies. Transferable water rights and trading have emerged as a potential solution to these challenges. By equalizing water prices and providing insights into the value of water across different uses, these mechanisms aim to optimize resource allocation and promote sustainable water management practices. However, implementing water markets introduces complexities related to transaction costs and third-party impacts. This article discusses these challenges, critiques of water markets, and provides examples of water trading, concluding with suggestions for future research directions (Chong and Sunding, 2006).

The western United States and other semi-arid agricultural regions have historically relied heavily on government-subsidized water supply augmentation to meet their water needs. This approach, characterized by the construction of dams, reservoirs, and canals, has enabled the capture and transportation of surface water to agricultural and urban users, while also generating hydropower and providing flood control. Despite the significant costs associated with long-distance water transport, water prices have remained artificially low. This has contributed to a steady increase in water demand, driven by expanding agricultural practices and population growth. Agriculture currently accounts for approximately 80% of water use in the western United States, with the remaining portion allocated to municipal and industrial uses. A major challenge in this system is the discrepancy in water pricing. Urban users often pay significantly higher rates than agricultural users, sometimes up to ten times more, even after accounting for delivery and treatment costs. This disparity violates the principle of a single market price, creating an imbalance in water allocation and potentially hindering efficient resource management (Calif. Dept. Water Resour., 1998; Fischhendler and Zilberman, 2005).

A water right refers to the privilege of utilizing or benefiting from the flowing water in a stream. This right can arise from owning land adjacent to the stream (known as riparian ownership) or from actually using the stream's water. The right may be regulated and managed by a government agency, or it may not be formally governed and instead be enforced through the court system. Additionally, a water right can be acquired through a contractual agreement with an existing rights holder. There are different types of water rights, including those that specify a fixed amount of water based on flow

DOI: 10.1201/9781003599593-9

rate, and others that have no limits as long as the holder doesn't harm the water flow or quality for others. Some water rights remain in effect only as long as the holder continues to use the water, while others persist regardless of whether the water is used or not (Scott and Coustalin, 1995).

6.1.1 Introduction to Blockchain Technology

The internet has enabled the sharing of information among various parties, as seen in online banking, e-commerce, digital currency management, and financial transactions. Traditionally, these transactions required a trusted intermediary. However, blockchain technology allows parties to interact directly without the need for a middleman. Blockchain is a decentralized, distributed public ledger system that records transactions across multiple computers, ensuring that multiple copies of the same ledger are maintained in different locations. If a change is made to the ledger in one location, it is reflected in all other locations. The applications of blockchain technology are vast, spanning industries such as healthcare, transportation, finance, and energy management. The decentralized, transparent, and fault-tolerant nature of blockchain makes it a versatile technology. It also has potential applications in communication networks, such as the Internet of Things (IoT). Blockchain technology offers a revolutionary approach to IoT infrastructure. By eliminating the need for central authorities or intermediaries, it creates a decentralized network where devices can interact directly and securely. This fosters trust and transparency, allowing for robust identity and data management, enhanced privacy, and secure communication channels. Furthermore, blockchain empowers users to monetize their IoT data and resources by establishing clear ownership and facilitating secure data exchange through decentralized markets and smart contracts (Rehmani, 2021).

A blockchain is essentially a unique type of database where data, once recorded, is immutable. This means that information cannot be altered or deleted after it's been added to the blockchain, ensuring its integrity and permanence. Instead, new data is added in a sequential manner, appended to the end of the blockchain, which makes the entire system highly resistant to tampering and alteration. This design ensures the integrity and permanence of the data stored within the blockchain (Rehmani, 2021).

6.2 THE CURRENT STATE OF WATER TRADING

In the Western United States, particularly in Arizona, a long-standing tradition in water law has hindered the efficient allocation of water resources. This tradition prohibits the separate sale of water rights from the land they are attached to, forcing entities like cities and municipal water districts to purchase land solely to acquire the accompanying water rights. This practice, known as "water ranching," is considered inefficient by many economists as it unnecessarily bundles resources, limiting their potential uses and mobility. Arizona's legal structure prohibits water trading, leaving nonagricultural users with limited options. They must either buy land with attached water rights or navigate complex procedures to establish rights to unclaimed water and arrange for its transfer. A more efficient alternative would be to reform water law to allow the separation of water rights from land. This would enable farmers to sell excess water to various sectors, including industrial, municipal, recreational, environmental, and other agricultural interests. Examples of successful water markets in New Mexico and Colorado demonstrate the benefits of this approach. New Mexico's state engineer has overseen groundwater exchanges since 1952, ensuring that consumptive use remains unchanged while facilitating water reallocation. Similarly, Colorado's Big Thompson transmountain diversion project, operational since 1957, allows for the free trade of water shares, promoting efficient water reallocation along the northern front range (Swaney, 1988).

Throughout history, various water allocation mechanisms have been employed, influenced by factors such as traditions, laws, cultural beliefs, and water scarcity. Over the past 50 years, however, water scarcity and competition for water supplies have increased dramatically. Rising incomes and population growth in emerging economies, as well as climate change in some locations, have driven up demands on surface water and groundwater sources. As a result, traditional water allocation mechanisms that were once effective are now becoming increasingly problematic. Many river basins have become closed systems, and groundwater depletion in certain areas is unsustainable at current rates of use. Moreover, water security in many countries is being threatened by increasing water demands, supply constraints, and heightened supply variability due to climate change (Jiménez Cisneros et al., 2014).

According to a recent OECD publication, emerging trends are affecting water allocation mechanisms in OECD countries (Table 6.1). Despite widespread recognition of water misallocation and scarcity at all levels of government and civil society, limited progress has been made globally. However, "better planning for water resource allocation" is crucial for enhancing water security and future climate resilience (World Bank Group, 2016).

The literature on water allocation and water allocation mechanisms is vast and extensive, spanning many decades (e.g., Dinar et al., 1997). It encompasses general principles, case studies, and examinations of various problems and potential solutions to water misallocation.

TABLE 6.1
Drivers of Water Allocation Decisions

Trends	Consequences for Water Resource Management
Shifting Trends in Water Demand	• Alters the available water resource pool • Influences competition among and within different water use categories • Affects the type of water demanded (e.g., piped, level, or quality) and the desired level of supply reliability
The Devastating Effects of Climate Change on Freshwater Resources	• Alters the timing, location, form, and amount of precipitation, thereby affecting the available water resource pool • Increases uncertainty about the availability of freshwater, making it more challenging to predict and manage water resources • With the increased frequency of extreme events, what were once considered 'exceptional circumstances' become more commonplace, further complicating water management
Decline of Water Quality: A Growing Concern	• Alters the available water resource pool, affecting the overall supply of water • Increases the cost of water use, making it more expensive for various sectors and individuals to access and utilize water resources
Enhancing Water Productivity: Strategies for Optimal Water Use	• Alters the available water resource pool, affecting the overall supply of water • Influences consumption rates, changing the way water is used and allocated • May reduce flows, reducing the availability of water for subsequent uses and users • Highlights the importance of specifying return flows in entitlements, ensuring that water is allocated and managed sustainably
Evolution of Societal Values: Changing Attitudes Toward Water Management	• Shifts the value placed on in situ and diverted water uses, altering the way water is valued and prioritized • Influences the value placed on water for environmental purposes, which in turn affects the definition and implementation of environmental flows
Advancing Scientific Knowledge: Uncovering the Complexities of Water Resources and Environmental Flow Needs	• Expands our understanding of the water resource and its interactions with other water bodies, fostering a more comprehensive appreciation of the water cycle • Refines the definition of environmental flows or other in situ flow requirements, ensuring that water is managed in a way that sustains ecosystem health • Also influences our understanding of the value of environmental flows, recognizing their crucial role in supporting key species, habitats, and ecosystem services

Source: OECD (2015, p. 21)

6.2.1 Case Studies

6.2.1.1 Water Market Transactions in South Africa: Examining Two Case Studies

This research investigated the emergence of water markets in South Africa's Lower Orange and Mhlatuze River regions through surveys conducted in 1997, 1998, and 2003/2004. The findings revealed a thriving water market in the Lower Orange River region, primarily driven by water scarcity and the growing demand from table grape farmers. The Department of Water Affairs and Forestry (DWAF) played a significant role in facilitating this market by managing water transfers and reducing transaction costs. The study suggests that further improvements in market efficiency could be achieved by delegating transfer approval authority to the regional DWAF office and extending support to market transfers of canal water (Nieuwoudt and Armitage, 2004).

The analysis revealed that water rights were preferentially transferred to farmers who demonstrated the most efficient water use, as measured by their estimated return per unit of water applied. This suggests that the market mechanism effectively allocated water to those who could maximize its value. Furthermore, the analysis indicated a bias toward farmers with the potential to cultivate high-value crops like table grapes, particularly those with significant amounts of undeveloped arable land. This suggests that the water market incentivized the development of high-value agricultural production, potentially contributing to economic growth in the region (Eslamian and Maleki, 2022).

The study's investment model highlighted key factors influencing future irrigation investment: expected profits, liquidity, risk aversion, and the security of water use rights. The analysis revealed that irrigators in the study area exhibited high risk aversion, and the insecurity of water rights had a significant negative impact on their willingness to invest. Interestingly, despite water scarcity, no water rights trading market developed along the Mhlatuze River. While 41% of respondents expressed interest in purchasing water rights, there were no willing sellers. This lack of a market can be attributed to several factors: farmers utilizing their full water allocation, significant capital investments in irrigated land, and a preference for retaining excess water as a safety buffer.

While achieving equity in water access for previously disadvantaged groups necessitates government intervention in water allocation, it is crucial to avoid stifling existing and potential future water markets for irrigation. These markets offer significant benefits in improving water use efficiency and allocation. To facilitate market performance, overcoming institutional and legal barriers is essential. This includes specifying water use allocations as perpetual or for extended periods, ideally with an expiry date closer to 40 years, and ensuring their inherent security. Additionally, legislative frameworks should explicitly permit water trading.

The experience of the western United States demonstrates that private use rights to water can be established and transferred through sale or rental, even in jurisdictions where the state claims ownership of water resources. This model holds relevance for South Africa, where the government acts as the custodian of the nation's water resources. It suggests that a similar system could be implemented in South Africa, allowing for the creation of a more dynamic and efficient water market.

South Africa's current water transfer system, based on diverted use rather than consumptive use, presents a unique challenge. While it allows farmers to irrigate larger areas by adopting water conservation technologies, this approach may have unintended consequences. Since the transfer of diverted use does not directly account for consumptive use, it may incentivize increased water consumption instead of releasing excess water. This is because there's no financial incentive to reduce consumptive use, as the price of water is not tied to its actual consumption (Nieuwoudt and Armitage, 2004).

6.2.1.2 California's Informal Agricultural Water Market: Trading Dynamics

While some water markets, like the Colorado–Big Thompson Project, have established market prices and anonymous trading, others, such as the Central Valley Project in California, lack these formal structures. This absence of established institutions in the CVP leads to significant transaction costs for buyers and sellers who must actively search for trading partners, negotiate prices, and navigate complex legal processes to transfer water rights. Despite these hurdles, true arms-length transactions still occur in these informal water markets, demonstrating the resilience of market forces even without posted prices or a centralized exchange.

Water-trading data from Westlands Water District between 1993 and 1996 is used to examine how seasonal and annual changes in climate and the crop cycle interact with existing institutional mechanisms to shape trading behavior. The analysis also explores how farmers of different sizes utilize the market and the role of individual loyalty in selecting trading partners in this informal market setting. Furthermore, the potential impact of changes in market structure on Westlands farmers and the potential distribution of these changes is discussed (Brozovic et al., 2002).

The Westlands Water District, situated in California's Central Valley, is the largest district served by the Central Valley Project, encompassing over 600,000 acres and supporting nearly 800 farms. However, due to its junior water rights, Westlands often receives less than its full allocation of 1.5 million acre-feet of water annually. This water scarcity has spurred the creation of the largest agricultural water market in the United States. Within Westlands, water allocation is determined by a priority rights system, dividing farms into two main priority areas. Priority Area One, encompassing 337,000 acres, holds the most senior water rights. This land, part of the original Westlands Water District, is entitled to 900,000 acre-feet of water in a full-delivery year, translating to 2.6 acre-feet per acre.

Priority Area Two, covering 187,000 acres, holds a more junior water right. This land, originally part of the former Westplains Water District, was later annexed to Westlands. It is entitled to 250,000 acre-feet of water in a full-delivery year, equivalent to 1.3 acre-feet per acre. A smaller Priority Area Three exists with the most junior water rights, but it is not included in this analysis due to its limited scope (Brozovic et al., 2002).

6.3 BLOCKCHAIN TECHNOLOGY: AN OVERVIEW

Blockchain technology, often associated with cryptocurrencies like Bitcoin, is a decentralized, distributed database that records transactions. Unlike traditional databases controlled by a central authority, blockchain records are validated and maintained by a global network of computers. This distributed nature ensures that no single entity has control, making it impossible to alter or erase transaction history. Instead of being stored on a single server like a conventional database, blockchain information is spread across the network's users. This decentralized structure makes it impossible for one entity to manipulate the data. When a transaction occurs, it is broadcast to the network, where algorithms verify its authenticity. Once verified, the new transaction is linked to the previous one, forming a chain of transactions known as the blockchain. This decentralized, peer-to-peer network structure is the foundation of blockchain technology, ensuring transparency, security, and immutability of data. (Sarmah, 2018).

6.3.1 Applications of Blockchain beyond Cryptocurrency

Blockchain technology, initially designed to support cryptocurrencies like Bitcoin, has expanded its utility across various industries, transforming traditional processes with its decentralized, immutable, and transparent features. In the finance sector, blockchain facilitates smart contracts that automate and enforce agreements without intermediaries, accelerates cross-border payments, and enhances identity verification to reduce fraud. Supply chain management benefits from blockchain through improved traceability, which ensures product authenticity and streamlines operations by reducing paperwork and enhancing efficiency.

In healthcare, blockchain ensures secure storage and sharing of patient records, tracks pharmaceuticals to prevent counterfeit drugs, and enhances the transparency of clinical trials. The real estate industry leverages blockchain to simplify property transactions and maintain tamper-proof land registries, reducing fraud and increasing transaction speed. Blockchain technology is transforming the energy sector by enabling peer-to-peer energy trading and decentralized grid management, which in turn promotes the adoption of renewable energy sources. Additionally, blockchain enhances voting systems by providing secure and transparent platforms that ensure accurate vote recording and reduce electoral fraud.

Other significant applications include intellectual property protection, where blockchain time-stamps and records creations to verify ownership and automate royalty payments. In education, blockchain simplifies credential verification and securely tracks learning histories, providing an immutable record of educational achievements. Blockchain's inherent ability to provide secure, transparent, and efficient solutions is driving its adoption across a wide range of sectors, leading to innovative improvements and paving the way for future advancements.

6.4 IMPLEMENTING BLOCKCHAIN IN WATER TRADING RIGHTS

Implementing blockchain technology in water trading rights offers a transformative solution to manage and allocate water resources more efficiently and transparently. Blockchain can create a decentralized ledger that records all water rights transactions in real time, ensuring immutability and reducing the risk of fraud. This blockchain-based system offers stakeholders, including government agencies, water users, and environmental organizations a transparent means of tracking water usage and ownership with clarity and precision. Smart contracts can automate the enforcement of water trading agreements, ensuring compliance with regulatory frameworks and facilitating quicker, more efficient transactions. By providing a clear and unalterable record of water rights, blockchain helps optimize water distribution, supports sustainable water management practices, and enhances trust among users in water-scarce regions.

6.4.1 Advantages of Blockchain in Water Trading

Blockchain technology offers significant advantages for water trading, revolutionizing water resource management by enhancing transparency, efficiency, and security. A key benefit is the creation of an immutable and transparent record of all water rights transactions, ensuring accountability and trust. This ensures that all stakeholders, including regulators, water users, and environmental groups, have access to accurate and real-time information, reducing disputes and fostering trust. Additionally, blockchain facilitates the use of smart contracts, which automate and enforce the terms of water trading agreements. This automation reduces administrative overhead, speeds up transactions, and ensures compliance with legal and regulatory frameworks.

Moreover, blockchain strengthens security in water trading by establishing a decentralized system that is highly resistant to manipulation and fraud. Each transaction is cryptographically secured and linked to previous transactions, making it practically impossible to alter records without detection. This level of security is crucial in preventing unauthorized access and ensuring that water rights are correctly allocated and transferred. Another significant advantage is the increased efficiency in water trading operations. Blockchain reduces the need for intermediaries, cutting down on costs and administrative burdens. This streamlined process enables quicker and more cost-effective transactions, which is particularly beneficial in regions facing water scarcity, where timely and efficient water distribution is critical.

Lastly, the use of blockchain in water trading supports better resource management and sustainability. By providing a comprehensive and transparent record of water usage, blockchain helps stakeholders make informed decisions about water allocation and conservation. This can lead to more sustainable practices, as users can track and verify the amount of water being used and traded, ensuring that resources are utilized responsibly. Overall, the integration of blockchain technology in water trading offers a robust solution to the challenges of managing water rights, promoting transparency, efficiency, security, and sustainability.

6.4.2 Smart Contracts for Water Rights

Smart contracts offer a powerful tool for managing water rights by automating the execution and enforcement of water trading agreements. Operating on a blockchain network, these self-executing contracts, with their terms directly embedded in code, guarantee transparency, security, and efficiency. By eliminating the need for intermediaries, smart contracts streamline the transfer of water rights, reducing administrative costs and minimizing delays. Once predefined conditions are met, such as the transfer of a specified amount of water at an agreed-upon price, the smart contract automatically executes the transaction, ensuring that all parties comply with the terms without the need for manual intervention.

One of the significant advantages of using smart contracts for water rights is the enhanced security and trust they provide. Every transaction is permanently recorded on the blockchain, creating an immutable and verifiable ledger that is accessible to all relevant stakeholders. This transparency reduces the risk of fraud and disputes, as all parties can independently verify the details of each transaction. Additionally, the cryptographic security inherent in blockchain technology ensures that water rights data is protected against unauthorized access and tampering, further safeguarding the integrity of water trading activities.

Moreover, smart contracts facilitate better regulatory compliance and resource management. By automating the enforcement of legal and regulatory requirements, smart contracts ensure that water trading activities adhere to established rules and standards. This automation not only simplifies the compliance process but also enhances accountability, as every transaction is permanently recorded on the blockchain. Furthermore, the use of smart contracts can promote more sustainable water management practices by providing real-time data on water usage and trading activities. This data can inform policy decisions and help stakeholders optimize water distribution, ensuring that resources are allocated efficiently and responsibly.

6.5 EXPLORING BLOCKCHAIN IMPLEMENTATION: CASE STUDIES AND PILOT PROJECTS

Several case studies and pilot projects around the world have demonstrated the potential of blockchain technology to revolutionize water trading. These initiatives showcase the advantages of blockchain in improving transparency, efficiency, and security in water management.

6.5.1 Case Study 1: South Australia's Water Ledger Pilot

In South Australia, a blockchain-based water trading platform called Water Ledger was piloted to streamline the allocation and trading of water rights. This project, supported by the government, aimed to enhance transparency and reduce administrative overhead in water trading. The Water Ledger platform used smart contracts to automate transactions, ensuring that trades were executed only when specific conditions were met. This not only sped up the process but also reduced the risk of disputes and errors. The pilot demonstrated how blockchain could provide real-time visibility into water usage, helping regulators and stakeholders make more informed decisions.

6.5.2 Case Study 2: IBM and SweetSense in California

In California, IBM and SweetSense partnered on a blockchain project to monitor groundwater usage. The initiative aimed to improve sustainability and compliance with water regulations. Sensors placed in wells collected real-time data on water extraction, which was then recorded on a blockchain. This data was accessible to all stakeholders, ensuring transparency and reducing the potential for over-extraction. The use of blockchain allowed for more efficient and secure tracking of water usage, promoting responsible water management practices and aiding in the enforcement of water rights.

6.5.3 Case Study 3: Colorado River Basin Project

The Colorado River Basin project is another significant example, involving a consortium of stakeholders including universities, government agencies, and technology firms. This pilot project explored the use of blockchain to manage water rights and trades across the basin. By using a blockchain-based platform, the project aimed to create a more transparent and efficient system for water trading. The immutable nature of blockchain records helped ensure that all transactions were accurately documented and easily verifiable. The project also explored the potential for smart contracts to automate water rights transfers, reducing the need for manual processing and minimizing delays.

6.6 TECHNICAL AND REGULATORY CHALLENGES IN WATER TRADING USING BLOCKCHAIN

The implementation of blockchain technology in water trading is promising, but it is not without its challenges. These challenges fall into two main categories: technical hurdles and regulatory and legal considerations. Addressing both is crucial to fully unlock the potential of blockchain in this sector.

6.6.1 Technical Hurdles

One of the primary technical hurdles is scalability. Blockchain networks, particularly those using proof-of-work consensus mechanisms, often struggle to handle a high volume of transactions quickly. This can be a significant drawback for water trading systems that need to process numerous transactions in real time. Scalability issues can lead to slower transaction times and higher costs, undermining the efficiency benefits of using blockchain.

Interoperability is another critical challenge. Water trading systems must integrate with various existing water management and monitoring systems, which often use different standards and protocols. Ensuring that blockchain platforms can seamlessly interact with these diverse systems is essential for creating a cohesive water trading environment. Without effective interoperability, the implementation of blockchain could be hampered by fragmentation and inefficiencies.

Data privacy is a significant concern in blockchain applications, as the decentralized and transparent nature of the technology means that all transactions are visible to network participants. In the context of water trading, this transparency must be balanced with the need to protect sensitive information about water rights holders and usage patterns. Implementing privacy-preserving techniques, such as zero-knowledge proofs or private blockchains, can help address these concerns, but they also add complexity to the system.

Security is another major issue. While blockchain is inherently secure against certain types of fraud, it is not immune to all threats. Potential vulnerabilities include attacks on consensus mechanisms, smart contract bugs, and phishing attacks on users. Ensuring robust security measures, such as regular audits, strong encryption, and secure coding practices is crucial to maintaining the integrity of water trading systems.

6.6.2 Regulatory and Legal Considerations

The legal and regulatory landscape for blockchain-based water trading is still evolving. Current legal frameworks often do not adequately address the unique characteristics and requirements of blockchain technology. Existing water rights regulations, for instance, may not accommodate the decentralized and automated nature of blockchain transactions, leading to legal ambiguities and potential conflicts.

Furthermore, the lack of standardization in blockchain technology can create challenges in ensuring compliance with regulations. Different jurisdictions may have varying requirements for data protection, transaction verification, and record-keeping, complicating the implementation of a unified blockchain-based water trading system.

To fully integrate blockchain technology into water trading, regulatory frameworks will need to evolve. Future trends may include the development of specific regulations for blockchain applications in water management, focusing on issues such as data privacy, transaction validation, and dispute resolution. Governments and regulatory bodies may also establish standards for interoperability and security to ensure consistent and reliable implementation across different regions.

Additionally, there may be a shift toward more flexible and adaptive regulatory approaches that can keep pace with rapid technological advancements. This could involve the creation of regulatory sandboxes, where blockchain-based water trading systems can be tested in a controlled environment before broader implementation. Such approaches would allow for the identification and resolution of potential issues while ensuring that regulations support innovation and sustainability in water management.

6.7 CONCLUSION

Addressing the technical and regulatory challenges of implementing blockchain in water trading is crucial for the successful adoption and integration of this technology. By overcoming scalability and interoperability issues, enhancing data privacy and security, and adapting regulatory frameworks to accommodate blockchain's unique features, stakeholders can leverage blockchain to create more transparent, efficient, and secure water trading systems. These efforts will pave the way for more sustainable and equitable water resource management.

REFERENCES

Brozovic, N., Carey, J. M. and Sunding, D. L., 2002. Trading activity in an informal agricultural water market: An example from California. *Journal of Contemporary Water Research and Education*, *121*(1), p. 2.

Calif. Dept. Water Resour. 1998. California Water Plan Update, Bull. 60–98, Sacramento, CA, USA.

Chong, H. and Sunding, D., 2006. Water markets and trading. *The Annual Review of Environment and Resources*, *31*, pp. 239–264.

Dinar, A., Rosegrant, M. W. and Meinzen-Dick, R. S. 1997. *Water Allocation Mechanisms: Principles And Examples*. Washington, DC USA World Bank Publications.

Eslamian, S. and Maleki, M., 2022, Development of an Earthquake-Induced Landslide Hazard Assessment. In Eslamian, S. and F. Eslamian (eds.) *Disaster Risk Reduction for Resilience: Disaster Economic Vulnerability and Recovery Programs*. Switzerland: Springer International Publishing, 179–189.

Fischhendler, I. and Zilberman, D. 2005. Packaging policies to reform the water sector: The case of the Central Valley Project Improvement Act. *Water Resources Research*, *41*, p. W0702.

Jiménez Cisneros, B. E., Oki, T., Arnell, N. W., Benito, G., Cogley, J. G., Döll, P., Jiang, T. and Mwakalila, S. S. 2014. Freshwater Resources. In Field, C. B., Barros, V. R., Dokken, D. J., Mach, K. J., Mastrandrea, M. D., Bilir, T. E., Chatterjee, M., Ebi, K. L., Estrada, Y. O., Genova, R. C., Girma, B., Kissel, E. S., Levy, A. N., Maccracken, S., Mastrandrea, P. R. and White, L. L. (eds.) *Climate Change 2014: Impacts, Adaptation, and Vulnerability. Part A: Global and Sectoral Aspects. Contribution of Working Group II to the Fifth Assessment Report of the Intergovernmental Panel of Climate Change*. Cambridge, United Kingdom and New York, NY USA Cambridge University Press.

Nieuwoudt, W. L. and Armitage, R. M., 2004. Water market transfers in South Africa: Two case studies. *Water Resources Research*, *40*(9). https://doi.org/10.1029/2003WR002840

OECD, 2015. *Water Resources Allocation: Sharing Risks and Opportunities*. OECD Studies on Water, Paris, France: OECD Publishing.

Rehmani, M. H., 2021. Blockchain fundamentals and working principles. In Rehmani, M. H. (ed.) *Blockchain Systems and Communication Networks: From Concepts to Implementation*, pp. 23–59. Cham: Germany, Springer International Publishing.

Sarmah, S. S., 2018. Understanding blockchain technology. *Computer Science and Engineering*, *8*(2), pp. 23–29.

Scott, A. and Coustalin, G., 1995. The evolution of water rights. *Natural Resources Journal*, *35*, p. 821.

Swaney, J. A., 1988. Trading water: Market extension, social improvement, or what?. *Journal of Economic Issues*, *22*(1), pp. 33–47.

World Bank Group. 2016. *High and Dry: Climate Change, Water, and the Economy*. Washington, DC USA World Bank.

Part IV

Blockchain Application in Ecological Protection and Water Quality

Part IV

7 Blockchain Systems for Ecological Sustainability

Advancing Resource Governance and Climate Resilience

Elena Gurgu

7.1 INTRODUCTION

A decentralized, secure method of transaction recording, blockchain technology has become widely acknowledged and has achieved traction for its transformative potential in everything from finance to governance (Haber & Stornetta, 1997; Nakamoto, 2008). The technology was originally designed to support cryptocurrencies, but it has quickly begun to grow and expand its uses, with potential applications for water trading and disaster recovery (Zheng et al., 2018). These areas are riddled with matters of inefficiency with resource allocation, transparency, and possible fraudulent activities that blockchain can help solve by confirming traceability, security, and automation with smart contracts (Albrecht et al., 2021; Savelyev, 2018).

A growing worldwide crisis, water scarcity affects billions of people and fuels disputes over water distribution (Mekonnen et al., 2022). In response, a market-driven way to maximize resource distribution is water trading, which is the exchange of water rights between users (Grafton et al., 2019). But conventional water trading systems are frequently unreliable, ineffective, and vulnerable to fraud (Albrecht et al., 2021).

Blockchain solves these problems by offering a decentralized ledger that transparently and irrevocably documents water transactions (Zwitter & Boisse-Despiaux, 2018). Blockchain ensures regulatory compliance and minimizes disputes by automating transactions based on predetermined conditions through the integration of smart contracts (Chen et al., 2021a).

According to Mekonnen et al. (2022), a blockchain-based water market has the potential to enforce sustainable usage limits, track water rights in real-time, and enable safe peer-to-peer transactions without the need for middlemen. These developments have already been investigated in nations such as Australia, where blockchain technology has the potential to improve market efficiency and lower administrative costs in water trading systems (Grafton et al., 2019). Additionally, blockchain improves water resource management accountability and traceability. Blockchain guarantees that decisions about water allocation are founded on precise, impenetrable data by connecting transactions to real-time hydrological data and Internet of Things (IoT)–enabled sensors (Zheng et al., 2018). This lowers the possibility of over-extraction and unsustainable practices while promoting fair and data-driven water distribution (Albrecht et al., 2021).

Significant logistical and financial obstacles frequently face disaster recovery efforts, such as mismanagement of resources, corruption, and delays in aid distribution (Gonzalez & Smith, 2017). By offering a safe and auditable system for monitoring donations, aid distribution, and logistics, blockchain can improve humanitarian response's transparency, effectiveness, and trust (Zwitter & Boisse-Despiaux, 2018).

DOI: 10.1201/9781003599593-11

Donation tracking is one of the most significant uses of blockchain in disaster relief. Due to inadequate oversight, traditional charitable donations frequently suffer from a lack of accountability, where money may be misappropriated or siphoned off (Savelyev, 2018). Real-time contribution tracking is made possible by blockchain-based platforms, which guarantee that money reaches its intended recipients without the need for middlemen (Tapscott & Tapscott, 2017).

Blockchain has already shown how to safely handle food distribution and humanitarian funds through projects like AidCoin and the World Food Programme's Building Blocks initiative (Zwitter & Boisse-Despiaux, 2018).

Blockchain enhances supply chain management and disaster logistics in addition to financial assistance. Humanitarian organizations can keep an eye on deliveries, stop fraud, and guarantee the effective distribution of necessities by tracking the movements of relief supplies on a blockchain (Gonzalez & Smith, 2017). Furthermore, displaced people can obtain aid and services more rapidly without depending on misplaced or unreachable documentation by using blockchain-powered identity verification (Zheng et al., 2018).

Blockchain is a game-changing tool for addressing pressing global issues, not just a technical advancement. Blockchain technology facilitates the safe, effective, and transparent distribution of limited water resources in water trading, guaranteeing equitable and sustainable distribution.

It helps reduce corruption and inefficiencies in crisis management by improving accountability and trust in the distribution of aid during disaster recovery. Blockchain's incorporation into these industries has enormous potential to enhance governance, lessen systemic inefficiencies, and promote more robust and equitable systems as it develops further (Hassan & De Filippi, 2021).

The historical evolution of blockchain, its fundamental workings, and the particular applications it is being used for to improve water trading and disaster response globally will all be covered in the sections that follow.

7.2 THE HISTORICAL EVOLUTION OF BLOCKCHAIN TECHNOLOGY FOR WATER AND ENVIRONMENT SYSTEMS

7.2.1 Early Foundations of Cryptography (1980s–1990s)

The development of cryptography in the 1980s and 1990s laid the groundwork for blockchain technology. Modern blockchain systems were made possible by a number of significant innovations, especially in the areas of digital signatures, secure timestamping, and cryptographic data structures. Decentralized and impenetrable systems that are currently being used for water and environmental management applications were made possible in large part by these early advancements.

Cryptographic timestamping was introduced by Haber and Stornetta (1997) to guarantee the immutability and integrity of digital documents. Their system suggested a way to time-stamp documents so that doing so would render the timestamp invalid. Blockchain technology, which is currently being used for environmental monitoring and compliance verification in water governance, was developed later as a result of this idea of chaining cryptographic proofs (Haber & Stornetta, 1997).

Ralph Merkle developed Merkle trees, which offered a quick and safe method of confirming the accuracy of data in big datasets (Merkle, 1988). This structure enables effective information authentication without necessitating access to the complete dataset—a concept that subsequently became crucial for blockchain verification procedures. Furthermore, secure identity verification in distributed systems was made possible by digital signatures, which were first introduced by Diffie and Hellman in 1976 and subsequently improved by RSA encryption (Rivest et al., 1978). These cryptographic methods are now essential for protecting access control and sharing of environmental data in decentralized water systems.

7.2.2 The Emergence of Bitcoin (2008–2013)

An important turning point in the development of blockchain technology was the publication of Satoshi Nakamoto's white paper (Nakamoto, 2008). Using a proof-of-work (PoW) consensus mechanism, the paper, "Bitcoin: A Peer-to-Peer Electronic Cash System," described a decentralized, trustless, and tamper-resistant system (Nakamoto, 2008). Blockchain solutions for tracking carbon credits, tracking pollution, and allocating water resources optimally were later developed as a result of this model.

The first decentralized blockchain network, Bitcoin, was introduced in 2009. In order to secure transactions and provide incentives to participants, this system introduced mining. According to Narayanan et al. (2016), Bitcoin proved that blockchain technology could be used to accomplish distributed consensus without requiring a central authority. Although the primary purpose of Bitcoin is finance, its underlying blockchain technology has been modified for use in sustainability applications like distributed energy management and transparent water usage tracking.

7.2.3 Smart Contracts and Programmable Blockchains (2014–2017)

Ethereum, a blockchain that introduced smart contracts and increased the functionality of Bitcoin, was proposed by Vitalik Buterin in 2014. Complex decentralized applications (DApps) are made possible by these self-executing contracts, which function according to predetermined rules (Buterin, 2014). Smart contracts enable automated transactions in water and environmental systems, including real-time monitoring of pollution discharge limits and pay-for-performance programs in water conservation projects.

Numerous decentralized applications in the fields of finance, gaming, governance, and supply chain management were developed as a result of Ethereum's launch (Wood, 2015). According to Zheng et al. (2018), smart contracts automated a number of transactional procedures and decreased the need for middlemen. DApps have been used in the environmental sector to improve carbon credit exchanges, peer-to-peer water trading, and ecosystem service monitoring.

7.2.4 Blockchain 3.0 and Industry Adoption (2018–Present)

Scalability issues emerged as blockchain adoption increased. To solve network congestion and transaction speed, innovations like Layer 2 solutions (such as the Lightning Network for Bitcoin and Rollups for Ethereum) were developed (Chen et al., 2021b). These solutions are crucial for managing massive sensor data from IoT devices that track hydrological cycles, climate impacts, and water quality in the context of water and environmental systems.

Applications of blockchain technology have spread from the financial industry to the fields of water and environmental management (Pathak et al., 2022). According to recent studies, blockchain can help decentralized water trading systems, improve supply chain transparency, and manage water resources more effectively (Xie et al., 2022; Zhang & Li, 2023). These applications show how blockchain technology can be used to solve issues with sustainability and legal compliance (Bastos et al., 2022). Blockchain is being investigated by governments and businesses to promote equitable water distribution, stop illicit water extraction, and produce unchangeable records for monitoring pollution.

7.3 THE BLOCKCHAIN STACK: LAYERS AND THEIR FUNCTIONS

Each layer of the multi-layered architecture of blockchain technology contributes differently to the efficiency, security, and integrity of the system. Applying blockchain solutions to environmental and water systems requires an understanding of how these layers work.

7.3.1 Hardware and Network Layer

The physical infrastructure (computers, servers) and the decentralized network that connects all participants make up the hardware and network layer, which serves as the framework for a blockchain network. When it comes to decentralization and energy consumption issues, it is essential.

Regarding the function of dispersed nodes in decentralization, we are aware that nodes in a blockchain network are dispersed throughout various geographic locations, each of which possesses a copy of the complete ledger. This decentralization guarantees that no one entity can control the system and does away with the need for a central authority (Shen et al., 2022). Additionally, distributed nodes boost the network's resilience and offer redundancy. Decentralized nodes can be used to gather and validate data from various sources for environmental monitoring, guaranteeing the accuracy and reliability of environmental data like pollution levels, water quality, and climate monitoring (Vladucu et al., 2024).

Because of the high processing power required for transaction validation and network security, blockchain networks—particularly those that use proof-of-work (PoW)—consume a lot of energy. In environmental applications, this problem is especially troubling. But new consensus systems like proof-of-stake (PoS) choose validators based on the quantity of cryptocurrency they own, not on computationally demanding tasks, in an effort to save energy (Wadhwa et al., 2022). PoS and other low-energy alternatives offer a more energy-efficient solution for environmental sustainability applications.

7.3.2 Data Layer

The data that will be stored on the blockchain must be organized and stored by the data layer. It guarantees transparency and data integrity. Merkle trees, a cryptographic structure at the heart of the data layer, effectively confirm the consistency and legitimacy of the data stored on the blockchain, thereby guaranteeing data integrity. Data is hashed into a tree-like structure using a Merkle tree, and the integrity of the entire dataset can be checked using the root hash (Zohar et al., 2021). For environmental monitoring, where data must be reliable and verifiable, this feature is essential. Stakeholders can verify environmental data in real time using Merkle trees, which guarantees its dependability and tamper-proofness.

Because it guarantees that recorded data cannot be changed or tampered with, blockchain's immutable ledger is perfect for applications like environmental monitoring where data integrity and transparency are crucial. Data on the blockchain is permanent and transparent once it is recorded; it cannot be changed or removed (Tapscott & Tapscott, 2017). This is especially crucial for monitoring environmental indicators like pollution, carbon emissions, and water use, where precise historical data is necessary for sustainability initiatives and policy decisions.

7.3.3 Consensus Layer

Validating transactions and upholding the consensus of the entire blockchain network are the responsibilities of the consensus layer. It guarantees that everyone involved agrees on the blockchain's current state.

The two most talked about consensus methods, PoW and PoS, take different tacks when it comes to securing the blockchain and validating transactions. To validate transactions and add them to the blockchain, PoW requires participants, or miners, to solve challenging mathematical puzzles. However, PoW uses a lot of energy because it uses a lot of computational resources (Narayanan et al., 2016). On the other hand, PoS drastically lowers the network's energy requirements by using a staking system to choose validators according to the quantity of cryptocurrency they own and are prepared to risk (Wadhwa et al., 2022).

PoS is a more energy-efficient and sustainable choice for environmental applications.

Ensuring data integrity and preventing unauthorized changes are crucial in environmental data management because they preserve the precision and dependability of transactions involving environmental data. By verifying the data before it is added to the blockchain, consensus mechanisms aid in the security of transactions. This guarantees the accuracy and immutability of environmental data, including water usage records, offering a solid foundation for decision-making (Shen et al., 2022).

7.3.4 Smart Contract Layer

By using self-executing contracts to automate procedures and transactions, the smart contract layer eliminates the need for middlemen. By enforcing predetermined terms, like water rights and conservation regulations, smart contracts can enable automated water trading and sustainability while guaranteeing effective and transparent resource allocation (Satilmisoglu, 2024). Water resources are distributed effectively and sustainably thanks to these contracts, which only carry out transactions when requirements are satisfied. Blockchain can simplify procedures like water distribution, quality control, and regulatory compliance by incorporating environmental laws into smart contracts. Smart contracts are vulnerable to security flaws, and errors in logic or code can result in exploitation or unexpected consequences, despite their benefits. This underscores the importance of careful design and testing (Miers et al., 2021).

Strict testing, official verification, and auditing are required to guarantee the security of these contracts in delicate applications like environmental monitoring and water trading.

7.3.5 Application Layer

Through decentralized applications (dApps), users engage with blockchain technology at the application layer. It offers the interfaces and tools needed to apply blockchain solutions in practical settings. Tracking water consumption, encouraging transparency in carbon emissions, and guaranteeing effective and transparent resource distribution during emergencies are some of the main uses of blockchain technology in water management, climate action, and disaster relief.

In water management, blockchain can be used to track water rights, monitor and manage water resources, and guarantee transparent distribution and use, all of which support accountability and sustainability (Satilmisoglu, 2024; Lezzi et al., 2024). By offering up-to-date information on water quality and availability, this can aid in addressing problems related to water scarcity.

By enabling transparent tracking of carbon credits and emissions and guaranteeing adherence to environmental regulations, blockchain technology can aid in climate action initiatives while promoting sustainability and accountability (Tapscott & Tapscott, 2017). Additionally, it can be used to confirm carbon offset initiatives and guarantee the accuracy of data pertaining to climate change.

By guaranteeing the transparent and effective distribution of resources, cutting down on delays, and improving coordination, blockchain can expedite the distribution of humanitarian aid in disaster relief. By automating the distribution of money or supplies according to verified needs, smart contracts can cut down on fraud and delays (Wadhwa et al., 2022). The application layer demonstrates how blockchain can be used to improve transparency, efficiency, and accountability in a range of sustainability and environmental-related fields.

7.4 BLOCKCHAIN APPLICATIONS ACROSS LAYERS IN WATER AND ENVIRONMENTAL SYSTEMS

Originally created for digital currencies, blockchain technology has since advanced to provide game-changing solutions for a number of industries, including environmental and water management. Its decentralized, transparent, and unchangeable features offer special chances to tackle difficult problems in these domains. Water trading and resource management, disaster recovery and humanitarian aid distribution, decentralized identity and IoT integration for water monitoring, and carbon credit trading for climate change mitigation are some of the many uses of blockchain technology that are examined in this section.

7.4.1 Water Trading and Resource Management

By bringing transparency, efficiency, and security to water transactions, blockchain technology has the potential to completely transform resource management and water trading. Water rights disputes, delayed transactions, and information asymmetry are some of the problems that traditional water markets frequently face. These problems can be successfully resolved by putting blockchain-based solutions into practice. Blockchain ensures trust and efficiency in water transactions by enabling the development of decentralized platforms where water rights can be traded safely and openly. To improve trust and transparency in water markets, for example, IBM is working with Australia to develop a blockchain-enabled water trading platform. The platform reduces the complexity and inefficiencies associated with traditional water trading systems by offering a shared, immutable ledger that documents all transactions. This ensures that market participants can audit trades and obtain comprehensive information (IBM, 2023). Blockchain can improve integrated water resource management beyond trading by guaranteeing precise water usage tracking, encouraging fair distribution, and boosting overall resource allocation efficiency. Data reliability is improved and well-informed decision-making in water resource management is supported by a study by Li et al. (2022) that suggests a blockchain-based method for tracking water abstraction permits and guaranteeing water quality.

7.4.2 Disaster Recovery and Humanitarian Aid Distribution

Humanitarian aid must be distributed effectively and transparently in disaster-affected areas. Blockchain technology provides ways to guarantee that resources reach the right people, cut down on fraud, and expedite the delivery of aid. By ensuring that all transactions pertaining to the distribution of aid are transparently recorded, blockchain's immutable ledger lowers the possibility of fraud and misallocation while simultaneously increasing accountability. Stakeholders can ensure that aid reaches those in need as soon as possible by monitoring the flow of resources. In order to ensure that recipients receive aid without the need for middlemen, the World Food Program, for instance, has investigated blockchain solutions for food assistance distribution. This has reduced costs and increased efficiency (World Food Program, 2019).

Because blockchain technology is decentralized, it facilitates better coordination between different humanitarian organizations by offering a common platform for information sharing, which improves cooperation and effectiveness in disaster relief operations. This lessens effort duplication and guarantees that aid is disbursed in accordance with current needs assessments. The integration of blockchain and IoT for disaster management is covered in a study by Singh et al. (2023), which emphasizes how these technologies can improve the effectiveness of aid distribution networks.

7.4.3 Decentralized Identity and IoT Integration for Water Monitoring

Blockchain technology combined with decentralized identity systems and IoT devices provides creative approaches to water management and monitoring.

IoT sensors allow for safe, real-time tracking of water conditions for better management by continuously monitoring water quality parameters like pH, turbidity, and contaminant levels. Blockchain integration with these sensors makes data easily accessible to stakeholders and impenetrable to tampering. For efficient management of water resources, this integration guarantees data integrity and makes real-time monitoring easier. Nododile and Nyirenda (2025) show how blockchain and the

Internet of Things can be used in water monitoring applications by investigating a hybrid blockchain-InterPlanetary File System (IPFS) solution for safe and scalable data collection from smart water meters.

Blockchain-based decentralized identity systems give people more control over their personal information, improving security and privacy while guaranteeing that sensitive data is better protected. Such systems can authenticate users gaining access to water resources or services in the context of water management, guaranteeing that only authorized individuals engage in water trading or obtain assistance. This method lowers the possibility of fraud while boosting trust. Singh et al. (2023) discuss how blockchain and IoT integration for water management also helps ensure safe and effective data management in water monitoring systems.

7.4.4 Carbon Credit Trading and Climate Change Mitigation

In order to improve the efficiency and transparency of carbon credit trading systems—which are crucial for international efforts to mitigate climate change—blockchain technology is essential.

Blockchain ensures more accuracy in carbon markets while lowering the risk of fraud and double-counting by offering an improved, transparent, and unchangeable record of carbon credit transactions. Increased involvement in carbon markets is encouraged by this transparency, which also fosters trust among stakeholders. Vladucu et al.'s 2024 study examines how blockchain is being used in environmental sustainability initiatives, emphasizing how it can be used to track greenhouse gas emissions and assist carbon credit schemes.

Effective real-time tracking of emissions is made possible by combining blockchain technology with Io T sensors. This enables prompt reporting, verification, and precise data management.

Accurate accounting of emissions reductions is made easier by this integration, which is essential for carbon credit schemes to remain credible. The potential of blockchain and IoT in emissions monitoring and reporting is highlighted by Singh et al.'s (2023) study on these technologies for environmental management.

Blockchain technology provides revolutionary solutions for environmental systems and water layers. Its uses in carbon credit trading, water trading, disaster recovery, and water monitoring show how adaptable it is and how well it can handle difficult problems in these fields. To fully reap the benefits of blockchain in environmental sustainability, it is imperative to tackle technical, organizational, and regulatory obstacles as adoption increases.

7.5 CHALLENGES AND FUTURE TRENDS IN BLOCKCHAIN DEVELOPMENT

A promising answer to many problems in water and environmental systems is blockchain technology. Notwithstanding its benefits, the technology has a number of important issues that need to be resolved if it is to be widely used and effective. Three main issues are examined in this section: performance and scalability optimization, energy consumption in relation to sustainability, and policy and regulatory issues. It also emphasizes upcoming developments in blockchain technology and how they might affect water and environmental systems.

7.5.1 Scalability and Performance Optimization

Scalability is one of the main issues with blockchain technology. It gets harder to maintain high performance and low latency as blockchain networks expand and process more transactions. This problem is particularly significant in the context of water and environmental systems since effective processing of vast amounts of data is necessary for real-time resource monitoring and management.

The ability of a blockchain to manage a growing volume of transactions without sacrificing performance is known as scalability, and it is a crucial issue for preserving effectiveness as the network grows. Blockchain needs to be able to handle transactions at scale in order to be used in applications like disaster recovery, carbon credit trading, and water management.

However, throughput issues plague many blockchain networks, especially those that use PoW consensus mechanisms. This problem is particularly noticeable in public blockchains such as Ethereum and Bitcoin, where network congestion can result in higher transaction costs and slow transaction times (Zohar et al., 2022).

Recent developments in blockchain technology have concentrated on enhancing performance and scalability, providing fixes and optimization strategies to boost effectiveness as networks expand. The creation of second-layer solutions, like the Lightning Network for Bitcoin or Ethereum's switch to PoS, which attempts to increase scalability and lower energy consumption, is one promising remedy.

Off-chain transactions are made possible by Layer 2 protocols, which improve throughput and lessen the strain on the primary blockchain. Another strategy, called sharding, increases the total transaction capacity by splitting the blockchain into smaller units, or "shards," that can process transactions concurrently (Cohen & Kim, 2021). These solutions will be crucial for enabling extensive environmental applications, such as real-time environmental data management and IoT–based water monitoring, as blockchain technology develops further.

7.5.2 Energy Consumption vs. Sustainability

When it comes to blockchain adoption, energy consumption is a major concern, especially when considering sustainability. The environmental impact of PoW–based blockchains, like Ethereum and Bitcoin (before switching to PoS), has been a major concern due to their energy-intensive nature. Blockchain's energy footprint and sustainability potential must be weighed in relation to water and environmental systems. Concerns regarding the environmental impact of blockchain networks are raised by PoW consensus mechanisms, which use a lot of energy and require miners to solve challenging mathematical puzzles in order to validate transactions.

According to a study by De Vries (2021), Bitcoin is one of the most energy-hungry technologies in use today, with energy consumption estimated to be higher than in some nations. This raises questions regarding blockchain networks' carbon footprint, particularly in light of the need for sustainable technologies in the environmental sector and the increased emphasis on mitigating climate change worldwide. Blockchain developers have been investigating more energy-efficient consensus mechanisms, like PoS, which uses significantly less processing power and provides a sustainable substitute for PoW, in response to energy concerns. It is anticipated that Ethereum will use over 99% less energy after switching to PoS (Kudzin, A., 2022).

Furthermore, there is growing interest in using renewable energy sources to power blockchain mining operations. According to studies, blockchain technology and renewable energy networks can work together to generate synergies that support sustainability while retaining the decentralized nature of blockchain (Sridhar et al., 2023). This trend is essential for environmental applications, where the technology's sustainability must match the environmental objectives it supports.

7.5.3 Regulatory and Policy Considerations

Blockchain technology's regulatory and policy environment is still mostly unclear, especially when it comes to its use in environmental and water systems. Effectively regulating blockchain-based applications while encouraging innovation is a challenge for governments and regulatory agencies. Blockchain adoption in regulated sectors like environmental monitoring and water management may be severely hampered by unclear regulations and legal frameworks.

Traditional regulatory frameworks that depend on centralized authority and control face difficulties due to blockchain's decentralized and unchangeable nature, which raises legal questions and compliance problems for its adoption. For instance, regulatory agencies may find it difficult to monitor adherence to environmental standards or enforce water rights on a blockchain in the water sector.

It is becoming more and more critical to have a clear regulatory framework that facilitates the incorporation of blockchain technology with current environmental regulations (Naqash et al., 2023).

Regulatory organizations are starting to investigate blockchain-specific regulations in an effort to address these issues, with the goal of developing frameworks that can facilitate the integration of the technology while maintaining security and compliance. In order to provide safe, international blockchain-based public services, the European Union, for instance, has taken the initiative to create legal frameworks for blockchain technology. One such framework is the European Blockchain Services Infrastructure (EBSI) (European Commission, 2022). Additionally, the creation of environmentally compliant smart contracts can aid in the implementation of legislation pertaining to pollution control, water rights, and carbon emissions (Lezzi et al., 2024).

Collaboration between technology developers, regulatory agencies, and legislators will be crucial as blockchain continues to gain traction in environmental applications in order to guarantee legal compliance and promote broad adoption.

7.5.4 Future Trends in Blockchain Development

A number of upcoming developments are anticipated to influence blockchain technology's function in water and environmental systems as it develops further.

Blockchain technology combined with artificial intelligence (AI) and IoT devices has the potential to completely transform water resource management and monitoring by facilitating more effective and transparent resource distribution. AI algorithms can evaluate real-time data on water quality, usage, and environmental conditions gathered by IoT sensors to maximize resource management. To ensure openness and confidence in environmental decision-making, blockchain can offer a safe, decentralized platform for storing and exchanging this data (Cheng et al., 2022).

Interoperability between various blockchain networks and legacy systems will be increasingly important as blockchain adoption increases in order to facilitate smooth integration and cooperation. Future blockchain platforms will probably concentrate on developing standardized frameworks and protocols that enable smooth integration across various industries and geographical areas. This will make it possible for blockchain to be used in more extensive ways in global environmental systems, including water management, disaster recovery, and carbon trading (Bendale et al., 2023). Because PoS and hybrid blockchain models offer more scalable and energy-efficient solutions, their adoption will continue to grow in light of concerns about scalability and energy consumption.

These models preserve the security and decentralization provided by blockchain technology while providing more energy-efficient solutions. Blockchain will be more appropriate for applications that need real-time data processing, like resource management and environmental monitoring, as it advances toward these models.

In conclusion, blockchain technology has a lot of potential to enhance water and environmental systems by providing answers for environmental monitoring, disaster recovery, and transparent resource management. However, a number of issues, such as scalability, energy consumption, and regulatory uncertainty, need to be resolved before its full potential can be achieved. Future developments in energy-efficient consensus techniques, scalability solutions, and regulatory frameworks will be crucial in determining how widely blockchain is adopted in environmental systems as it develops.

7.6 CONCLUSIONS

Over the past ten years, blockchain technology—which was first created as the foundation for cryptocurrencies—has undergone significant development. Nowadays, its uses go well beyond the financial industry, as there is growing interest in using its decentralized, transparent, and safe features to solve some of the most important problems facing the planet, such as those pertaining to environmental and water systems. The development of blockchain technology and its application to water and environmental systems are summarized in this chapter, which is followed by future projections regarding the technology's potential for sustainable resource management.

The history of blockchain started when it was linked to Bitcoin and the larger cryptocurrency community. Nonetheless, blockchain's intrinsic qualities—transparency, immutability, and decentralization—have made it extremely pertinent to industries that need effective and safe data management. For water and environmental systems, which depend on precise, unchangeable data for resource management, decision-making, and regulatory compliance, these characteristics are especially crucial (Kudzin, 2022).

Blockchain has demonstrated promise in water management by enabling the safe and transparent allocation of water rights, guaranteeing fair access, and stopping fraudulent water transaction practices (Satilmisoglu, 2024).

Furthermore, blockchain's immutability guarantees data integrity and helps prevent fraud in environmental monitoring applications like tracking carbon credits and emissions (Tapscott & Tapscott, 2017). Blockchain can offer real-time data on water quality, usage, and other environmental parameters through the integration of IoT devices, improving the management of these valuable resources even more (Cheng et al., 2022).

Blockchain adoption in water and environmental systems has not been without difficulties, though, despite its potential. The broad use of blockchain technology is still hampered by problems like scalability, energy consumption, and regulatory uncertainty. To fully utilize blockchain in these industries, these obstacles must be removed.

In the future, blockchain integration with other cutting-edge technologies like artificial intelligence (AI) and the Internet of Things (IoT) will be crucial in revolutionizing environmental and water management. As blockchain technology advances, new avenues for sustainable resource management will become possible due to its capacity to automate procedures via smart contracts, produce transparent and auditable records, and facilitate decentralized decision-making (Sridhar et al., 2023).

The scalability of blockchain networks is one important area of development. The scalability and energy efficiency issues with conventional Proof-of-Work (PoW) systems are being investigated with solutions such as Proof-of-Stake (PoS) and hybrid blockchain models (Zohar et al., 2022).

Large-scale blockchain applications will be made possible by these developments, especially in water management systems that need to integrate data from various sources and process data in real time.

Additionally, blockchain will remain a crucial weapon in the battle against climate change and environmental degradation due to its capacity to offer safe and open platforms for the trading of carbon credits, the distribution of water rights, and emissions monitoring. Transparency in transaction tracking and verification will enhance resource management and boost confidence in regulatory frameworks that are essential for overseeing climate action and sustainability objectives (Lezzi, 2024).

The regulatory environment will also need to change as blockchain technology advances. Wider adoption will be made possible by well-defined frameworks for integrating blockchain technology into water and environmental systems, which will guarantee that it can be utilized in accordance with environmental laws and regulations (Naqash et al., 2023). To accomplish these objectives and make sure that blockchain technology supports sustainable environmental management, cooperation between legislators, regulators, and blockchain developers will be essential.

In summary, blockchain technology has the potential to significantly influence water and environmental systems by providing creative answers to problems like accountability, efficiency, and transparency. Although there are still challenges to be solved, mainly in the areas of energy consumption, scalability, and regulatory frameworks, blockchain technology has a bright future in sustainable resource management.

We can imagine a future where resource management is not only more sustainable but also more efficient by combining blockchain with IoT, AI, and other cutting-edge technologies. Blockchain could completely transform the way we manage water and environmental resources with continued development and cross-sector cooperation, paving the way for a more just and sustainable future for everybody.

REFERENCES

Albrecht, C., Burzlaff, B., & Steinbrink, J. (2021). Blockchain technology for water trading: A solution for sustainable water management? *Sustainability*, 13(8), 4123. https://doi.org/10.3390/su13084123

Bastos, M. S., Oliveira, L. F., & Santos, P. R. (2022). Blockchain applications in sustainable water management: A systematic review. *Environmental Research Letters*, 17(3), 1–15.

Bendale, S. P., Sharma, S. V., Shinde, S. B., Sakunde, S. S., & Kokane, A. S. (2023). Blockchain and IoT-driven carbon offsetting for a decentralized carbon economy. *Materials Today: Proceedings*.

Buterin, V. (2014). A next-generation smart contract and decentralized application platform. *Ethereum Whitepaper*.

Chen, Y., Bellavitis, C., & Wang, M. (2021a). Blockchain and smart contracts in resource management: Innovations in water trading and supply chains. *Journal of Environmental Management*, 295, 113043.

Chen, Y., Xu, H., & Liu, Z. (2021b). A survey on blockchain scalability: Challenges and solutions. *Journal of Blockchain Research*, 5(1), 45–68.

Cheng, D., Li, Y., & Zhang, Y. (2022). Blockchain and IoT for sustainable water resource management: A comprehensive review. *Journal of Environmental Management*, 301, 113759. https://doi.org/10.1016/j.jenvman.2021.113759

Cohen, T., & Kim, Y. (2021). Sharding in blockchain: The key to scalability. *Journal of Cryptography and Security*, 12(3), 143–157.

De Vries, A. (2021). Bitcoin's energy consumption: Environmental impact and mitigation. *Environmental Economics*, 18(1), 3–25.

European Commission. (2022). The European Blockchain Services Infrastructure (EBSI). Retrieved from https://ec.europa.eu/digital-strategy/our-policies/european-blockchain-services-infrastructure

Gonzalez, R., & Smith, C. (2017). Transparency and accountability in disaster relief funding: The role of blockchain technology. *Journal of Humanitarian Logistics and Supply Chain Management*, 7(3), 324–342.

Grafton, R. Q., Libecap, G. D., & McGlennon, S. (2019). Water markets and the role of blockchain in improving water governance. *Water Resources Research*, 55(2), 1651–1662.

Haber, S., & Stornetta, W. S. (1997). How to time-stamp a digital document. *Journal of Cryptology*, 3(2), 99–111.

Hassan, S., & De Filippi, P. (2021). Decentralized governance in blockchain systems: A critical analysis of DAOs. *New Media & Society*, 23(8), 2369–2392.

IBM. (2023). *How blockchain enables trust in water trading*. IBM. Retrieved from https://www.ibm.com/think/insights/how-blockchain-enables-trust-in-water-trading

Kudzin, A. (2022). Scaling Ethereum 2.0's cross-shard transactions with refined data structures. *Cryptography*, 6(4), 57. https://doi.org/10.3390/cryptography6040057

Lezzi, M., Del Vecchio, V., & Lazoi, M. (2024). Using blockchain technology for sustainability and secure data management in the energy industry: Implications and future research directions. *Sustainability*, 16(18), 7949. https://doi.org/10.3390/su16187949

Li, Y., Zhang, H., & Liu, J. (2022). A framework of blockchain technology in intelligent water resource management. *Frontiers in Environmental Science*, 10, 909606. https://doi.org/10.3389/fenvs.2022.909606

Mekonnen, M. M., Hoekstra, A. Y., & Gerbens-Leenes, W. (2022). Blockchain for water sustainability: Opportunities and challenges. *Environmental Research Letters*, 17(4), 045021.

Merkle, R. C. (1988). A digital signature based on a conventional encryption function. In A. M. Odlyzko (Ed.), *Advances in Cryptology — CRYPTO '87* (Vol. 293, pp. 369–378). Springer. https://doi.org/10.1007/3-540-48184-2_32

Miers, I., Andrews, S., & Wiemer, K. (2021). The security of blockchain-based smart contracts. *Journal of Cryptography and Information Security*, 45(5), 39–53.

Naqash, M. T., Syed, T. A., Alqahtani, S. S., Siddiqui, M. S., Alzahrani, A., & Nauman, M. (2023). A blockchain-based framework for efficient water management and leakage detection in urban areas. *Urban Science*, 7(4), 99. https://doi.org/10.3390/urbansci7040099

Nakamoto, S. (2008). Bitcoin: A peer-to-peer electronic cash system. *Bitcoin Whitepaper*.

Narayanan, A., Bonneau, J., Felten, E., Miller, A., & Goldfeder, S. (2016). *Bitcoin and cryptocurrency technologies: A comprehensive introduction*. Princeton University Press.

Nododile, T., & Nyirenda, C. (2025). A hybrid blockchain-IPFS solution for secure and scalable data collection and storage for smart water meters. *arXiv preprint arXiv:2502.03427*. Retrieved from https://arxiv.org/abs/2502.03427

Pathak, S., Huda, M. B., Lone, M. A., & Eslamian, S. (2022). Reduction of Water Leakage by Smart Water Technologies, Ch. 9. In Eslamian et al. (Ed.), *Handbook of Eurasian Forecasting*, Nova Science Publishers, Inc., USA, 225–232.

Rivest, R. L., Shamir, A., & Adleman, L. (1978). A method for obtaining digital signatures and public-key cryptosystems. *Communications of the ACM*, 21(2), 120–126. https://doi.org/10.1145/359340.359342

Satilmisoglu, T. K. (2024). Blockchain opportunities for water resources management: A comprehensive review. *Sustainability*, 16(6), 2403. https://doi.org/10.3390/su16062403

Savelyev, A. (2018). Contract law 2.0: "Smart" contracts as the beginning of the end of classic contract law. *Information & Communications Technology Law*, 27(2), 133–154.

Shen, X. S., Liu, D., Huang, C., Xue, L., Yin, H., Zhuang, W., & Sun, R. (2022). Blockchain for transparent data management toward 6G. *Engineering*, 8, 74–85. https://doi.org/10.1016/j.eng.2021.07.010

Singh, S., Singh, P., & Singh, R. (2023). Blockchain and IoT for drinking water: A game-changing opportunity. *Sustainable Development Goals Report*. Retrieved from https://sdgs.un.org/sites/default/files/2023-05/A46%20-%20Singh%20-%20Blockchain%20and%20IoT%20for%20water%20A%20Game-Changing%20Opportunity.pdf

Sridhar, R., Singh, P., & Zhang, T. (2023). Renewable energy and blockchain integration: A sustainable future. *Renewable Energy Journal*, 45(2), 119–136.

Tapscott, D., & Tapscott, A. (2017). *Blockchain revolution: How the technology behind bitcoin is changing money, business, and the world*. Penguin Books.

Vladucu, M.-V., Wu, H., Medina, J., Salehin, K. M., Dong, Z., & Rojas-Cessa, R. (2024). Blockchain in environmental sustainability measures: A survey. *arXiv preprint arXiv:2412.15261*. Retrieved from https://arxiv

Wadhwa, S., Rani, S., Kavita, V. S., Shafi, J., & Wozniak, M. (2022). Energy efficient consensus approach of blockchain for IoT networks with edge computing. *Sensors*, 22(10), 3733. https://doi.org/10.3390/s22103733

Wood, G. (2015). *Ethereum: A secure decentralized generalized transaction ledger*. Ethereum Yellow Paper.

World Food Programme. (2019). *Year in Review 2019*. Rome, Italy: World Food Programme.

Xie, F., Zhang, Y., & Li, M. (2022). Leveraging blockchain for water resource management: A systematic review. *Water Science and Technology*, 85(4), 567–590.

Zhang, L., & Li, X. (2023). Decentralized water trading via blockchain: A case study. *Journal of Environmental Economics*, 12(1), 67–83.

Zheng, Z., Xie, S., Dai, H., Chen, X., & Wang, H. (2018). Blockchain challenges and opportunities: A survey. *International Journal of Web and Grid Services*, 14(4), 352–375.

Zohar, M., Reichman, M., & Jaffe, J. (2021). Merkle trees and their role in blockchain security. *Journal of Cryptographic Engineering*, 10(1), 89–101.

Zohar, T., Goldberg, A., & Lee, S. (2022). Blockchain scalability: Solutions and challenges. *Journal of Blockchain Technology*, 19(4), 219–231.

Zwitter, A., & Boisse-Despiaux, M. (2018). Blockchain for humanitarian action and development aid. *Journal of International Humanitarian Action*, 3(1), 17.

8 Blockchain for Water Quality Monitoring
Enhancing Transparency, Security, and Accountability

Harpreet Kaur Channi, Pulkit Kumar, and Saeid Eslamian

8.1 INTRODUCTION

The world has recently witnessed a growing concern for environmental conservation and sustainability. As the global population continues to grow and human activities place increasing pressure on our planet's finite resources, there is an urgent need for innovative solutions to address environmental challenges. One such innovative technology that has gained significant attention is blockchain. Originally developed as the underlying technology for cryptocurrencies like Bitcoin, blockchain technology has evolved beyond its financial roots and found applications in various sectors. Environmental conservation is one of the most promising and impactful domains where Blockchain is making its mark.

Environmental conservation involves the protection, preservation, and sustainable management of natural resources and ecosystems. It encompasses many efforts, from biodiversity preservation and climate change mitigation to waste reduction and sustainable agriculture. Blockchain technology offers unique capabilities that can revolutionize how we approach these challenges, providing transparency, trust, and traceability in environmental initiatives. This introductory guide explores the intersection of blockchain technology and environmental conservation. We will delve into the fundamental concepts of blockchain, its key features, and how it can be harnessed to address environmental issues.

Additionally, we will examine specific use cases and examples of how blockchain is already being employed to promote environmental sustainability and conservation efforts worldwide. As we navigate through the following sections, it will become clear that blockchain technology has the potential to empower individuals, organizations, and governments to make more informed decisions, track environmental impact, and drive positive change for the planet. Whether combating illegal logging, promoting sustainable supply chains, or incentivizing renewable energy production, blockchain is playing a pivotal role in shaping the future of environmental conservation.

8.1.1 Literature Review

To this end, a proof-of-concept of a blockchain-based wildlife conservation performance payments scheme in Namibia: the habitat integrity of an elephant corridor is assessed by remote sensing algorithms, which trigger fictitious blockchain innovative contract payments to surrounding communities (Oberhauser 2019). This commentary explores how blockchain is being leveraged to address the fundamental problems with market-based forest protection globally. Despite the increasing importance of blockchain in improving supply chain efficiencies and bringing societal changes, research investigating its potentialities from the lens of sustainability is scarce (Howson et al. 2019). The

DOI: 10.1201/9781003599593-12

primary aim is to fill this knowledge gap and synthesize the literature from leading journals on blockchain and its relation to supply chain sustainability (Rejeb and Rejeb 2020). The author addresses the sustainable microgrid design problem by leveraging blockchain technology to provide real-time–based demand response programs. This study investigates the impact of blockchain technology on green supply chain practices to promote pro-environmental settings in supply chains of manufacturing firms (Mubarik et al. 2021, Tsao et al. 2021). The author has provided an interconnection diagram of organizational, technological, and environmental blockchains. Future research should explore whether blockchain initiatives may translate into concrete environmental outcomes and contribute to the well-being of natural resource managers (Abramyan et al. 2021, Stuit et al. 2022).

In response to rising environmental concerns, blockchain technology (BCT) is transforming green innovation, culminating in green economic practices and well-established business models. Recognizing this, the author investigates the role of blockchain technology in green innovation practices and its impact on green economic sustainability (Polas et al. 2022). The author‘s primary purpose is to explore the literature relevant to blockchain for sustainable supply chain management, proposing an environmental accounting system model based on artificial intelligence blockchain and embedded sensors (Munir et al. 2022, Zhang and Zhu 2022).

In his study, the author shows blockchain technology and the sustainable supply chain, theoretically exploring adoption barriers. The technology-organization-environment framework and force field theories are utilized to investigate blockchain adoption barriers (Kouhizadeh et al. 2019, Kouhizadeh et al. 2021). Another author provides a comprehensive survey on the literature on blockchain technology applied to smart cities (Xie et al. 2019). The researcher aims to investigate the effects of relative advantage, complexity, upper management support, cost, market dynamics, competitive pressure, and regulatory support on blockchain adoption for operations and supply chain management among small-medium enterprises (SMEs) in Malaysia (Wong et al. 2020). In this study, the author examines the government intervention choices to desocialize negative environmental externalities caused by high-energy consuming blockchain technology designs (Truby 2018). The subject is to provide an overview of blockchain technology and Industry 4.0 for advancing supply chains toward sustainability (Esmaeilian et al. 2020). In another article, the author focuses on a specific distributed ledger IT called blockchain, which has been heralded as possessing the capability to transform many industries radically. The aim is to elucidate the impact of organizational factors on the adoption of blockchain in companies based in Ireland (Clohessy and Acton 2019).

8.1.2 Background and Motivation

The background for exploring the synergy between blockchain technology and environmental conservation stems from the urgent need to address the escalating environmental crises facing our planet. Climate change, biodiversity loss, resource depletion, and pollution have reached critical levels, demanding innovative solutions. Meanwhile, integrating blockchain into environmental conservation stems from the technology's remarkable attributes. Blockchain offers transparency, decentralization, security, and the capability to execute smart contracts, making it a powerful tool to enhance environmental efforts. By leveraging these features, we can foster trust, accountability, and traceability in environmental initiatives, ultimately working toward a sustainable and greener future for our planet.

8.2 WATER QUALITY MONITORING: CHALLENGES AND TRADITIONAL APPROACHES

8.2.1 Challenges in Water Quality Monitoring

Monitoring and maintaining water quality is critical for ensuring safe and sustainable access to clean water (Shah Naqvi et al., 2015). However, several challenges persist in water quality monitoring, as shown in Figure 8.1

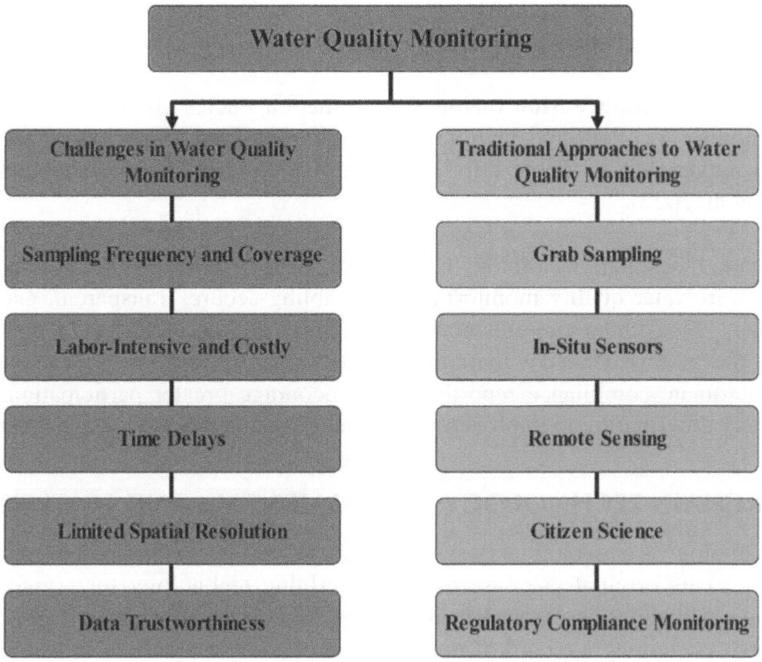

FIGURE 8.1 Challenges and Traditional approaches to water quality monitoring.

- **Sampling Frequency and Coverage**: Traditional monitoring methods rely on infrequent, sporadic sampling at specific locations. This approach may miss sudden changes in water quality or fail to capture a broader picture of water conditions (Xia, et al. 2022).
- **Labor-Intensive and Costly**: Collecting, transporting, and analyzing water samples can be labor-intensive and expensive. It requires trained personnel, specialized equipment, and significant resources.
- **Time Delays**: Traditional lab-based analysis introduces delays between sample collection and obtaining results, hindering timely response to water quality issues.
- **Limited Spatial Resolution**: Traditional methods often provide data for specific points or regions, making it challenging to accurately assess water quality on a larger scale or identify pollution sources.
- **Data Trustworthiness**: The potential for human error, data manipulation, or inaccuracies in manual record-keeping can raise concerns about data trustworthiness and reliability (Bodkhe et al. 2022).

8.2.2 Traditional Approaches to Water Quality Monitoring

- **Grab Sampling**: Water samples are collected manually at specific locations and times in this method. These samples are then transported to a laboratory for analysis. It is a widely but infrequently used method.
- **In-Situ Sensors**: In-situ sensors are deployed directly in the water body to continuously measure temperature, pH, dissolved oxygen, and turbidity. These sensors can provide real-time data but may require regular maintenance (Miloudi et al. 2020).
- **Remote Sensing**: Remote sensing technologies, such as satellites and drones, can capture data on water quality parameters from a distance. While they offer broader spatial coverage, they may lack the precision of in-situ measurements.

- **Citizen Science**: Citizen science programs involve volunteers in collecting water quality data. While they can enhance spatial coverage, ensuring data accuracy and consistency can be challenging.
- **Regulatory Compliance Monitoring**: Government agencies often conduct routine water quality monitoring to enforce environmental regulations and ensure compliance by industries and municipalities. These efforts tend to focus on specific pollutants and locations (Ahmed et al. 2022).

In recent years, blockchain technology has emerged as a promising solution to address some of these challenges in water quality monitoring. By enabling secure, transparent, and decentralized data management, blockchain can enhance the reliability and accessibility of water quality data while reducing the risks associated with traditional methods. This technology can facilitate real-time data sharing, automate compliance reporting, and encourage greater participation in monitoring efforts, ultimately contributing to improved water quality management and conservation.

8.3 BLOCKCHAIN TECHNOLOGY: FUNDAMENTALS AND FEATURES

Blockchain technology is a revolutionary system characterized by its distributed ledger architecture. It has transcended its original use case as the underlying technology for cryptocurrencies like Bitcoin, finding applications across various industries. At its core, Blockchain consists of a chain of data blocks, each containing a record of multiple transactions or pieces of information. What sets blockchain apart are its fundamental features (Ølnes and Jansen 2017).

8.3.1 Understanding Blockchain Technology

Originally conceived as the underlying architecture for cryptocurrencies like Bitcoin, blockchain technology has evolved into a transformative and versatile tool with applications across various industries. At its core, blockchain is a decentralized, distributed ledger system that records transactions or data transparently and immutable. To comprehend blockchain technology fully, it's essential to grasp its fundamental components and how they function. Unlike traditional centralized systems, where a central authority holds control, blockchain operates on a distributed computer network (nodes). Each node maintains a copy of the entire blockchain (Jaradat et al. 2022).

This decentralization enhances security by eliminating a single point of failure and reducing the risk of manipulation. The data in a blockchain is organized into blocks, each containing a set of transactions or information. These blocks are linked chronologically to form a chain called "blockchain." Once a block is added to the chain, its data is virtually impossible to alter, ensuring the immutability of the historical record. One of blockchain's defining features is transparency. Every transaction recorded on the blockchain is visible to all participants in the network. This transparency fosters trust and accountability, as anyone can independently verify the validity of transactions and data. Immutability is achieved through cryptographic hashing (Mukherjee and Pradhan 2021).

Each block contains a unique cryptographic hash of the previous block, creating a secure and tamper-resistant connection. Changing the data in one block would require altering the data in all subsequent blocks, a computationally infeasible task. To add new transactions to the blockchain, network participants must reach a consensus. Various consensus mechanisms validate and verify transactions, such as proof of work (PoW) and proof of stake (PoS). These mechanisms ensure that only legitimate transactions are added to the blockchain. Blockchain relies on cryptographic algorithms to secure transactions and data. Each participant has a private key to sign transactions, while public keys verify ownership. This cryptographic security makes blockchain highly resistant to fraud and unauthorized access. Smart contracts are self-executing programs with predefined rules and conditions. They automate and enforce agreements, eliminating the need for intermediaries. Smart contracts extend the utility of blockchain technology beyond simple transactions, enabling complex business processes and applications (Elangovan et al. 2020).

Blockchain technology has applications in diverse sectors, including finance, supply chain management, healthcare, and real estate. It offers solutions for secure data sharing, provenance tracking, supply chain transparency, and decentralized finance (DeFi). Understanding blockchain technology is pivotal as it represents a fundamental shift in how data is stored, verified, and transacted in the digital age. Its potential to provide transparent, secure, and efficient solutions has the potential to disrupt industries and transform various aspects of our lives. As the technology continues to evolve, its implications are far-reaching, making it an area of growing interest and innovation (Xu, et al. 2021).

8.3.2 Key Features and Advantages of Data Management

Blockchain technology offers several key features and advantages for data management that make it a compelling choice for various industries and applications, as depicted in Figure 8.2:

- **Security**: Blockchain's security is paramount. Each transaction is cryptographically linked to the previous one, forming an immutable chain of blocks. Once data is recorded, altering or deleting it is exceedingly tricky, providing high data integrity and protection against fraud or unauthorized access (Aoun et al. 2021).
- **Transparency**: Transparency is a fundamental feature of blockchain. All participants in a blockchain network have access to the same ledger, and transactions are visible to all. This transparency enhances trust among parties and allows for independent data verification, reducing the need for intermediaries.
- **Decentralization**: Blockchain operates on a decentralized network of nodes. This decentralization eliminates the reliance on a central authority and mitigates the risk of a single point of failure. It enhances the resilience and availability of data, even in the face of network disruptions or attacks (Alam 2022).
- **Immutability**: Data stored on a blockchain is tamper resistant. Once recorded, it cannot be changed without consensus from most network participants. This feature is invaluable for maintaining a reliable historical record.

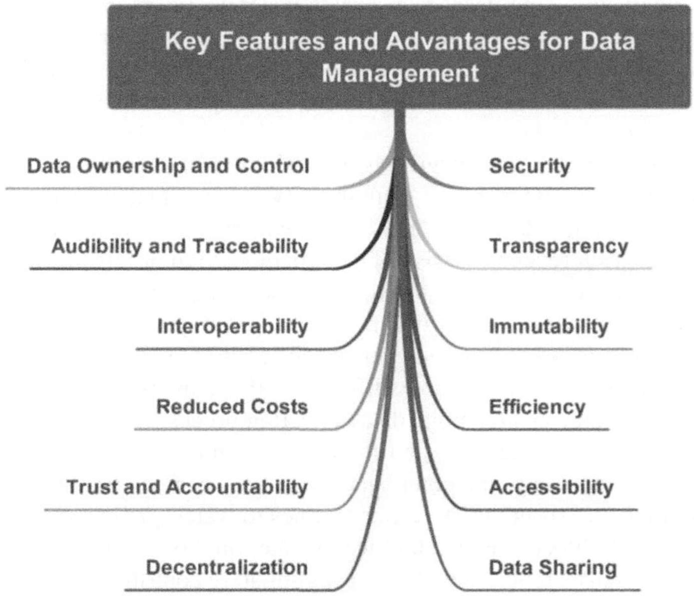

FIGURE 8.2 Advantages and key features for data management.

- **Efficiency**: Blockchain technology can streamline data management processes. Smart contracts, self-executing agreements with predefined rules, automate tasks and enforce agreements, reducing the need for manual intervention and paperwork. This efficiency can lead to cost savings and faster transaction processing times.
- **Accessibility**: Blockchain data is accessible 24/7 from anywhere with an internet connection. This accessibility enables real-time data sharing and collaboration among stakeholders, regardless of geographical location.
- **Data Ownership and Control**: Participants in a blockchain network have control over their data. They can grant or revoke access, enhancing data ownership and privacy.
- **Audibility and Traceability**: Blockchain's transparent and immutable ledger allows for comprehensive auditing and traceability. Every transaction is recorded, providing a complete and auditable history of data changes.
- **Interoperability**: Blockchain technology can be integrated with existing systems and technologies, enabling seamless data exchange and interoperability between different parties and platforms (Alam 2022).
- **Reduced Costs**: By eliminating intermediaries, reducing the risk of errors, and automating processes, blockchain can lead to cost savings in data management and transactions.
- **Trust and Accountability**: The combination of security, transparency, and immutability instills trust among participants. It ensures that data is accurate and any changes are traceable, fostering greater accountability in data management.
- **Data Sharing**: Blockchain facilitates secure data sharing between parties who may not fully trust each other. It enables controlled access to data while maintaining data integrity (Alam 2022).

These key features and advantages make blockchain technology attractive for a wide range of data management applications, from supply chain tracking and healthcare record management to financial transactions and voting systems. As blockchain continues to evolve and gain broader adoption, its potential to revolutionize data management practices becomes increasingly evident.

8.3.3 Relevance of Blockchain in Water Quality Monitoring

Blockchain technology holds significant relevance in water quality monitoring because it addresses several critical challenges in this field. By providing a secure, transparent, and decentralized framework for data management, blockchain can enhance the accuracy and reliability of water quality data, leading to more effective conservation efforts and improved access to clean water (Frikha et al. 2023).

In water quality monitoring, data integrity is paramount, as it directly impacts public health and environmental preservation. Blockchain's immutability ensures that it remains tamper-proof and trustworthy once water quality data is recorded. This feature is particularly vital when dealing with regulatory compliance, as it reduces the risk of data manipulation or falsification. Moreover, blockchain's transparency fosters trust among stakeholders, including government agencies, environmental organizations, and the public. Real-time access to water quality data from multiple sources ensures that information is up to date and openly accessible. This transparency can incentivize greater collaboration and data sharing among different entities involved in water quality monitoring.

Additionally, blockchain can streamline data collection processes by enabling the integration of data from various sensors, IoT devices, and monitoring stations. Through smart contracts, automated data verification, reporting, and compliance processes can be established, reducing the need for manual intervention and speeding up response times to water quality issues (Frikha et al. 2023, Zhai et al. 2023). Overall, blockchain's relevance in water quality monitoring lies in its potential to enhance data security, transparency, and efficiency, ultimately contributing to better-informed decisions, improved environmental conservation, and the sustainable management of water resources.

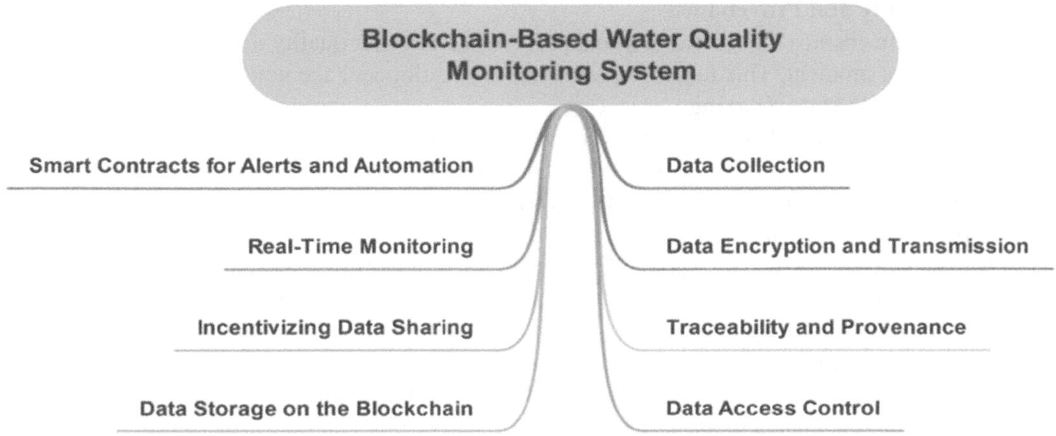

FIGURE 8.3 Blockchain-based water quality monitoring system.

8.4 BLOCKCHAIN-BASED WATER QUALITY MONITORING SYSTEM

A blockchain-based water quality monitoring system represents a cutting-edge approach to addressing water quality challenges with enhanced transparency, security, and efficiency. This system leverages the core features of blockchain technology to create a robust and reliable framework for monitoring, managing, and ensuring the quality of water resources (Jin and Chang 2023). Figure 8.3 shows the blockchain-based water quality monitoring system.

- **Data Collection**:
  Sensors and IoT devices are deployed at various water monitoring points, such as rivers, lakes, reservoirs, and water treatment facilities. These sensors continuously collect data on key water quality parameters, including pH levels, turbidity, dissolved oxygen, temperature, and the presence of contaminants (Chen 2023).
- **Data Encryption and Transmission**:
  Data collected by the sensors are encrypted and transmitted securely to the blockchain network. Blockchain's cryptographic features ensure the integrity and confidentiality of the data during transmission.
- **Data Storage on the Blockchain**:
  The data is recorded as transactions on the blockchain. Each transaction is stored in a block linked to the previous block, ensuring immutability. These transactions include the water quality data and metadata, such as the timestamp, location, and information about the sensors used (Chen 2023).
- **Real-Time Monitoring**:
  Stakeholders, including government agencies, environmental organizations, water treatment facilities, and the public, can access the blockchain-based platform in real time to monitor water quality data. Transparency is a key feature, allowing anyone to verify the accuracy and authenticity of the data (Cai et al. 2023).
- **Smart Contracts for Alerts and Automation**:
  Smart contracts are deployed on the blockchain to automate specific actions based on predefined criteria. For instance, if water quality data exceeds safety thresholds, an intelligent contract could alert relevant authorities for immediate intervention. Automated compliance reporting can also be implemented, simplifying regulatory processes.

- **Traceability and Provenance**:
  Blockchain ensures the traceability and provenance of water quality data from its source to the present moment. This feature is valuable for auditing and accountability purposes.
- **Incentivizing Data Sharing**:
  To encourage data sharing among various stakeholders, blockchain can facilitate token-based incentives. Those who contribute high-quality data or participate in monitoring efforts could receive tokens as rewards (Midhila et al. 2023).
- **Data Access Control**:
  Blockchain allows for fine-grained access control, ensuring that sensitive data is only accessible to authorized parties. This is crucial for privacy and data protection compliance.

  A blockchain-based water quality monitoring system offers several advantages, including enhanced data accuracy, reduced risk of fraud or manipulation, and streamlined data management processes. It empowers stakeholders to make informed decisions regarding water resource management, pollution control, and public health. Furthermore, the transparency and accountability provided by blockchain technology can strengthen public trust and encourage collaboration among government agencies, environmental organizations, and the broader community, ultimately contributing to the preservation and sustainable management of precious water resources (Zeng et al. 2023).

8.4.1 Architecture and Components of the System

The architecture of a blockchain-based water quality monitoring system is designed to integrate blockchain technology with data collection, transmission, storage, and user interfaces. The system comprises several key components to ensure accurate, secure, and transparent water quality monitoring. Here are the primary components and their roles within the system:

- **Data Collection Sensors and IoT Devices**:
  These are physical devices deployed in water bodies, such as rivers, lakes, reservoirs, and treatment facilities. Sensors measure various water quality parameters, including pH levels, turbidity, dissolved oxygen, temperature, and the presence of contaminants. IoT devices transmit sensor data to the system's data-processing component (Zeng et al. 2023).
- **Data Processing and Encryption**:
  This component is responsible for receiving and processing data from IoT devices. Data is encrypted to ensure confidentiality and integrity during transmission. Any preprocessing or filtering of data to remove noise or anomalies can occur at this stage.
- **Blockchain Network**:
  The heart of the system is the blockchain network, which comprises a distributed ledger of interconnected nodes. The blockchain records water quality data that is tamper-resistant and immutable (Xu et al. 2023).
  It consists of multiple components:
  1. **Consensus Mechanism**: This ensures that network participants agree on the validity of transactions before they are added to the blockchain. Common mechanisms include PoW or PoS.
  2. **Smart Contracts**: These self-executing contracts contain predefined rules and conditions. They can automate triggering alerts or compliance reporting based on data thresholds.
  3. **Transactions**: Each transaction represents a set of water quality data, including measurements, timestamps, sensor information, and location data (Faridi et al., 2023).
- **User Interfaces and Access Control**:
  The system provides user interfaces for various stakeholders, including government agencies, environmental organizations, and the public. Access control mechanisms ensure users have appropriate permissions to view specific data based on their roles and responsibilities.

- **Alerts and Notifications**:
  Smart contracts within the blockchain can trigger real-time alerts and notifications to relevant authorities or stakeholders when water quality data exceeds safety thresholds or triggers predefined conditions.
- **Data Storage**:
  The blockchain itself serves as the primary data storage mechanism. Data is stored in blocks, each linked to the previous one, ensuring immutability. In addition to the blockchain, off-chain or external storage solutions may be used for large data sets or historical records (Faridi et al. 2023).
- **Tokenization (Optional)**:
  Token-based incentives or reward systems may be integrated to encourage data sharing and participation. Users who contribute valuable data or engage in monitoring efforts can receive tokens as incentives.
- **External Integration and Reporting**:
  The system can integrate with external databases, reporting tools, or regulatory systems to facilitate compliance reporting and data sharing with government agencies or environmental bodies. The architecture and components of a blockchain-based water quality monitoring system provide a holistic approach to addressing water quality challenges. This system ensures data accuracy, transparency, and security while enabling real-time monitoring and timely response to water quality issues. Additionally, it promotes collaboration among stakeholders, fosters trust, and contributes to the sustainable management of water resources and environmental conservation efforts (Satılmıoğlu and Keskin 2023).

8.4.2 Data Storage and Encryption

Data storage and encryption are critical components of a blockchain-based water quality monitoring system, ensuring the security, integrity, and privacy of the water quality data collected and transmitted. Here's a closer look at how data storage and encryption work within the system:

8.4.2.1 Data Storage

The primary data storage mechanism in this system is the blockchain itself. Each water quality data entry is recorded as a transaction on the blockchain. Data is organized into blocks, and each block is linked to the previous one through cryptographic hashing, forming a continuous data chain. This linking ensures data immutability, as altering data in one block would require changing data in all subsequent blocks, which is computationally infeasible. Water quality data transactions typically include parameters like pH levels, turbidity, dissolved oxygen, temperature, sensor information, timestamps, and location data. These transactions are appended to the blockchain in a tamper-resistant manner. The blockchain ledger provides a transparent, publicly accessible, and auditable record of water quality data, making it available for real-time monitoring, historical analysis, and regulatory compliance (Martínez-Castañeda and Feijoo 2023).

8.4.2.2 Data Encryption

Data encryption is crucial in securing water quality data at various stages, including during transmission and storage. Encryption During Transmission: the IoT devices and sensors collect water quality data and transmit it to the data processing component over networks. Encryption protocols, such as Secure Sockets Layer (SSL) or Transport Layer Security (TLS), encrypt data in transit, safeguarding it from eavesdropping or interception during transmission. Encryption During Storage: While data is on the blockchain, it benefits from the blockchain's inherent security features. Each transaction is cryptographically hashed and linked to the previous one, making it extremely difficult to alter or manipulate. This process ensures data integrity and immutability.

Moreover, blockchain technology employs advanced cryptographic techniques to protect data at rest. Public and private keys secure access to the bockchain, ensuring that only authorized users can read and write data. Users' private keys should be confidential to prevent unauthorized access (Ismail et al. 2023).

By combining data storage on the blockchain with encryption techniques during transmission and storage, the system provides a multilayered approach to data security. This ensures that water quality data is protected from tampering, unauthorized access, and interception, maintaining its accuracy and reliability for monitoring, analysis, and decision-making processes (Vats and Sharma 2023).

8.5 ADVANTAGES AND BENEFITS OF BLOCKCHAIN INTEGRATION

Integrating blockchain technology into water quality monitoring systems represents a significant leap forward in our ability to safeguard one of our most vital resources: clean water. The advantages and benefits of blockchain integration in this context are multifaceted and far-reaching. Firstly, blockchain's core features, including immutability, transparency, and cryptographic security, ensure that water quality data is accurate and tamper resistant (Leong et al. 2023). This level of data integrity is paramount when addressing critical issues related to public health and environmental preservation. Real-time monitoring capabilities further enhance our capacity to respond swiftly to water quality issues. By providing instant access to data from numerous monitoring points, blockchain enables rapid intervention, reducing the risk of contamination and its associated consequences. Moreover, the system's transparency fosters trust among stakeholders, creating a collaborative ecosystem where government agencies, environmental organizations, and the public can actively participate in monitoring efforts and make informed decisions.

Decentralization, another critical feature of blockchain, reduces the reliance on centralized authorities and intermediaries. This enhances the system's resilience and empowers local communities to take ownership of their water quality data. Smart contracts, automated through predefined rules, streamline compliance reporting and response mechanisms, reducing administrative burdens and costs. Data sharing and collaboration, facilitated by blockchain, promote a collective effort to ensure clean water. Blockchain's ledger's traceability and provenance features enable comprehensive auditing, while fine-grained access control respects individual privacy and data protection regulations (Iranmanesh et al. 2023). The potential for cost savings in data management and compliance reporting cannot be understated. By eliminating intermediaries, automating processes, and reducing the risk of errors or data manipulation, blockchain technology optimizes resource allocation to maintain and improve water quality. Integrating blockchain technology in water quality monitoring systems marks a watershed moment in our commitment to environmental conservation and public health. It offers a secure, transparent, and efficient approach to data management, fostering collaboration, trust, and accountability among stakeholders. As we grapple with water quality challenges in an increasingly interconnected world, Blockchain provides a powerful tool for ensuring the sustainable management of this precious resource for current and future generations (Rehali et al. 2023).

8.6 CHALLENGES AND MITIGATION STRATEGIES

Integrating blockchain technology into water quality monitoring systems offers numerous advantages. Still, it is not without its share of challenges, which must be thoughtfully addressed to ensure the system's effectiveness and sustainability. One of the primary challenges is the initial cost of deploying the necessary infrastructure, including sensors, IoT devices, and blockchain networks. To mitigate this challenge, organizations can explore collaborative funding models such as public-private partnerships, seek government grants, or tap into existing environmental conservation budgets. Over time, the operational efficiencies and cost savings facilitated by the blockchain system can justify the initial investment (Liu et al. 2023).

Another critical challenge is ensuring the privacy and security of sensitive water quality data. Unauthorized access or data breaches can compromise public trust and environmental integrity. Robust cybersecurity measures, including strong encryption, access controls, and regular security audits, are essential to mitigate these risks. Moreover, the accuracy and reliability of sensors and IoT devices can vary, leading to potential data discrepancies (Karthick et al. 2023). Calibration and maintenance routines should be rigorously upheld to ensure data accuracy, and redundancy through multiple sensors for critical measurements can provide added reliability. Network connectivity can be a stumbling block for data transmission in remote or underserved areas. To overcome this challenge, alternative connectivity options such as satellite or mesh networks should be explored, and data storage solutions should be equipped to handle intermittent connectivity. As the monitoring network expands, scalability concerns may arise. Selecting a blockchain platform capable of handling growing data volumes and transaction loads is crucial (Bhattu and Singh 2023). Standardizing data formats, units, and metadata conventions is essential for streamlined data integration and analysis. Public participation in data sharing and monitoring efforts can be challenging to incentivize, but education and outreach programs, coupled with token-based incentives, can foster community engagement. Regulatory compliance is another multifaceted challenge, requiring close collaboration with regulatory bodies and smart contracts to automate compliance reporting. Technological evolution demands regular system updates and maintenance to align with emerging standards and technologies. By proactively addressing these challenges through careful planning, organizations can harness the full potential of blockchain technology to revolutionize water quality monitoring, improve data accuracy, enhance transparency, and contribute to the preservation of clean water resources and the environment for generations to come as shown in Figure 8.4 (Utami et al. 2023).

8.7 FUTURE TRENDS AND RESEARCH DIRECTIONS

The future of blockchain technology in environmental conservation holds immense promise, with exciting possibilities for addressing global ecological challenges. Blockchain's ability to enhance transparency, security, and trust in environmental data and transactions is set to revolutionize conservation efforts. One of the most compelling trends is the emergence of carbon markets and emissions-tracking systems. Blockchain will enable precise monitoring and verification of carbon credits and emissions reductions, supporting climate change mitigation, as shown in Figure 8.5 Supply chain sustainability will see further adoption of blockchain to trace the origins and environmental impact of products, empowering consumers to make eco-conscious choices (Yadav et al. 2023). Wildlife conservation will benefit from blockchain's capacity to combat illegal trade by enabling the tracking of endangered species and verifying the legitimacy of transactions. Environmental impact tokens will incentivize green actions, offering tangible rewards for sustainable practices. Standardization and interoperability efforts will enhance data sharing among diverse stakeholders, while decentralized autonomous organizations will automate environmental governance. Privacy-preserving technologies will secure sensitive data, and global collaboration will promote standardized approaches to conservation. The future of blockchain in environmental conservation is not just a technological evolution; it's a catalyst for a more sustainable and harmonious coexistence with our planet (Charles et al. 2023).

8.8 CASE STUDIES AND REAL-WORLD IMPLEMENTATIONS

Blockchain technology has found compelling applications in environmental conservation through various case studies and real-world implementations. For instance, the OpenSC Blockchain, a collaboration between the WILD Foundation, Viant, and Conservation International, tracks the journey of sustainably caught fish, combating illegal fishing practices. IBM Food Trust employs blockchain to enhance supply chain transparency, allowing consumers to make informed choices and reduce food waste (Sreenivasan and Suresh 2023). Climate Vault uses blockchain to purchase and retire

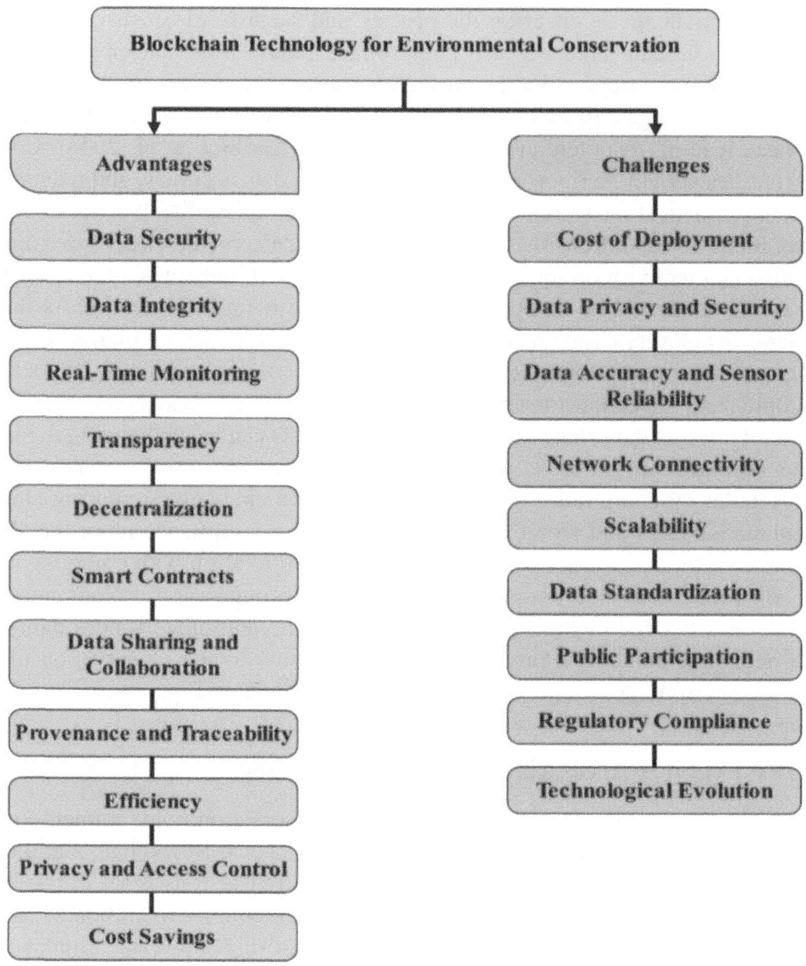

FIGURE 8.4 Advantages and challenges of blockchain for environmental conservation.

carbon allowances, contributing to emissions reduction efforts. In wildlife conservation, organizations like WildTrack employ blockchain to monitor endangered species by tracking footprints and collecting data through blockchain-powered apps. Renewable energy trading via blockchain, as seen in the Brooklyn Microgrid project, promotes green energy adoption and peer-to-peer energy trading. Blockchain also supports clean water initiatives, exemplified by the WaterChain Project in California, which incentivizes water conservation and efficient use through intelligent contracts. Forest management and protection against illegal logging are facilitated by the Rainforest Foundation's blockchain-based land titles (Alqudah et al. 2023).

The Plastic Bank initiative combats plastic pollution by incentivizing plastic waste collection through blockchain-based mobile apps. These diverse case studies demonstrate blockchain's versatility in addressing environmental conservation challenges, from supply chain transparency to wildlife protection and sustainable energy practices. Here are some case studies and implementations for environmental conservation using BlockChain:

- **Green Energy Certification**:
  The Power Ledger platform, operating in multiple countries, enables peer-to-peer renewable energy trading. It uses blockchain to certify the origin of green energy and allows users to buy and sell excess energy directly, promoting renewable energy adoption.

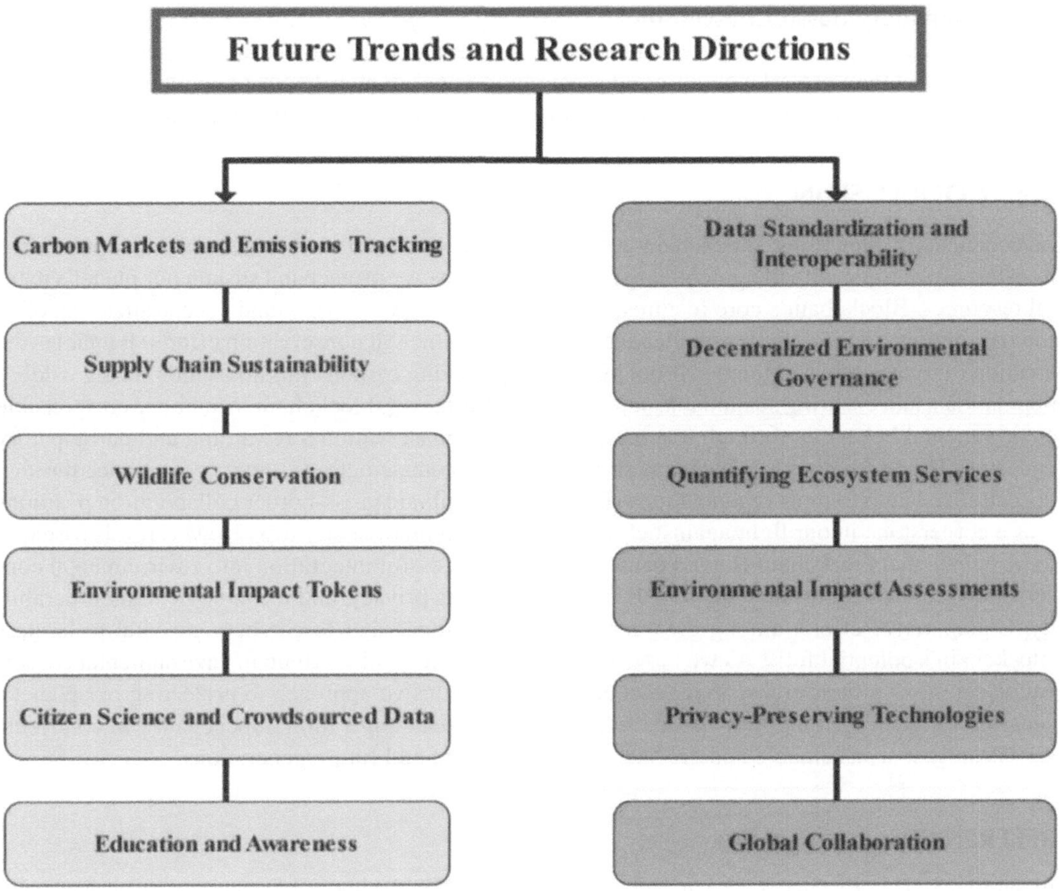

FIGURE 8.5 Research directions and future trends of blockchain in environmental conservation.

- **Plastic Waste Tracking**:
  The Circularise project uses blockchain to trace and verify plastics' origins and recycling paths. It helps reduce plastic waste by ensuring transparency and accountability in the recycling industry.
- **Emission Reduction and Offset**:
  The AirCarbon Exchange is a blockchain-based platform that allows companies to trade carbon offset tokens. It provides a transparent marketplace for buying and selling carbon credits, fostering emissions reductions.
- **Forestry and Carbon Credits**:
  The Nori platform focuses on reversing climate change by incentivizing carbon sequestration through reforestation and afforestation. Farmers and landowners can earn carbon removal certificates recorded on the blockchain (Junaid et al. 2023).
- **Sustainable Fisheries**:
  Fishcoin, operating in Southeast Asia, employs blockchain to trace the seafood supply chain. It ensures the authenticity of sustainably sourced seafood products, reducing overfishing and promoting responsible practices.
- **E-Waste Management**:
  The VH Certifier platform utilizes blockchain to verify the ethical disposal and recycling of electronic waste (e-waste). It certifies the responsible handling of e-waste, reducing environmental hazards.

- **Community-Based Conservation**:
  The Reserve platform empowers local communities and indigenous tribes by enabling them to protect their natural resources and ecosystems. Blockchain helps these communities manage land rights and receive direct financial support for conservation efforts (Junaid et al. 2023).

8.9 CONCLUSION

In conclusion, integrating blockchain technology into environmental conservation represents a transformative force with the potential to revolutionize how we protect and sustain our planet's natural resources. Blockchain's core features, such as transparency, security, and decentralization, offer robust solutions to the complex challenges facing environmental conservation efforts. It enables the creation of transparent and tamper-proof systems for tracking carbon emissions, monitoring wildlife populations, and ensuring sustainable supply chains. Moreover, blockchain incentivizes responsible environmental behaviors through tokenization, making conservation a rewarding and participatory endeavor. The technology's capacity to streamline data management, automate governance through decentralized autonomous organizations (DAOs), and facilitate cross-border collaboration positions it as a cornerstone in our fight against climate change and biodiversity loss. However, it is essential to recognize that blockchain is not a panacea, and its successful integration into environmental conservation requires addressing challenges such as cost, data privacy, and technological interoperability. Collaborative efforts among governments, organizations, and researchers are vital to harness Blockchain's potential fully. As we move forward, the future of blockchain in environmental conservation promises a more transparent, accountable, and inclusive approach to protecting our planet's ecosystems. With ongoing innovation, research, and global cooperation, blockchain can contribute significantly to a sustainable and greener world for current and future generations.

REFERENCES

Abramyan, S. G., E. Klenina, A. Peskov, & A. Burlachenko (2021). Monitoring of organizational, technological and environmental solutions in construction with blockchain technologies. *IOP Conference Series: Materials Science and Engineering, 1079, Paper 052060. Proceedings of the International Scientific Conference "FarEastCon 2020"* (6–9 October 2020, Vladivostok, Russian Federation). https://doi.org/10.1088/1757-899X/1079/5/052060

Ahmed, R. A., E. E. D. Hemdan, W. El-Shafai, Z. A. Ahmed, E. S. M. El-Rabaie and F. E. Abd El-Samie (2022). "Climate-smart agriculture using intelligent techniques, blockchain and Internet of Things: Concepts, challenges, and opportunities." *Transactions on Emerging Telecommunications Technologies* 33(11): e4607.

Alam, A. (2022). Platform Utilising Blockchain Technology for eLearning and Online Education for Open Sharing of Academic Proficiency and Progress Records. In R. Asokan, D. P. Ruiz, Z. A. Baig, & S. Piramuthu (Eds.), *Smart Data Intelligence* (Algorithms for Intelligent Systems, pp. 307–320). Springer, Singapore. https://doi.org/10.1007/978-981-19-3311-0_26

Alqudah, M., L. Ferruz, E. Martín, H. Qudah and F. Hamdan (2023). "The sustainability of investing in cryptocurrencies: A bibliometric analysis of research trends." *International Journal of Financial Studies* 11(3): 93.

Aoun, A., A. Ilinca, M. Ghandour and H. Ibrahim (2021). "A review of Industry 4.0 characteristics and challenges, with potential improvements using blockchain technology." *Computers & Industrial Engineering* 162: 107746.

Bhattu, M., & J. Singh. (2023). Recent advances in nanomaterials based sustainable approaches for mitigation of emerging organic pollutants. *Chemosphere*, 321, 138072, pp. 1–22. https://doi.org/10.1016/j.chemosphere.2023.138072

Bodkhe, U., S. Tanwar, P. Bhattacharya and N. Kumar (2022). "Blockchain for precision irrigation: Opportunities and challenges." *Transactions on Emerging Telecommunications Technologies* 33(10): e4059.

Cai, L., Q. Gong, F. Jiang, M. Yuan, Z. Xiao, S. Zhang, C. Zheng and Y. Wu (2023). "The use of the intelligent Bayesian network method combined with blockchain technology in the optimisation of tunnel construction quality control." *IET Software, The Institution of Engineering and Technology*, https://doi.org/10.1049/sfw2.12109

Charles, V., A. Emrouznejad and T. Gherman (2023). "A critical analysis of the integration of blockchain and artificial intelligence for supply chain." *Annals of Operations Research*, 327: 7–47.

Chen, Y. (2023). "How blockchain adoption affects supply chain sustainability in the fashion industry: A systematic review and case studies." *International Transactions in Operational Research*, 31(6), Special Issue on Operational Research Models for Blockchain: 3592–3620

Clohessy, T. and T. Acton (2019). "Investigating the influence of organizational factors on blockchain adoption: An innovation theory perspective." *Industrial Management & Data Systems*, 119(7): 1457–1491.

Elangovan, D., C. S. Long, F. S. Bakrin, C. S. Tan, K. W. Goh, Z. Hussain, Y. M. Al-Worafi, K. S. Lee, Y. W. Kassab, & L. C. Ming (2020). Application of Blockchain Technology in Hospital Information System. In *Mathematical Modeling and Soft Computing in Epidemiology* (pp. 231–246). CRC Press, Boca Raton, FL. https://doi.org/10.1201/9781003038399-12

Esmaeilian, B., J. Sarkis, K. Lewis and S. Behdad (2020). "Blockchain for the future of sustainable supply chain management in Industry 4.0." *Resources, Conservation and Recycling* 163: 105064.

Faridi, A. R., A. Hafeez, F. Masood and M. Y. Salmony (2023). Stubble Burning Detection and Prevention using Machine Learning and Blockchain. *2023 10th International Conference on Computing for Sustainable Global Development (INDIACom)*, IEEE.

Frikha, T., J. Ktari, N. B. Amor, F. Chaabane, M. Hamdi, F. Denguir and H. Hamam (2023). "Low Power Blockchain in Industry 4.0 Case Study: Water Management in Tunisia." *Journal of Signal Processing Systems* 96(3), 257–271.

Howson, P., S. Oakes, Z. Baynham-Herd and J. Swords (2019). "Cryptocarbon: The promises and pitfalls of forest protection on a blockchain." *Geoforum*, 100: 1–9.

Iranmanesh, M., P. Maroufkhani, S. Asadi, M. Ghobakhloo, Y. K. Dwivedi and M.-L. Tseng (2023). "Effects of supply chain transparency, alignment, adaptability, and agility on blockchain adoption in supply chain among SMEs." *Computers & Industrial Engineering* 176: 108931.

Ismail, S., H. Reza, K. Salameh, H. Kashani Zadeh and F. Vasefi (2023). "Toward an Intelligent Blockchain IoT-Enabled Fish Supply Chain: A Review and Conceptual Framework." *Sensors* 23(11): 5136.

Jaradat, A., O. Ali and A. AlAhmad (2022). "Blockchain technology: A fundamental overview." In: Muthu, S.S. (eds) *Blockchain Technologies for Sustainability. Environmental Footprints and Eco-design of Products and Processes*. Springer, Singapore. https://doi.org/10.1007/978-981-16-6301-7_1

Jin, S. and H. Chang (2023). "The trends of blockchain in environmental management research: a bibliometric analysis." *Environmental Science and Pollution Research* 30(34): 81707–81724.

Junaid, M., S. Nazir, & M. Ali (2023). Role of Blockchain Technology Adoption between Sustainability-Related Supply Chain Risks and Triple Bottom Line Performance. In *Blockchain Driven Supply Chain Management* (Management for Professionals Series, pp. 181–199). Springer Nature Singapore Pte Ltd. https://doi.org/10.1007/978-981-99-0699-4_11

Karthick, S., S. Kermanshachi, A. Pamidimukkala and M. Namian (2023). "A review of construction workforce health challenges and strategies in extreme weather conditions." *International Journal of Occupational Safety and Ergonomics* 29(2): 773–784.

Kouhizadeh, M., S. Saberi and J. Sarkis (2021). "Blockchain technology and the sustainable supply chain: Theoretically exploring adoption barriers." *International Journal of Production Economics* 231: 107831.

Kouhizadeh, M., J. Sarkis and Q. Zhu (2019). "At the nexus of blockchain technology, the circular economy, and product deletion." *Applied Sciences* 9(8): 1712.

Leong, L.-Y., J.-J. Hew, V.-H. Lee, G. W.-H. Tan, K.-B. Ooi and N. P. Rana (2023). "An SEM-ANN analysis of the impacts of Blockchain on competitive advantage." *Industrial Management & Data Systems* 123(3): 967–1004.

Liu, F., Y. Khan and M. Marie (2023). "Carbon neutrality challenges in Belt and Road countries: what factors can contribute to CO2 emissions mitigation?" *Environmental Science and Pollution Research* 30(6): 14884–14901.

Martínez-Castañeda, M. and C. Feijoo (2023). "Use of blockchain in the agri-food value chain: State of the art in Spain and some lessons from the perspective of public support." *Telecommunications Policy* 47(6): 102574.

Midhila, P., M. Manjushree, D. Raajasubramanian, S. Srinivasan, H. Chalwadi, R. Devi, A. J. Adur and K. Narendra (2023). "Importance and Applications of Internet of Things (IoT)." Internet of Things: Applications for Sustainable Development: In book: *Internet of Things* (pp.1–16), Chapter: 1, Publisher: Taylor & Francis Group. doi:10.1201/9781003226888-1

Miloudi, L., K. Rezeg, O. Kazar and M. K. Miloudi (2020). Smart sustainable farming management using integrated approach of IoT, blockchain & geospatial technologies. *Advanced Intelligent Systems for Sustainable Development (AI2SD'2019) Volume 2-Advanced Intelligent Systems for Sustainable Development Applied to Agriculture and Health*, Springer.

Mubarik, M., R. Z. Raja Mohd Rasi, M. F. Mubarak and R. Ashraf (2021). "Impact of blockchain technology on green supply chain practices: Evidence from emerging economy." *Management of Environmental Quality: An International Journal* 32(5): 1023–1039.

Mukherjee, P. and C. Pradhan (2021). *Blockchain 1.0 to blockcTechnology: Applications and Challenges*, Springer: 29–49.

Munir, M. A., M. S. Habib, A. Hussain, M. A. Shahbaz, A. Qamar, T. Masood, M. Sultan, M. Mujtaba, S. Imran and M. Hasan (2022). "Blockchain adoption for sustainable supply chain management: Economic, environmental, and social perspectives." *Frontiers in Energy Research* 10: 899632.

Oberhauser, D. (2019). "Blockchain for environmental governance: Can smart contracts reinforce payments for ecosystem services in Namibia?" *Frontiers in Blockchain* 2: 21.

Ølnes, S. and A. Jansen (2017). Blockchain technology as s support infrastructure in e-government. *Electronic Government: 16th IFIP WG 8.5 International Conference, EGOV 2017*, St. Petersburg, Russia, September 4–7, 2017, Proceedings 16, Springer.

Polas, M. R. H., A. I. Kabir, A. S. M. Sohel-Uz-Zaman, R. Karim and M. I. Tabash (2022). "Blockchain technology as a game changer for green innovation: Green entrepreneurship as a roadmap to green economic sustainability in Peru." *Journal of Open Innovation: Technology, Market, and Complexity* 8(2): 62.

Rehali, M., Elhassani, M. E., Jaouhari, A. E., & Berrada, M. (2023, January). The use of artificial intelligence and blockchain in healthcare applications: introduction for beginning researchers. In *International Conference on Digital Technologies and Applications* (pp. 986–997). Cham: Springer Nature Switzerland.

Rejeb, A. and K. Rejeb (2020). "Blockchain and supply chain sustainability." *Logforum* 16(3), 363–372,

Satılmıoğlu, T. K., & H. Keskin (2023). Smart water chain: Immutable, distributed and decentralized water transaction ledgers. In *IOP Conference Series: Earth and Environmental Science, 1136, 012014. Proceedings of the 14th International Conference on Hydroinformatics (HIC 2022)*, held 4–8 July 2022, Bucharest, Romania. IOP Publishing, Bristol, England, UK. https://doi.org/10.1088/1755-1315/1136/1/012014

Shah Naqvi, S. A. A., A. Sultan, and S. Eslamian, (2015). Water Quality Issues in Urban Water, In *Urban Water Reuse Handbook, Ch. 8*, Eslamian, S. (Ed.), Taylor &Francis Group, CRC Group, USA, 99–112.

Sreenivasan, A. and M. Suresh (2023). "Start-up sustainability: Does blockchain adoption drives sustainability in start-ups? A systematic literature reviews." *Management Research Review*, 47(2). doi:10.1108/MRR-07-2022-0519

Stuit, A., D. Brockington and E. Corbera (2022). "Smart, commodified and encoded." *Conservation & Society* 20(1): 12–23.

Truby, J. (2018). "Decarbonizing Bitcoin: Law and policy choices for reducing the energy consumption of Blockchain technologies and digital currencies." *Energy Research & Social Science* 44: 399–410.

Tsao, Y.-C., V.-V. Thanh and Q. Wu (2021). "Sustainable microgrid design considering blockchain technology for real-time price-based demand response programs." *International Journal of Electrical Power & Energy Systems* 125: 106418.

Utami, B., W. Widjayanti and K. Sukmawati (2023). "Technology-based financial risk management strategy: Overview of recent developments." *Endless: International Journal of Future Studies* 6(2): 316–328.

Vats, G. and P. K. Sharma (2023). An Exhaustive Analysis on Security Issues concerning IoT using Blockchain. *2023 International Conference on Disruptive Technologies (ICDT)*, IEEE.

Wong, L.-W., L.-Y. Leong, J.-J. Hew, G. W.-H. Tan and K.-B. Ooi (2020). "Time to seize the digital evolution: Adoption of blockchain in operations and supply chain management among Malaysian SMEs." *International Journal of Information Management* 52: 101997.

Xia, W., X. Chen and C. Song (2022). "A framework of blockchain technology in intelligent water management." *Frontiers in Environmental Science* 10: 909606.

Xie, J., H. Tang, T. Huang, F. R. Yu, R. Xie, J. Liu and Y. Liu (2019). "A survey of blockchain technology applied to smart cities: Research issues and challenges." *IEEE Communications Surveys & Tutorials* 21(3): 2794–2830.

Xu, P., J. Lee, J. R. Barth and R. G. Richey (2021). "Blockchain as supply chain technology: Considering transparency and security." *International Journal of Physical Distribution & Logistics Management* 51(3): 305–324.

Xu, X., X. Bao, H. Yi, J. Wu and J. Han (2023). "*A Novel Resource-Saving and Traceable Tea Production and Supply Chain Based on Blockchain and IoT*." IEEE Access.

Yadav, N., S. Luthra and D. Garg (2023). "Blockchain technology for sustainable supply chains: A network cluster analysis and future research propositions." *Environmental Science and Pollution Research*, 30(24), 64779–64799.

Zeng, H., G. Dhiman, A. Sharma, A. Sharma and A. Tselykh (2023). "An IoT and Blockchain-based approach for the smart water management system in agriculture." *Expert Systems* 40(4): e12892.

Zhai, Q., Q. Li, A. Sher and C. Chen (2023). "The role of information heterogeneity in blockchain-based traceability systems: Evidence from fresh fruits buyers in China." *International Food and Agribusiness Management Review*, 26(3), 489–518.

Zhang, W., & M. Zhu (2022). Environmental accounting system model based on artificial intelligence, blockchain, and embedded sensors. *Computational Intelligence and Neuroscience*, 2022(1), Article 3803566.

Part V

Blockchain Technology for Environmental Innovation, Conservation and Sustainability

9 Blockchain for Environmental Innovation

Balancing Sustainability and Global Challenges

Erginbay Uğurlu, Irena Jindrichovska, and Burak Arslan

9.1 INTRODUCTION

Environmental conservation has become a crucial topic that is widely debated across many different sciences. The necessity for environmental conservation arises from the observation of human degradation of the environment. Given the importance of natural resources, humans have long discussed boundaries for economic progress. It is well known that economic expansion picked up steam mainly following the Industrial Revolution, and that growth led to a rise in the use of natural resources and energy. Scientists have viewed natural resources as the primary barrier to perpetually unrestricted economic progress, at least since the late 1700s.

Most discussions regarding the ultimate boundaries of economic growth are predicated on the idea that natural resources are the primary constraints on the continuous expansion of economic output. Based on this discussion on the effect of economic growth and development on natural resources, governments have increasingly focused on sustainable development and climate change mitigation. Researchers also began to study sustainability and sustainable development, resulting in a vast literature on the subject.

At this point, sustainable development has emerged as an important concept. The concrete results of international initiatives on environmental degradation and climate change launched in the 1980s were the United Nations Conference on Environment and Development held in Rio de Janeiro in 1992, and the Kyoto Protocol adopted in 1997. The Intergovernmental Panel on Climate Change (IPCC) has reported on the importance of climate change in human life (IPCC, 2014). The United Nations Sustainable Development Goals (SDGs) were set in 2015 as part of this process. Many different tools, technologies, strategies, and efforts have been used to enable sustainable development and environmental conservation. The motivation and urgency to address environmental issues are increasing; therefore, there is hope for innovative technologies to address the challenges and find sustainable solutions. New technologies are becoming increasingly important (Kouhizadeh & Sarkis, 2018). Blockchain technology is one of the infant technologies used to provide environmental conservation. Blockchain is not a widely used and well-known technology, but the role of blockchain in environmental conservation is increasing as the global community faces unprecedented environmental challenges. Blockchain technology can be useful for environmental sustainability in a number of ways, including the development of new green manufacturing techniques, the tracking and archiving of data-related activities, the real-time gathering and analysis of low-carbon or green data, and numerous other uses that have yet to be discovered (Mendoza Tovar, 2023).

As noted, several challenges are quickly becoming imminent threats, for example, climate change, excessive pollution, rapid urbanization, and rising urban problems, including criminality and excessive traffic, noise, and pollution (Abdelkrim & Eslamian, 2024). There is also the negative

DOI: 10.1201/9781003599593-14

effect of industrial activity in terms of deforestation and a decrease in biodiversity. However, new technologies such as digitalization, big data solutions, and blockchain technology provide a broad opportunity to use innovative solutions to help overcome these alarming issues.

Blockchain technology provides a broad range of options for transforming the landscape of environmental conservation (Alhasan & Hamdan, 2023; Balzarova, 2020). Disruptive innovations play a pivotal role in addressing environmental challenges by introducing groundbreaking technologies that revolutionize traditional practices. These innovations often provide sustainable solutions, fostering conservation and reducing the ecological footprint. Disruptive innovations, such as renewable energy breakthroughs and efficient waste management systems, pave the way for a more environmentally conscious and resilient future.

Blockchain technology is well-suited for improving environmental areas due to its strengths in supporting transparent data sharing, optimizing business processes, and reducing operating costs.

The intersection of blockchain and environmental conservation introduces a novel approach to addressing longstanding challenges in the field. The transparency and security inherent in blockchain technology can be leveraged to enhance various aspects of environmental conservation efforts (Kouhizadeh & Sarkis, 2018; Parmentola et al., 2022).

In this chapter, our main aim is to provide blockchain technology use cases in environmental conservation. Section 9.2 presents a general framework for environmental conservation and its relationship with the Sustainable Development Goals. Section 9.3 connects blockchain technology and environmental conservation by presenting some cases of blockchain technology use in environmental conservation. In Section 9.4, we conclude the chapter.

9.2 ENVIRONMENTAL CONSERVATION AND SDGs

Tisdell (2005) gave very broad information about environmental conservation and its economic base. He states that there are three conservation strategies: the first uses macro-variables, the second uses micro-adjustments, and the last is a mixture of the first and the second strategies. These strategies for conservation are used in every aspect of life, including human economic welfare, agriculture, forests, marine and fresh water, genetic diversity, wildlife, soil, and more.

The international conservation movement underwent a fundamental change with the publication of the World Conservation Strategy (IUCN et al., 1980), which was prepared by the International Union for the Conservation of Nature and Natural Resources, the United Nations Environment Program, and the World Wide Fund for Nature. The World Conservation Strategy, Caring for the Earth: A Strategy for Sustainable Living (IUCN et al., 1991), was published to extend and emphasize the message. The strategy provided (IUCN et al., 1980) was the key to a sustainable society and shifted the topic from the traditional focus on cure to prevention (McCormick, 1986). The starting point of concern for environmental conservation can be found in the 1972 United Nations Conference on the Human Environment (UNCHE), which marked the beginning of concerns regarding natural environmental degradation and the need for sustainable development (Uğurlu, 2019). These concerns increased with new conferences. One of the important conferences was the United Nations Conference on Environment and Development (UNCED) held in Rio de Janeiro between June 3 and 14, 1992. The declarations following these conferences emphasized the importance of the environment in relation to development and promoted an action plan to preserve it. In the following years, there were many efforts to advance the topic. In 2015, the United Nations developed its sustainable efforts, placing 17 SDGs on its agenda (United Nations, 2015). These SDGs are summarized in Table 9.1. With these goals, the UN has 169 targets.

These goals have two key features: economic security and environmental sustainability; unfortunately, achieving these goals will be very hard for countries to succeed by 2030 (Arora & Mishra, 2019). One of the goals of this chapter is to investigate environmental conservation, which is related to one of the afore-stated concerns, which is environmental sustainability. Environmental sustainability directly relates to SDG6, SDG7, SDG12, SDG13, SDG14 and SDG15, which are clean water

TABLE 9.1
Sustainable Development Goals and Targets

SDG	Name	Number of Targets
SDG1	No poverty	7
SDG2	Zero hunger	8
SDG3	Good health and well-being	13
SDG4	Quality education	10
SDG5	Gender equality	9
SDG6	Clean water and sanitation	8
SDG7	Affordable and clean energy	5
SDG8	Decent work and economic growth	12
SDG9	Industrial innovation and infrastructure	8
SDG10	Reduced inequalities	10
SDG11	Sustainable cities and communities	10
SDG12	Responsible consumption and production	11
SDG13	Climate action	5
SDG14	Life below water	10
SDG15	Life on land	12
SDG16	Peace, justice and strong institutions	12
SDG17	Partnerships for the goals	19

Source: Uğurlu, 2019.

and sanitation, affordable and clean energy, responsible consumption and production, climate action, life below water and life on land, respectively. Moreover, clean water and sanitation are related to the conservation of water sources, and responsible consumption and production include the conservation of resources by changing consumption habits to conserve natural resources.

Investigating the conservation of all the environmental issues in one chapter is impossible. Still, we will give some examples to show the relationship between environmental conservation and the Sustainable Development Goals. For example, severe deterioration of land and soil has been observed throughout history but continues nowadays because of unsustainable land use, deforestation, and climate change. That is why scientists and governments are developing strategies for soil conservation. These strategies are related to the SDGs, which are SDG 1 (no poverty), SDG 2 (zero hunger), SDG 6 (clean water and sanitation), SDG 13 (climate action) and SDG 15 (life on land) (Albaladejo et al., 2021). They are related to SDG1, SDG2, and SDG6 because soil conservation maintains or improves the land's productivity and has positive effects on water management based on its agricultural role (Bouma, 2019).

Biodiversity conservation is one of the parts of environmental conservation, and it is particularly related to SGD15. To understand this issue, we have to understand the term "built environment", which is defined in Kaklauskas and Gudauskas (2016) as: "human-made surroundings that provide the setting for human activity, ranging in scale from buildings and parks or green space to neighborhoods and cities that can often include their supporting infrastructure, such as water supply or energy networks". Human activities in the built environment do contribute to the loss of biodiversity (Zari, 2014). Moreover, one of the targets of SDG15 is to ensure the conservation of mountain ecosystems, including their biodiversity.

Many other factors affect environmental conservation, such as small ruminants. Small ruminants have both positive and negative effects on the environment. While they enable greater species diversity by grazing, they also contribute to desertification. Overgrazing can cause water and wind erosion, negatively impacting environmental conservation (El Aich & Waterhouse, 1999). Modifications to small ruminant systems are having a negative influence on existing ecosystems and habitats. If small ruminants are integrated, any negative impacts could be reduced, and environmental quality could be increased.

Sea-level conservation is another important type of environmental conservation. Arctic sea ice melting and rising sea levels are two important indicators of climate change. The melting of Arctic Sea ice causes rising sea levels, impacting land and human life. One of the effects of rising sea levels is increased coastal erosion (Nummedal et al., 1987), increased coastal flooding (Michener et al., 1997), and saltwater intrusion into estuaries, deltas, and aquifers (Titus, 1990), resulting in changes in water quality, salinity, and biological activity (Feenstra et al., 1998). It is related to several SDGs; because it affects land, it is related to SDG1, SDG2, and SDG15; because it affects the sea, water, and water quality, it is related to SDG6 and SDG14; because ice melting is caused by climate change, it is related to SDG13. It also indirectly affects the economy and economic growth. Thus, it implies the existence of other related SDGs.

Many countries have coastlines, and coastal and marine ecosystem conservation is an important part of environmental conservation. It has a direct relation to SDG14 and an indirect relation to other SDGs. These indirectly related SGDs are SDG2, SDG6, SDG7, SDG8, SDG10, SDG11, SDG12, SDG13, and SDG15 (Islam & Shamsuddoha, 2018). One of the tools for managing marine social–ecological systems is marine protected areas (MPAs). The MPA system prevents some of the fishable stock from being caught. To achieve this, it is necessary to restrict or outlaw extractive activities like fishing in a defined area (Cinner & David, 2011). While areas of importance for biodiversity have made moderate progress toward ecological representativeness, the connectivity of MPAs has not yet been evaluated, according to the most recent Global Assessment on Biodiversity and Ecosystem Services (Gannon et al., 2017).

9.3 THE BLOCKCHAIN AND ENVIRONMENTAL CONSERVATION

9.3.1 The Blockchain

Blockchain is widely recognized as the underlying technology of Bitcoin; however, its applications extend significantly beyond cryptocurrencies. Using cryptographic techniques to ensure the security of transactions, Blockchain offers a distributed database, commonly known as a "digital ledger" that is accessible to every participant on the network. This network essentially consists of a sequence of interconnected computers, which necessitate a unanimous consensus for the validation and registration of each transaction (Hutt, 2016).

Blockchain is a disruptive innovation initially known for its links with cryptocurrencies (Bai et al., 2019). It is a decentralized and distributed ledger technology. It securely records and authenticates transactions across a network of computers. Unlike conventional centralized systems, blockchain functions on a peer-to-peer (P2P) network, providing transparency, security, and persistence. Each block in the chain holds a timestamped record of transactions, and once added, any attempt to retroactively alter it would necessitate changes to subsequent blocks. Additionally, blockchain technology directly relates to SDG 9 (industry, innovation, and infrastructure) (Fraga-Lamas & Fernández-Caramés, 2020). Figure 9.1 shows the historical evolution of blockchain technology.

The process of verifying ownership and transactions relies on the utilization of search puzzles based on hash functions. The utilization of blockchain technology encompasses a decentralized digital ledger system that enhances transparency, traceability, and security as its primary focus (Manullang et al., 2021). Network participants are required to solve these puzzles to append valid blocks to the chain. The complexity of these puzzles is regularly adjusted to accommodate variations in the computing power of connected devices and to ensure that there is a consistent interval of approximately ten minutes between the addition of each block (Stoll et al., 2019). Consensus algorithms, as an integral aspect of blockchain technology, assume a crucial function. Consensus

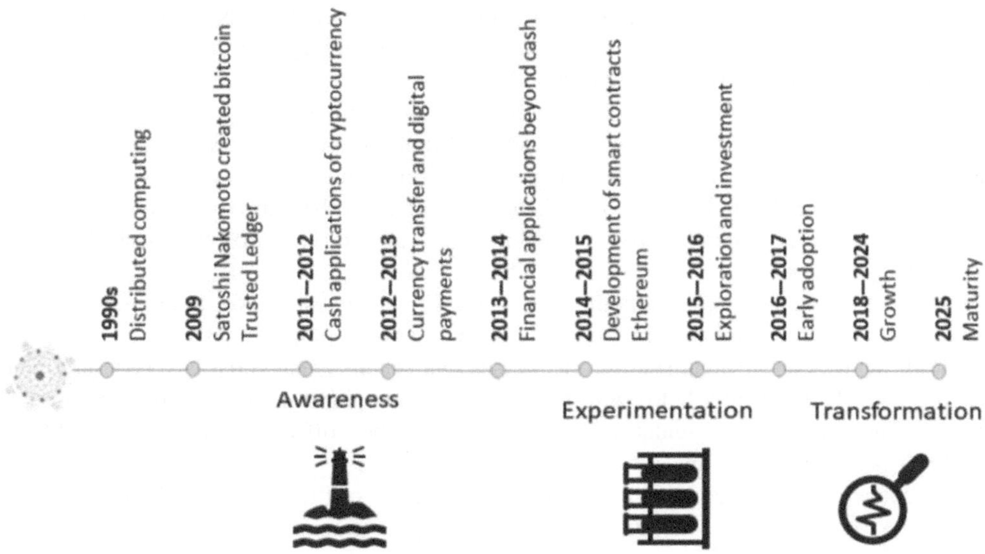

FIGURE 9.1 Historical Evolution of Blockchain.

Source: Aysan et al., 2021, p. 5.

mechanisms are necessary for decision-making, enabling all participants to achieve mutual understanding. The objective of the consensus process is to authenticate any data appended to the ledger. This procedure guarantees that the information enclosed in each block accurately represents the latest series of transactions occurring within the network. With the progression of blockchain technologies, diverse modes of consensus models have been formulated (Arshad et al., 2023).

Bitcoin employs a consensus mechanism known as proof of work to prevent instances of double spending and manipulation. Various consensus algorithms exist within the realm of blockchain technology. Blockchain algorithms can be divided into two primary classifications. The initial classification is based on proof, where the nodes participating in the validation network must demonstrate their superior qualifications compared to others to carry out the act of appending. The subsequent classification is centered around voting, necessitating the exchange of verification outcomes for a new block or transaction among network nodes before reaching a conclusive determination (Nguyen & Kim, 2018). With the advance of blockchain technologies, various methods of consensus models have been formulated (ESCAP, 2022).

Proof of work (PoW) refers to the phenomenon where multiple nodes compete to achieve a consensus by successfully resolving an intricate mathematical problem. With each resolution obtained, the complexity of the mathematical problem and the computational resources required for its resolution increase. Consequently, due to its cost-intensive nature, PoW effectively deters any attempts at fraudulent activities within the network.

Proof of stake (PoS) enables consensus through the exposure of capital. Nodes deposit their cryptocurrencies to be granted the privilege of validating novel transactions and appending them to the blockchain. If a node fails to adhere to the prescribed regulations, it forfeits its stake.

Delegated proof of stake (DPOS) corresponds to the concept of PoS. Individuals place their coins as collateral and entrust their decision-making authority to a higher-level governing body. In certain instances, real-time voting and social reputation are employed to attain consensus.

Proof of space (PoSp) permits individuals to allocate computational capacity. Consensus is reached when an individual demonstrates that a small chunk of data truly utilizes the assigned capacity at the minimum level of affirmation. Blockchain operates as a decentralized ledger, meaning it does not rely on a central authority for control. This inherent decentralization ensures that information is distributed across a network of computers, reducing the risk of data manipulation or fraud. In the context of environmental conservation, this transparency is instrumental in tracking and validating activities related to sustainable practices.

One of the notable applications of blockchain in environmental conservation is supply chain management. The technology enables the creation of transparent and traceable supply chains, particularly in industries notorious for their negative environmental impact, such as forestry, fishing, and agriculture. By utilizing blockchain, stakeholders can track the journey of products from origin to consumer, ensuring adherence to sustainability and ethical standards (Gurtu & Johny, 2019; Moosavi et al., 2021).

Blockchain technology is also suitable for creating reliable systems for tracking and managing other negative features in today's environment, such as carbon footprints. By using smart contracts to track all devices in an organization, businesses can securely and reliably record and measure emissions data. This not only helps with better understanding and monitoring their environmental impact but also provides a foundation for developing carbon offset initiatives and enables more accurate sustainability reporting.

Smart contracts – self-executing contracts with the terms of the agreement directly written into code – play a crucial role in automating and enforcing environmental agreements. For instance, stakeholders involved in land conservation can utilize smart contracts to ensure that land use adheres to predefined sustainable practices. This not only streamlines the enforcement process but also reduces the need for intermediaries.

After decentralized blockchain technology was conceptualized with the use of cryptocurrencies in 2009, it became quite popular in a really short period of time. Aside from cryptocurrencies, its technology finds usage in different sectors with different purposes. Blockchain technology provides a wide range of benefits by employing a smooth and seamless procedure of unchangeable and open records and by transforming assets into digital tokens (also known as tokenization) with the assistance of smart contracts (Aysan et al., 2021).

Smart contracts can be described as digital contracts which are stored on a blockchain and are programmed to be automatically executed upon meeting specific predetermined terms and conditions. Smart contracts function by adhering to simplistic "if/when... then..." statements that are encoded into the blockchain. Once predetermined conditions have been fulfilled and validated, a network of computers will execute the designated actions. These actions have the potential to incorporate the disbursement of funds to appropriate entities, registering a motor vehicle, transmitting notifications, or providing a citation. Upon the completion of the transaction, the blockchain is subsequently updated. Consequently, this signifies that the transaction retains an unalterable state, and only individuals who have been given authorization possess the capability to perceive the outcomes (Chen et al., 2022).

Blockchain technology and environmental sustainability are becoming a focus (Arshad et al., 2023), as witnessed in a recent numeric summary publication, which is summarized in Table 9.2. The most important topics are summarized in the following table according to journals and keywords (see: Arshad et al., 2023). In the table, IJEEP, Sus, AE, IJEM, RSER, and BSE are the *International Journal of Energy Economics and Policy*, *Sustainability*, *Applied Energy*, *International Journal of Energy Sector Management*, *Renewable and Sustainable Energy Reviews*, and *Business Strategy and Environment*, respectively.

The authors have characterized recent literature according to progressive journals using keyword search. Blockchain technology and energy are the most often cited issues. That is why the recent technology of the blockchain is promising in fulfilling SDG goals, specifically SDG 7 (Affordable and clean energy) and SDG13 (Climate action).

Suggested blockchain technology solutions concentrate on four primary motives: (1) emissions-trading schemes, (2) voluntary carbon markets, (3) Article 6 of the Paris Agreement, and (4) the Carbon Offsetting and Reduction Scheme for International Aviation (CORSIA). If these and other new systems in climate change mitigation require data transfer, the blockchain can provide data

TABLE 9.2
Numeric Summary of the Review Sample

Search Strings	Scopus	IJEEP	Sus	AE	IJEM	RSER	BSE
Blockchain Technology AND Environment	921	3	1	112	3	57	63
Blockchain Technology AND Green Finance	2	2	9	29	8	37	13
Blockchain Technology AND Carbon Emissions	57	5	6	28	9	15	5
Blockchain Technology AND Climate Change	47	7	1	18	5	13	9
Blockchain Technology AND Greenhouse Gases	36	8	10	5	3	10	6
Blockchain Technology AND Energy	1585	2	14	4	8	2	9
Blockchain Technology AND SDGs	59	32	12	18	16	12	3

Source: Arshad et al. 2023, p. 2607.

transfer that corresponds to the current requirements of climate markets (Dorfleitner et al., 2021). Blockchain's adaptability to different technologies and environmental conservation areas is key. It can be used to implement smart contracts and achieve transactions in CO_2 reduction mechanisms and renewable energy markets. Additionally, in green production, users need to monitor and store data-related activities, and blockchain can be used in the green supply chain (Parmentola et al., 2022).

9.3.2 Blockchain-Enabled Solutions for Environmental Conservation and SDGs

Blockchain technology is used in many different sectors. Some uses relate to SDGs and environmental conservation, directly or indirectly; these include sustainable water management, sustainable energy, reducing inequality, and the blockchain tokenization of carbon emissions. However, the examples of blockchain technology in environmental conversation are not limited to these; all SDGs have at least one blockchain technology use, which is summarized in Table 9.3.

Microgrids are the next generation of power transmission systems that conserve energy and have economic, reliability, environmental, and social benefits. Additionally, microgrids offer deferred transmission and distribution benefits (Parag & Ainspan, 2019). Uğurlu & Muratoğlu (2019) give examples of microgrid usage in solar energy and list 24 companies that use P2P companies mainly focus on blockchain applications. The authors focus on the Enervalis company, which uses the consensus mechanism, encryption algorithms, smart contracts, and distributed data storage. Security and sustainability are the two most common difficulties in conventional P2P energy trading; blockchain technology solves this problem by using a sustainable microgrid (SMG) with secure P2P trading (Tsao & Thanh, 2021). SMG is a conflicting multi-objective optimization problem where social and environmental concerns, in addition to economic benefits, must be considered simultaneously (Hall et al., 2012).

Water is an inherently finite asset integral to humanity's survival and economic progress. SDG6 concerns clean water and sanitation, but the insufficiency of water poses a pivotal predicament in numerous regions across the globe (Naqash et al., 2023). As urban areas expand, the demand for water provision increases. This has led to a requirement for a centralized and synchronized water provision structure, which mandates the implementation of more sustainable and effective methods for managing infrastructure-related resources (Pahonţu et al., 2023). Here, the significance of blockchain technology lies in its capacity to facilitate the real-time processing of water resources data, thereby augmenting the efficacy of resource management and strengthening the stability of new energy economic development. This can promote the economy's and resources' sustainable progress (Zhang, 2023).

Naqash et al. (2023) illustrate a smart water dashboard and middleware blockchain system. The blockchain system is constructed for the real-time monitoring of water distribution networks.

TABLE 9.3
Main Blockchain Use Cases for SDGs

SDG	Case
SDG1	Efficient access to credit, crowdsourcing and crowdlending
SDG2	Sustainable/traceable food production systems along their life cycle
SDG3	Crowdsourcing health systems
SDG4	Education records and certificates
SDG5	Micro-credits, ICT learning
SDG6	Smart IoT management systems
SDG7	Micro-transactions, transparent local market
SDG8	Optimized financial infrastructures
SDG9	Supply chain traceability
SDG10	Financial inclusion, sharing economy
SDG11	Smart cities, collective decisions
SDG12	Traceability of the production process
SDG13	Accountability and transparency for rapid decarbonization
SDG14	Incentivized actions
SDG15	Accountability and transparency to terrestrial ecosystems
SDG16	Transparency in public spending
SDG17	Global competition

Source: Adopted from Fraga-Lamas and Fernández-Caramés (2020), p. 6, Figure 9.1.

Internet of Things (IoT) sensors facilitate the instantaneous observation and dissemination of information concerning urban water management. These sensors gather and transfer data about various aspects of water management, including water pressure, flow rates, and water quality. Blockchain technology guarantees the authenticity and reliability of data by preserving sensor data in a decentralized and tamper-proof manner. Each block of data is interconnected with the preceding block, rendering any attempt to manipulate the data impossible through computational means. IoT sensors, the blockchain, and a water distribution dashboard enable precise decision-making, proactive maintenance, and the effective identification of water leaks, ultimately optimizing water distribution networks in urban areas.

The global energy sector is experiencing an increasing trend toward rapid decarbonization and decentralization. Nevertheless, renewable energy sources often exhibit variability, signifying their reliance on intermittent weather conditions. Consequently, this phenomenon has given rise to fresh problems in the administration and operation of electricity systems. To effectively address these challenges, it becomes imperative to possess suitable monitoring instruments that guarantee the proficient management of demand and supply (Social Alpha Foundation, 2022). Utilizing blockchain technology to establish smart grid systems supports the creation and maintenance of sustainable local energy markets. These markets facilitate the efficient execution of micro-transactions while simultaneously removing the need for central intermediaries. Furthermore, they serve to enhance the overall reliability and equality of the various agents engaged in these markets (Fraga-Lamas & Fernández-Caramés, 2020).

Applying blockchain technology can streamline the pricing hierarchy and determine the source of renewable energy resources through peer-to-peer transactions within a communal power grid. The viability of blockchain technology in the realm of clean energy can be evaluated through economic, technological, and social perspectives, encompassing considerations about consumer demand and producer supply (Social Alpha Foundation, 2022).

Blockchain has the capacity to overcome several obstacles faced by the impoverished and possesses immense potential to offer scalable resolutions to the problems of poverty and inequality. This encompasses the ability to provide a lasting and unalterable record of personal identity within the blockchain, which individuals own and hold the power to revolutionize the situation. Furthermore, by utilizing smart contracts, blockchain can guarantee that the funds provided by donors are efficiently and transparently directed to their intended recipients, eliminating the need for intermediaries and preventing any loss or leakage along the way. Additionally, the delivery of aid can be monitored through transparently documented "way-stations", which display the location in the supply chain and the final delivery, thereby facilitating financial inclusion (Thomason, 2021).

In theory, the Blockchain's capabilities make it a catalyst for facilitating a sharing economy characterized by a democratic ownership structure, thereby bypassing unnecessary intermediaries. Certain scholars within the literature have analyzed the consequences of blockchain technology on income inequality and argue that, despite its potential to produce positive outcomes, it could also exacerbate existing imbalances in the allocation of wealth.

Blockchain introduces the concept of tokenization, which can be leveraged to create digital assets representing environmental indicators or conservation efforts. These tokens, often referred to as "conservation tokens", can be traded or used to attract funding for specific environmental projects. This financial incentive model encourages private and public participation in conservation efforts, enforcing a more sustainable and collaborative approach (Huseynov & Mitchell, 2023; Sun et al., 2023).

Blockchain technology presents a dependable, effective, all-encompassing, convenient, and accessible solution for carbon credit markets. The public blockchain is easily reachable for small- and medium-scale transactions, and the rate of entry into the carbon trading market is diminishing. The information disseminated by corporations is transparent and available to all. Recent advances include digitalized product instruments and automated market traders (AMTs) that operate independently, enabling direct trading on the blockchain without incurring algorithmic fees. Moreover, blockchain furnishes the necessary infrastructure to ensure the existence of all blocks and establish a digital carbon credit ecosystem (Saraji & Borowczak, 2021).

9.3.3 Cryptocurrencies

In the first decade of the new millennium, various cryptocurrencies emerged, starting with Bitcoin, the first peer-to-peer electronic cash system, introduced in 2009. After the success of this new investment opportunity, other cryptocurrencies followed, for example, Ethereum, Tether, BMB, and XRP. Nevertheless, this investment opportunity regarding energy consumption also has a very negative side. Cryptocurrencies consume a significant amount of electrical energy compared to their technical performance, for instance (Corbet et al., 2020; Kohli et al., 2023; Li et al., 2019; Nguyen et al., 2022).

Nevertheless, incorporating blockchain technology into other industrial branches and integrating the revolutionary methods of Industry 4.0 and corporate sustainability amplifies its impact by ensuring transparency and traceability throughout the supply chain in any branch of industrial activity. Blockchain's decentralized ledger enhances accountability, providing an irreversible record of every transaction and process. This not only promotes trust among stakeholders but also facilitates the verification of sustainable practices, such as responsible sourcing and ethical production. Combining Industry 4.0, sustainability initiatives, and blockchain methodology creates a powerful framework for building resilient and eco-friendly business models. This combination of groundbreaking methods fosters a future where innovation and environmental consciousness work well in a progressively enriching combination.

Since 2009, many cryptocurrencies have surfaced, commencing with Bitcoin, which served as the inaugural and widely recognized manifestation of Satoshi Nakamoto's blockchain technology that debuted in 2008. In due course, it became the foremost cryptocurrency globally, boasting a market value of $614.9 billion as of July 2021, surpassing 5,655 acknowledged cryptocurrencies (Kohli et al., 2023). The advent of Bitcoin, coupled with the utilization of blockchain technology, has substantially enhanced the conventional payment system. Normally, individuals must rely on a third party to authenticate the legitimacy of their transactions before their execution. Regrettably, skepticism surrounds such intermediaries as they may be susceptible to exploitation and deception. The problem with a formal centralized authority is that all operations center on this single organization, rendering trust insufficient. The resolution to this quandary entails the utilization of numerous autonomous entities to validate transactions, thereby changing from centralization to decentralization. The environmental impact of cryptocurrency markets is raising concern, with the level of discussion and engagement surrounding their energy consumption increasing. Despite the global push for a greener and more sustainable planet, the impact of cryptocurrency growth and its rising energy consumption has not been a focal point in high-level policy debates. The recent adoption of Bitcoin as an official currency in El Salvador signals the possible beginning of cryptocurrency legalization, emphasizing the need for a comprehensive assessment of its environmental impacts. Recently a new measure called the Cryptocurrency Environmental Attention (ICEA) index was created to assess sustainability concerns in cryptocurrency markets (Wang et al., 2022). Researchers worry about the carbon footprint of digital currencies. Most researchers consider cryptocurrencies as speculative assets with potential price bubbles. Furthermore, their societal value is questionable.

9.3.3.1 The Relation Between Cryptocurrency and the Environment

Blockchain technology holds great promise for reshaping the conditions for environmental improvements. Its decentralized nature, transparency, and ability to facilitate trust among stakeholders represent a powerful tool for encompassing the varied nature of sustainable practices. As we reconsider the challenges of preserving our planet, the integration of the blockchain into environmental initiatives represents a significant step toward a better structured and more transparent, accountable, and collaborative approach to conservation (Allena, 2020; Parmentola et al., 2022).

With the increase in environmental challenges, integrating innovative technologies has become crucial for effective conservation efforts. Blockchain is an innovative technology with practical applications in various fields. Blockchain has grown from its financial origins to find applications in multiple sectors, including environmental conservation (Ayan et al., 2022; Esmaeilian et al., 2020; Kohli et al., 2023).

Beyond finance, the blockchain, as a decentralized and transparent method, can revolutionize how we approach and manage environmental initiatives in most industrial segments (Parmentola et al., 2022). This system explores the multifaceted role of blockchain in environmental conservation, highlighting its key features, applications, and the positive impact it can have on sustainable practices.

In the face of escalating environmental challenges, integrating innovative technologies has become imperative for effective conservation efforts. Initially recognized for its association with cryptocurrencies, the blockchain has transcended its financial origins to find applications in various sectors, including environmental conservation. This decentralized and transparent technology has the potential to revolutionize the way we approach and manage various environmental initiatives. This chapter explores the multifaceted role of the Blockchain in environmental conservation, highlighting its key features, applications, and the positive impact it can have on sustainable practices.

At present, there are both positive as well as negative views regarding cryptocurrencies' potential environmental impact. Moreover, Andoni et al. (2019) and Gallersdörfer et al. (2020) assume that the energy consumption of cryptocurrencies, especially Bitcoin using ASIC algorithms, appears disproportionately large compared to the energy consumed in market capitalization. These digital currencies show promising technological advances, particularly in the widespread adoption of blockchain technology across various industries. Wang et al. (2020) discuss its application in supply chain management, while Kimani et al. (2020) investigate its role in corporate operations and taxation. Furthermore, blockchain technologies have found utility in business and infrastructure sectors

(Bai et al., 2019; Shojaei et al., 2019) and academia, notably in management (Centobelli et al., 2021). Singh and Kim (2019) emphasized the importance of the blockchain in the context of the fourth industrial revolution.

Yang and Hamori (2021) performed a study on the carbon footprint network that reveals varying percentiles, highlighting potential atmospheric risks. Abubakr Naeem et al. (2022) asserted that Bitcoin's clean energy hedging ratio and effectiveness make it a viable diversification option. Pham et al. (2022) noted a lack of strong connections between green cryptocurrencies and Bitcoin/Ethereum, emphasizing the need for more green financial assets. They stressed the necessity of a robust policy assessment due to the susceptibility of green assets in the USA to stock market volatilities (Abubakr Naeem et al., 2022).

A detailed exploration of cryptocurrencies' environmental impact becomes imperative amid these contrasting perspectives. Human-induced changes, such as global warming caused by the accumulation of CO_2 emissions, pose significant threats to the planet. These environmental changes affect various aspects of human life, including flooding, alterations in weather patterns, urbanization, agricultural productivity losses, and threats to human health. Moreover, they constrain economic growth and development, impacting overall quality of life. Recognizing the paramount importance of the environment and the potential consequences of cryptocurrencies on environmental quality prompts the need for this study. The existing literature on cryptocurrencies tends to emphasize their positive attributes, often overlooking the environmental implications of mining and transacting in these digital currencies.

From a technical point of view, cryptocurrencies rely on blockchain technology, requiring users to place trust in their cryptographic algorithms and peer-to-peer technologies. This task can be challenging to achieve and disseminate. In cryptocurrency mining, each node independently verifies a set of transactions before incorporating them into the blockchain.

Several authors provided case studies on cryptocurrency networks, specifically focusing on Ethereum 2.0 and Pi Network, and discussed their potential to address and resolve certain challenges within the cryptocurrency landscape.

The research employed the MAC (mining allocation cost) method to evaluate energy consumption, emphasizing renewable energy capabilities in key Bitcoin mining regions. The relative consumption ratio (RCR) revealed that major mining countries such as China, the USA, Germany, and Canada are more conducive to allocating renewable energy than Russia, Iran, Malaysia, Ireland, and Kazakhstan. Two case studies focusing on Ethereum 2.0 and the Pi Network were presented. These studies highlighted endeavors to implement consensus algorithms like PoS and SCP, as well as concepts such as sharding, to distribute the load and reduce redundancy in cryptocurrency networks. These initiatives aim to decrease overall energy consumption and carbon footprints, demonstrating a commitment to addressing the environmental impact of cryptocurrencies and enhancing their sustainability and widespread acceptance despite being in the developmental phase.

Some pros and cons of blockchain technology. As a decentralized and fraud-resistant ledger technology, blockchain highlights several positive features. Its transparency ensures trust among participants, reducing fraud and enhancing security. Its decentralized nature minimizes the risk of a single point of failure, increasing system reliability. Smart contracts automate and rationalize processes, reducing costs and improving efficiency. Furthermore, blockchain facilitates faster and more secure cross-border transactions, promoting positive financial effects and economic growth. Despite technological advances, cryptocurrencies have been scrutinized for their environmental impact and enormous carbon footprint. As seen in Bitcoin, proof-of-work consensus mechanisms consume an extensive amount of energy, contributing to carbon inefficiency. The mining process involves complex mathematical computations, demanding powerful hardware and causing high electricity consumption. The sustainability of cryptocurrencies and their alignment with global environmental goals have raised concerns. Additionally, their social impact is also questionable.

Blockchain is becoming an incentive for innovative solutions in response to environmental concerns. Blockchain technology enables transparent and traceable supply chain management, promoting sustainable practices. Carbon offset projects and renewable energy trading platforms are leveraging the blockchain to ensure transparency and efficiency. These applications enhance accountability and promote eco-friendly practices, demonstrating the potential of blockchain for driving environmental innovations.

Blockchain plays an important role in furthering SDG 7 (Affordable and Clean Energy) and SDG 13 (Climate Action) by facilitating transparent energy transactions and promoting decentralized energy grids. Blockchain contributes to the expansion of clean energy sources. Additionally, blockchain supports climate action by enhancing transparency in carbon markets and ensuring the credibility of emissions reduction initiatives. This vital combination with SDGs underscores the potential of the blockchain to drive positive environmental outcomes and foster a sustainable future.

9.4 CONCLUSIONS

Interest in climate change began to grow mainly after the 1980s. Climate change is one of the factors responsible for increasing temperatures, melting ice, ozone depletion, and much environmental degradation. Environmental degradation will affect all human activities, and because of this, the environment has to be protected from climate change and environmental degradation. These protection efforts and strategies began with environmental conservation strategies. Further steps have been taken along with the initiative of such organizations and institutions as the UN, IPCC, EU, and many others. Governments' sustainable development aims led to the creation of a new worldwide strategy, the 17 Sustainable Development Goals (SDGs), which are being implemented globally. To achieve the targets of the SDGs, many different projects/tools/policies are being applied using different technologies. In this decade, we have seen many developments in the digital sector. One of them is blockchain technology. Blockchain technology holds great promises for improvements in environmental conservation. Its decentralized nature, transparency, and ability to facilitate trust among stakeholders make it a powerful tool for addressing the complexities of sustainable practices. As we navigate the challenges of preserving our planet, integrating blockchain into environmental initiatives represents a significant step toward a more transparent, accountable, and collaborative approach to conservation.

In conclusion, while the positive features of blockchain offer transformative solutions in various sectors, the environmental impact of cryptocurrencies requires a careful balance. The synergy between blockchain and environmental innovations inspires viable expectations for sustainable development, as evidenced by their contributions to SDG 7 and SDG 13. As the technology evolves, a nuanced approach is crucial to enforce the positive aspects of the blockchain while mitigating its environmental footprint.

REFERENCES

Abdelkrim, Z., & Eslamian, S. 2024. Potential impacts of climate change, In S. Eslamian, M. B. Huda, N. A. Rather, & F. A. Eslamian (Eds.), *Handbook of Climate Change Impacts on River Basin Management, Ch. 7, Vol. 3: Technology, Innovations and Management Practices*. Taylor & Francis Group, CRC Group, USA.

Abubakr Naeem, M., Iqbal, N., Lucey, B. M., & Karim, S. (2022). Good versus bad information transmission in the cryptocurrency market: Evidence from high-frequency data. *Journal of International Financial Markets, Institutions and Money*, *81*, 101695. https://doi.org/10.1016/j.intfin.2022.101695

Albaladejo, J., Díaz-Pereira, E., & de Vente, J. (2021). Eco-holistic soil conservation to support land degradation neutrality and the sustainable development goals. *CATENA*, *196*, 104823. https://doi.org/10.1016/j.catena.2020.104823

Alhasan, H., Hamdan, A. (2023). Blockchain Technology and Environmental Sustainability. In: El Khoury, R., Nasrallah, N. (eds) *Emerging Trends and Innovation in Business and Finance. Contributions to Management Science*. Springer, Singapore. https://doi.org/10.1007/978-981-99-6101-6_54

Allena, M. (2020). Blockchain technology for environmental compliance: Towards a "Choral" approach. *Environmental Law*, *50*(4), 1055–1103.

Andoni, M., Robu, V., Flynn, D., Abram, S., Geach, D., Jenkins, D., McCallum, P., & Peacock, A. (2019). Blockchain technology in the energy sector: A systematic review of challenges and opportunities. *Renewable and Sustainable Energy Reviews*, *100*, 143–174. https://doi.org/10.1016/j.rser.2018.10.014

Arora, N. K., & Mishra, I. (2019). United nations sustainable development goals 2030 and environmental sustainability: Race against time. *Environmental Sustainability*, *2*(4), 339–342. https://doi.org/10.1007/s42398-019-00092-y

Arshad, A., Shahzad, F., Ur Rehman, I., & Sergi, B. S. (2023). A systematic literature review of blockchain technology and environmental sustainability: Status quo and future research. *International Review of Economics & Finance*, *88*, 1602–1622. https://doi.org/10.1016/j.iref.2023.07.044

Ayan, B., Güner, E., & Son-Turan, S. (2022). Blockchain technology and sustainability in supply chains and a closer look at different industries: A mixed method approach. https://doi.org/10.3390/logistics6040085

Aysan, A. F., Bergigui, F., & Disli, M. (2021). Using Blockchain-Enabled Solutions as SDG accelerators in the international development space. *Sustainability*, *13*(7), Article 7. https://doi.org/10.3390/su13074025

Bai, C., Cordeiro, J., & Sarkis, J. (2019). Blockchain technology: Business, strategy, the environment, and sustainability. *Business Strategy and the Environment*, *29*. https://doi.org/10.1002/bse.2431

Balzarova, M. A. (2020). Blockchain technology – a new era of ecolabelling schemes? *Corporate Governance*, *21*(1), 159–174. https://doi.org/10.1108/CG-08-2020-0328

Bouma, J. (2019). How to communicate soil expertise more effectively in the information age when aiming at the UN Sustainable Development Goals. *Soil Use and Management*, *35*(1), 32–38. https://doi.org/10.1111/sum.12415

Centobelli, P., Cerchione, R., Esposito, E., Passaro, R., & Shashi. (2021). Determinants of the transition towards circular economy in SMEs: A sustainable supply chain management perspective. *International Journal of Production Economics*, *242*(C). https://econpapers.repec.org/article/eeeproeco/v_3a242_3ay_3a2021_3ai_3ac_3as0925527321002735.htm

Chen, J., Xia, X., Lo, D., Grundy, J., Luo, X., & Chen, T. (2022). Defining smart contract defects on Ethereum. *IEEE Transactions on Software Engineering*, *48*(1), 327–345. https://doi.org/10.1109/TSE.2020.2989002

Cinner, J. E., & David, G. (2011). The human dimensions of coastal and marine ecosystems in the Western Indian Ocean. *Coastal Management*, *39*(4), 351–357. https://doi.org/10.1080/08920753.2011.589207

Corbet, S., Larkin, C. J., Lucey, B. M., Meegan, A., & Yarovaya, L. (2020). *The Impact of Macroeconomic News on Bitcoin Returns* (SSRN Scholarly Paper 3550842). https://doi.org/10.2139/ssrn.3550842

Dorfleitner, G., Muck, F., & Scheckenbach, I. (2021). Blockchain applications for climate protection: A global empirical investigation. *Renewable and Sustainable Energy Reviews*, *149*, 111378. https://doi.org/10.1016/j.rser.2021.111378

El Aich, A., & Waterhouse, A. (1999). Small ruminants in environmental conservation. *Small Ruminant Research*, *34*(3), 271–287. https://doi.org/10.1016/S0921-4488(99)00079-6

ESCAP. (2022). *Frontier ICTs for sustainable development for digital leaders: Overview module*. ESCAP. https://www.unescap.org/kp/2022/frontier-icts-sustainable-development-digital-leaders-overview-module

Esmaeilian, B., Sarkis, J., Lewis, K., & Behdad, S. (2020). Blockchain for the future of sustainable supply chain management in Industry 4.0. *Resources, Conservation and Recycling*, *163*, 105064. https://doi.org/10.1016/j.resconrec.2020.105064

Feenstra, J. F., Smith, J. B., & Tol, R. S. J. (1998). *Handbook on Methods for Climate Change Impact Assessment and Adaptation Strategies*. https://wedocs.unep.org/xmlui/handle/20.500.11822/32746

Fraga-Lamas, P., & Fernández-Caramés, T. M. (2020). *Leveraging Blockchain for Sustainability and Open Innovation: A Cyber-Resilient Approach toward EU Green Deal and UN Sustainable Development Goals*. https://doi.org/10.5772/intechopen.92371

Gallersdörfer, U., Klaaßen, L., & Stoll, C. (2020). Energy consumption of cryptocurrencies beyond bitcoin. *Joule*, *4*(9), 1843–1846. https://doi.org/10.1016/j.joule.2020.07.013

Gannon, P., Seyoum-Edjigu, E., Cooper, H., Sandwith, T., Ferreira de Souza Dias, B., Palmer, C., Lang, B., Ervin, J., & Gidda, S. (2017). Status and prospects for achieving Aichi Biodiversity Target 11: Implications of national commitments and priority actions. *Parks*, *23*, 13–26. https://doi.org/10.2305/IUCN.CH.2017.PARKS-23-2PG.en

Gurtu, A., & Johny, J. (2019). Potential of blockchain technology in supply chain management: A literature review. *49*, 881–900. https://doi.org/10.1108/IJPDLM-11-2018-0371

Hall, J., Matos, S., & Silvestre, B. (2012). Understanding why firms should invest in sustainable supply chains: A complexity approach. *50*(5), 1332–1348.

Huseynov, F., & Mitchell, J. (2023). Blockchain for environmental peacebuilding: Application in water management. *Digital Policy, Regulation and Governance*, ahead-of-print (ahead-of-print). https://doi.org/10.1108/DPRG-06-2023-0080

Hutt, R. (2016, June 17). All you need to know about blockchain, explained simply. *World Economic Forum*. https://www.weforum.org/agenda/2016/06/blockchain-explained-simply/

IPCC. (2014). *AR5 Synthesis Report: Climate Change 2014*. https://www.ipcc.ch/report/ar5/syr/

Islam, M. M., & Shamsuddoha, M. (2018). Coastal and marine conservation strategy for Bangladesh in the context of achieving blue growth and sustainable development goals (SDGs). *Environmental Science & Policy*, *87*, 45–54. https://doi.org/10.1016/j.envsci.2018.05.014

IUCN, UNEP, & WWF. (1980). *World Conservation Strategy: Living Resource Conservation for Sustainable Development/: Prepared by the International Union for Conservation of Nature and Natural Resources (IUCN)*. IUCN. https://digitallibrary.un.org/record/91329

IUCN, UNEP, WWF (1991), *Caring for the Earth: A Strategy for Sustainable Living*, IUCN, Gland, Switzerland

Kaklauskas, A., & Gudauskas, R. (2016). Intelligent decision-support systems and the Internet of Things for the smart built environment. In F. Pacheco-Torgal, E. Rasmussen, C.-G. Granqvist, V. Ivanov, A. Kaklauskas, & S. Makonin (Eds.), *Start-Up Creation* (pp. 413–449). Woodhead Publishing. https://doi.org/10.1016/B978-0-08-100546-0.00017-0

Kimani, D., Adams, K., Attah-Boakye, R., Ullah, S., Frecknall-Hughes, J., & Kim, J. (2020). Blockchain, business and the fourth industrial revolution: Whence, whither, wherefore and how? *Technological Forecasting and Social Change*, *161*, 120254. https://doi.org/10.1016/j.techfore.2020.120254

Kohli, V., Chakravarty, S., Chamola, V., Sangwan, K. S., & Zeadally, S. (2023). An analysis of energy consumption and carbon footprints of cryptocurrencies and possible solutions. *Digital Communications and Networks*, *9*(1), 79–89. https://doi.org/10.1016/j.dcan.2022.06.017

Kouhizadeh, M., & Sarkis, J. (2018). Blockchain practices, potentials, and perspectives in greening supply chains. *Sustainability*, *10*(10), Article 10. https://doi.org/10.3390/su10103652

Li, J., Li, N., Peng, J., Cui, H., & Wu, Z. (2019). Energy consumption of cryptocurrency mining: A study of electricity consumption in mining cryptocurrencies. *Energy*, *168*, 160–168. https://doi.org/10.1016/j.energy.2018.11.046

Manullang, S. O., Prasetya, A. B., Kusumadewi, Y., & Zakaria, O. P. (2021). Understanding relevancy of technological trends in environmental conservation efforts and business sustainability in Indonesia: A systematic review study. *Annals of the Romanian Society for Cell Biology*, *25*(5), 2230–2240.

McCormick, J. (1986). The origins of the world conservation strategy. *Environmental Review: ER*, *10*(3), 177–187. https://doi.org/10.2307/3984544

Mendoza Tovar, E. (2023). *Blockchain for environmental sustainability* [Masaryk University, Faculty of Economics and Administration]. https://is.muni.cz/th/hyrel/?lang=en

Michener, W. K., Blood, E. R., Bildstein, K. L., Brinson, M. M., & Gardner, L. R. (1997). Climate change, Hurricanes and Tropical Storms, and Rising Sea level in Coastal Wetlands. *Ecological Applications*, *7*(3), 770–801. https://doi.org/10.2307/2269434

Moosavi, J., Naeni, L. M., Fathollahi-Fard, A. M., & Fiore, U. (2021). Blockchain in supply chain management: A review, bibliometric, and network analysis. *Environmental Science and Pollution Research International*. https://doi.org/10.1007/s11356-021-13094-3

Naqash, M. T., Syed, T. A., Alqahtani, S. S., Siddiqui, M. S., Alzahrani, A., & Nauman, M. (2023). A blockchain based framework for efficient water management and leakage detection in Urban Areas. *Urban Science*, *7*(4), Article 4. https://doi.org/10.3390/urbansci7040099

Nguyen, C. P., Schinckus, C., Su, T. D., & Chong, F. H. L. (2022). Determinants of stock market returns in emerging markets: The linkage between institutional quality and macro liquidity. *International Journal of Finance & Economics*, *27*(4), 4472–4486. https://doi.org/10.1002/ijfe.2382

Nguyen, G.-T., & Kim, K. (2018). A Survey about Consensus Algorithms Used in Blockchain. *Journal of Information Processing Systems*, *14*(1), 101–128.

Nummedal, D., Pilkey, O. H., & Howard, J. D. (1987). *Sea-Level Fluctuation and Coastal Evolution. SEPM Society for Sedimentary Geology*. https://doi.org/10.2110/pec.87.41

Pahonţu, B.-I., Arsene, D.-A., Predescu, A., Mocanu, M., & Gheorghiţă, A. (2023). Blockchain-based decision support system for water management. *Studies in Informatics and Control*, *32*(3), 131–140. https://doi.org/10.24846/v32i3y202312

Parag, Y., & Ainspan, M. (2019). Sustainable microgrids: Economic, environmental and social costs and benefits of microgrid deployment. *Energy for Sustainable Development*, *52*, 72–81. https://doi.org/10.1016/j.esd.2019.07.003

Parmentola, A., Petrillo, A., Tutore, I., & De Felice, F. (2022). Is blockchain able to enhance environmental sustainability? A systematic review and research agenda from the perspective of Sustainable Development Goals (SDGs). *Business Strategy and the Environment*, *31*(1), 194–217. https://doi.org/10.1002/bse.2882

Pham, L., Karim, S., Naeem, M. A., & Long, C. (2022). A tale of two tails among carbon prices, green and non-green crypto currencies. *International Review of Financial Analysis*, *82*, 102139.

Saraji, S., & Borowczak, M. (2021). *A Blockchain-based Carbon Credit Ecosystem* (arXiv:2107.00185). arXiv. https://doi.org/10.48550/arXiv.2107.00185

Shojaei, A., Wang, J., & Fenner, A. (2019). Exploring the feasibility of blockchain technology as an infrastructure for improving built asset sustainability. *Built Environment Project and Asset Management*, *10*(2), 184–199. https://doi.org/10.1108/BEPAM-11-2018-0142

Singh, M., & Kim, S. (2019). Chapter Four: Blockchain technology for decentralized autonomous organizations. In S. Kim, G. C. Deka, & P. Zhang (Eds.), *Advances in Computers* (Vol. 115, pp. 115–140). Elsevier. https://doi.org/10.1016/bs.adcom.2019.06.001

Social Alpha Foundation. (2022, January 19). *Blockchain for sustainable energy and climate in the Global South*. UNEP - UN Environment Programme. http://www.unep.org/resources/report/blockchain-sustainable-energy-and-climate-global-south

Stoll, C., Klaaßen, L., & Gallersdörfer, U. (2019). The carbon footprint of Bitcoin. *Joule*, *3*(7), 1647–1661. https://doi.org/10.1016/j.joule.2019.05.012

Sun, L., Hua, G., Teunter, R. H., Cheng, T. C. E., & Shen, Z. J. M. (2023). The effects of tokenization on ride-hailing blockchain platforms. *Production and Operations Management*. https://doi.org/10.1111/poms.14007

Thomason, J. (2021). Blockchain for growth: Applying DLTs to the UN sustainable development goals. In E. Kaili & D. Psarrakis (Eds.), *Disintermediation Economics: The Impact of Blockchain on Markets and Policies* (pp. 93–110). Springer International Publishing. https://doi.org/10.1007/978-3-030-65781-9_6

Tisdell, C. (2005). Economics of environmental conservation. In *Economics of Environmental Conservation*, Second Edition. https://doi.org/10.4337/9781845428266

Titus, J. G. (1990). Greenhouse effect, sea level rise and land use. *Land Use Policy*, *7*(2), 138–153. https://doi.org/10.1016/0264-8377(90)90005-J

Tsao, Y.-C., & Thanh, V.-V. (2021). Toward sustainable microgrids with blockchain technology-based peer-to-peer energy trading mechanism: A fuzzy meta-heuristic approach. *Renewable and Sustainable Energy Reviews*, *136*, 110452. https://doi.org/10.1016/j.rser.2020.110452

Uğurlu, E. (2019). Renewable energy strategies for sustainable development in the European Union. In D. Kurochkin, E. V. Shabliy, & E. Shittu (Eds.), *Renewable Energy: International Perspectives on Sustainability* (pp. 63–87). Springer International Publishing. https://doi.org/10.1007/978-3-030-14207-0_3

Uğurlu, E., & Muratoğlu, Y. (2019). Blockchain technology in solar energy [Chapter]. In *Architectures and Frameworks for Developing and Applying Blockchain Technology*, IGI Global. https://doi.org/10.4018/978-1-5225-9257-0.ch006

United Nations. (2015). *Transforming our world: The 2030 Agenda for sustainable development (A/RES/70/1)*. m https://sdgs.un.org/2030agenda

Wang, M., Wu, Y., Chen, B., & Evans, M. (2020). Blockchain and supply chain management: A new paradigm for supply chain integration and collaboration. *Operations and Supply Chain Management: An International Journal*, *14*(1), 111–122. https://doi.org/10.31387/oscm0440290

Wang, Y., Lucey, B. M., Vigne, S. A., & Yarovaya, L. (2022). The effects of central bank digital currencies news on financial markets. *Technological Forecasting and Social Change*, *180*, 121715. https://doi.org/10.1016/j.techfore.2022.121715

Yang, L., & Hamori, S. (2021). The role of the carbon market in relation to the cryptocurrency market: Only diversification or more? *International Review of Financial Analysis*, *77*, 101864. https://doi.org/10.1016/j.irfa.2021.101864

Zari, M. P. (2014). Ecosystem services analysis in response to biodiversity loss caused by the built environment. *S.A.P.I.EN.S. Surveys and Perspectives Integrating Environment and Society*, *7.1*, Article 7.1. https://journals.openedition.org/sapiens/1684

Zhang, J. (2023). Application of blockchain technology in water management and new energy economic development. *Water Supply*, *23*(10), 4227–4241. https://doi.org/10.2166/ws.2023.258

Part VI

Blockchain for Sustainable Practices in Irrigation and Crop Systems

10 Empowering SMEs with Blockchain

Advancing Sustainability in Water Systems

Gautam Bapat, Purvi Pujari, Anuj Kumar, and Saeid Eslamian

10.1 INTRODUCTION

Blockchain technology, initially designed as the underlying architecture for cryptocurrencies like Bitcoin, has evolved into a transformative force with wide-ranging applications. At its core, a blockchain is a decentralized and immutable ledger that records transactions across a network of computers (Aste et al., 2017). What sets it apart is its ability to maintain trust and transparency in a trustless environment, making it particularly relevant for small and medium-sized enterprises (SMEs) (Avevor, 2016).

For SMEs, the adoption of blockchain technology represents a significant opportunity to drive sustainable growth while addressing unique challenges. One of the primary advantages is *transparency* (Badhwar et al., 2023). SMEs often operate in highly competitive markets where trust is paramount. Blockchain's transparent and tamper-resistant ledger ensures that all transactions are verifiable and traceable. This not only builds trust with customers and partners but also minimizes the risk of fraud (Bilal and Al Mqbali, 2015) (Figure 10.1).

Another crucial feature of blockchain is *decentralization*. SMEs typically have limited resources, making them vulnerable to centralized control or disruptions. Blockchain operates on a decentralized network of computers, reducing dependence on a single entity and enhancing resilience. This ensures that critical data, such as supply chain information or environmental impact metrics, remains secure and accessible even in the face of unforeseen challenges.

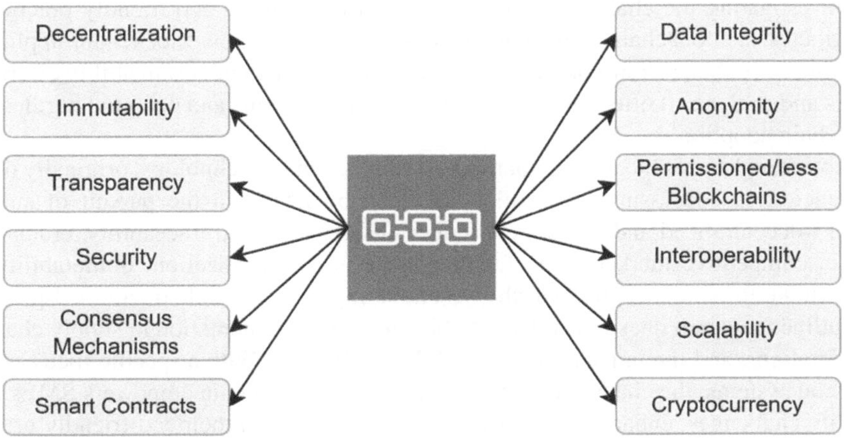

FIGURE 10.1 Features of blockchain (Authors' own).

DOI: 10.1201/9781003599593-16

Immutability is another key attribute. Once data is recorded on a blockchain, it cannot be altered or deleted. For SMEs focused on sustainability, this means that records related to resource consumption, emissions, or sustainable sourcing remain unaltered and trustworthy. This immutability is invaluable for proving compliance with sustainability standards and regulations.

Blockchain also promotes *efficiency*. In SMEs, streamlined processes are essential, and blockchain can eliminate intermediaries and automate complex tasks. This not only reduces costs but also accelerates processes, such as verifying the authenticity of eco-friendly products or tracking the provenance of sustainable materials (Bordel et al., 2019).

Blockchain technology offers SMEs a potent tool to navigate the sustainability landscape. Its features – transparency, decentralization, immutability, and efficiency – align perfectly with the needs and challenges faced by SMEs. By leveraging blockchain, SMEs can not only enhance their sustainability practices but also foster trust, reduce operational costs, and contribute to a more sustainable future (Cant, 2012). This chapter will answer the following research questions:

1. How can blockchain technology be effectively implemented by SMEs to enhance transparency and traceability in their supply chains, specifically in the context of sustainable sourcing and eco-friendly product verification?
2. What are the key challenges and regulatory considerations that SMEs may encounter when integrating blockchain technology to track and report their environmental impact and sustainability efforts, such as carbon emissions and resource consumption?
3. In what ways can blockchain-driven automation and decentralization contribute to the efficiency and effectiveness of sustainability initiatives within SMEs, particularly in circular economy models and renewable energy trading?

The growing concern for sustainability in business operations, especially within the orbit of water and environmental systems, is a prominent global trend. In recent years, there has been a profound shift in how companies perceive and address their environmental responsibilities. This transformation is propelled by escalating awareness of climate change, resource scarcity, and ecological degradation. Stakeholders, including consumers, investors, and governments, are increasingly demanding eco-conscious practices. Consequently, businesses are recognizing that sustainable operations not only mitigate environmental risks but also offer competitive advantages, cost savings, and enhanced reputations. This mounting concern underscores the urgency for innovative solutions, with blockchain technology emerging as a promising enabler of sustainability initiatives.

The purpose of this chapter is to explore the pivotal role of blockchain technology in aiding SMEs in their journey toward sustainability, with a specific focus on water and environmental systems. By investigating the challenges faced by SMEs in adopting eco-friendly practices and the unique attributes of blockchain, this chapter aims to showcase how blockchain applications can empower SMEs to enhance transparency, efficiency, and trust in their sustainability efforts. Through case studies and analysis, it offers insights into the tangible benefits and potential hurdles associated with blockchain adoption.

The chapter emphasizes the transformative role of blockchain technology, originally designed for cryptocurrencies, in addressing the challenges faced by SMEs in the pursuit of sustainability. Blockchain's decentralized and transparent ledger ensures trust and traceability, crucial for SMEs operating in competitive markets. Key features such as decentralization, immutability, and efficiency contribute to resilience, trustworthy sustainability records, and streamlined processes. The text also outlines research questions related to blockchain implementation in supply chains, regulatory considerations, and the impact on sustainability initiatives. With a specific focus on water and environmental systems, the chapter aims to showcase how blockchain empowers SMEs to navigate sustainability challenges, enhance transparency, and build trust in their eco-friendly practices. The urgency for innovative solutions, especially in the context of water and environmental concerns, underscores the growing importance of blockchain as a promising enabler of sustainability initiatives for businesses.

10.2 UNDERSTANDING THE SUSTAINABILITY CHALLENGES FOR SMEs

10.2.1 Challenges Faced by SMEs

Small and medium-sized enterprises (SMEs) face a distinct set of challenges when it comes to adopting sustainable practices (Centobelli et al., 2022). These challenges often stem from their size, limited resources, and the competitive business environment (Chapman, 2021). Here are some of the unique challenges (Figure 10.2):

- Resource Constraints: SMEs typically have limited financial, human, and technological resources compared to larger corporations. This can make it challenging to invest in sustainable technologies or implement comprehensive sustainability strategies (Daghighi and Shoushtari, 2023).
- Lack of Awareness: Many SMEs may not be fully aware of the environmental and social impacts of their operations or the potential benefits of sustainability. This lack of awareness can hinder the initiation of sustainable practices.
- Compliance Burdens: SMEs often struggle to keep up with evolving environmental regulations and standards. Compliance can be costly and complex, and noncompliance can lead to fines and reputational damage (Dasaklis et al., 2022).
- Supply Chain Challenges: SMEs often rely on complex supply chains that may span multiple countries. Ensuring sustainability throughout the supply chain can be difficult, especially when dealing with suppliers that are not yet committed to eco-friendly practices.
- Access to Capital: SMEs may face challenges in securing financing for sustainability initiatives. Traditional lenders may view such investments as risky, and SMEs may not have the collateral or track record to secure loans (Diegtiar et al., 2020).
- Competitive Pressures: In highly competitive markets, SMEs may perceive sustainability initiatives as a cost burden that could make them less competitive in terms of pricing. This can discourage investment in sustainability.
- Limited Expertise: SMEs may lack in-house expertise in sustainability practices and may not have the resources to hire dedicated sustainability professionals.
- Short-Term Focus: SMEs may be more focused on short-term survival and growth, which can lead to prioritization of immediate financial concerns over long-term sustainability goals (Dogo et al., 2019).
- Resistance to Change: Cultural resistance to change within SMEs can be a significant challenge. Employees and management may be hesitant to adopt new practices or technologies, even if they are environmentally friendly.

FIGURE 10.2 Challenges faced by SMEs (Authors' own).

Despite these challenges, SMEs also have certain advantages, such as agility and the ability to quickly implement changes when they commit to sustainability. Overcoming these obstacles often involves creative solutions, partnerships with larger organizations, and leveraging technology, like blockchain, to streamline and enhance sustainability efforts.

10.2.2 Significance of Environmental Issues to SMEs

The significance of water and environmental issues is paramount for both global sustainability and business operations, shaping the landscape of responsible and forward-thinking enterprises (Drăgulinescu et al., 2021).

Healthy environments are essential for sustaining life on Earth. Biodiversity, clean air, and water are foundational to ecosystem health (Vernekar AG., 2020). Environmental issues, including habitat destruction, pollution, and climate change, directly impact these elements. Additionally, environmental issues, particularly those related to carbon emissions, contribute significantly to climate change (Efanov D and Roschin P., 2018). Mitigating these issues is critical for limiting global temperature rise and avoiding catastrophic climate impacts.

Sustainable use of resources, including water, forests, and minerals, is vital for long-term global sustainability. Environmental issues like deforestation, overfishing, and resource depletion threaten the availability of these critical resources (Erol et al., 2022). Furthermore, environmental issues can exacerbate the impact of natural disasters, making communities and economies more vulnerable (Friedman and Ormiston, 2022). Sustainable practices can enhance resilience and reduce the harm caused by events like hurricanes, floods, and droughts.

Environmental issues also have a direct impact on business operations. Companies must navigate these challenges to ensure both their sustainability and profitability. Environmental regulations and standards are increasingly stringent (Frikha et al., 2023). Noncompliance can result in fines, legal liabilities, and damage to a company's reputation. Businesses rely on various natural resources, such as water and energy, for their operations. Managing these resources efficiently can reduce costs and enhance competitiveness (Furze et al., 2022).

Consumer expectations are shifting toward eco-conscious products and services. Companies that prioritize environmental issues can build a positive brand image and attract environmentally conscious customers (Garcia-Torres et al., 2022). Environmental issues, like extreme weather events or resource shortages, can disrupt supply chains. Businesses that integrate sustainability practices are often better prepared to navigate such disruptions (Harper and Snowden, 2017). Some markets, especially in the developed world, require adherence to sustainability standards. Access to these markets can be restricted for companies that do not meet environmental criteria.

Water and environmental issues are not only critical for the long-term health of our planet but also integral to the success and sustainability of businesses. Addressing these issues is not just an ethical imperative but a strategic necessity for businesses looking to thrive in an increasingly eco-conscious and regulated global marketplace (Jena and Dash, 2021).

10.3 OVERVIEW OF BLOCKCHAIN TECHNOLOGY

10.3.1 Role of Blockchain in Addressing Sustainability Challenges

Blockchain's principles of decentralization, immutability, and transparency make it particularly well-suited for addressing sustainability challenges in various ways (Khan FA et al., 2020).

Decentralization: Decentralization eliminates the need for a single central authority, such as a government or a corporation, to oversee and control transactions or data. In the context of sustainability, this reduces dependence on centralized bodies to monitor and enforce eco-friendly practices (Khan M et al., 2022). Decentralization enhances trust, as no single entity has disproportionate control or influence over sustainability data, making it harder for data manipulation or greenwashing. Furthermore, the distributed networks of nodes in blockchain make it resistant to single points of failure, ensuring that sustainability data remains accessible even in the face of disruptions, such as natural disasters or cyberattacks, contributing to business continuity in eco-conscious practices (Laurence T, 2011).

Immutability: Immutability means that once data is recorded on a blockchain, it cannot be altered or deleted without consensus from the network. In sustainability, this feature ensures the integrity of records related to resource consumption, emissions, and eco-friendly certifications (Lee K, Neff J. 2011). Stakeholders can trust that the sustainability data on a blockchain has not been tampered with, enhancing the credibility of eco-friendly claims. This is crucial for businesses seeking to prove their commitment to sustainability. Additionally, immutability allows for easy auditing of sustainability practices. Regulators, investors, and consumers can review historical data on the blockchain to verify adherence to environmental standards.

Transparency: Transactions and data on a blockchain are transparent and accessible to authorized participants. In sustainability efforts, this transparency enables stakeholders, including consumers and regulatory bodies, to monitor and verify environmental impact data. It provides visibility into the practices and actions of organizations, promoting greater accountability (Leng J et al., 2020). The transparent nature of blockchain holds organizations accountable for their sustainability claims, as false or misleading claims can be quickly identified and challenged. Furthermore, in supply chains, transparency is crucial for sustainability. Blockchain enables end-to-end traceability, allowing consumers to verify the origins and eco-friendly attributes of products, promoting responsible consumption (Maulekhi A et al., 2020).

Decentralization, immutability, and transparency are fundamental principles of blockchain that align closely with the objectives of sustainability initiatives. They foster trust, accountability, and data reliability, making blockchain a powerful tool for promoting and verifying sustainable practices, whether in resource management, supply chains, or carbon footprint tracking. As businesses and organizations increasingly prioritize sustainability, blockchain's role in ensuring the authenticity and effectiveness of these efforts is set to become even more significant (Mishra and Kaushik, 2023).

10.4 BLOCKCHAIN, SMES IN WATER AND ENVIRONMENTAL SUSTAINABILITY

10.4.1 Supply Chain Transparency and Traceability

Blockchain technology can significantly enhance transparency and traceability in supply chains, providing SMEs with the tools to verify the origins and environmental impact of their products (Mora et al., 2021).

Blockchain's immutability ensures that once data is recorded, it cannot be altered or deleted without consensus from the network. This feature makes supply chain records tamper-proof, guaranteeing the accuracy and integrity of information related to the production and distribution of products (Namasudra S, Akkaya K, 2023).

Every transaction or event in a supply chain can be recorded as a block on the blockchain. These blocks are linked chronologically to form a complete and transparent chain of events, from raw material sourcing to final product delivery (Naqash MT et al., 2023). SMEs can trace the journey of their products backward or forward in the supply chain, verifying where materials originated and how they were processed. This helps in identifying potential environmental concerns, such as unsustainable sourcing practices or excessive carbon emissions during transportation (Ogbokor and Ngeendepi, 2012).

Blockchain allows SMEs to securely verify the authenticity and sustainability credentials of their suppliers (Owen and Pickering., 2006). Smart contracts can automate checks against predefined criteria, ensuring that suppliers meet the required environmental standards before transactions are approved.

Participants in the supply chain can update the blockchain in real time, allowing for immediate visibility of any changes or delays. This transparency is invaluable for SMEs seeking to manage their supply chains efficiently and reduce waste (Paiva Sobrinho R et al., 2022).

Blockchain enables SMEs to share supply chain data directly with consumers. QR codes or product labels can be scanned to access information about the product's journey, including details about its environmental impact. This transparency fosters consumer trust and empowers eco-conscious buying decisions (Pant et al., 2015).

Counterfeit products can have detrimental environmental impacts due to their often-substandard quality and unethical production methods. Blockchain's transparency and traceability make it difficult for counterfeit products to enter the supply chain, protecting both consumers and the environment.

SMEs can more easily comply with sustainability regulations and reporting requirements by maintaining transparent and auditable records on the blockchain. This reduces the risk of noncompliance penalties and improves the company's overall sustainability reputation (Pimentel et al., 2004).

By streamlining supply chain operations and reducing the risk of errors or fraud, blockchain can result in cost savings for SMEs. These savings can be reinvested in sustainability initiatives or used to improve product quality.

Blockchain technology revolutionizes supply chain management for SMEs by enhancing transparency and traceability. It empowers businesses to verify the origins and environmental impact of their products, thereby promoting responsible sourcing, reducing waste, and meeting the growing demand for eco-conscious consumer choices (Qin et al., 2007).

10.4.2 Water Management and Conservation

Blockchain technology is a powerful tool for monitoring and managing water usage efficiently, minimizing waste, and promoting responsible consumption in various sectors, including industrial processes, agriculture, and urban water management (Saleh et al., 2008) (Figure 10.3).

Real-Time Monitoring: Integration of IoT devices like smart meters with blockchain networks enables continuous monitoring of water flow, quality, and consumption. This real-time data offers valuable insights into usage patterns (Saleh et al., 2008).

Data Transparency: All information from IoT devices is securely recorded on the blockchain, ensuring transparency and data integrity. Stakeholders, including water utilities and

FIGURE 10.3 Role of blockchain in water management and conservation (Authors' own).

consumers, can access this data, reducing the risk of tampering or fraudulent reporting (Shaikh et al., 2021).

Smart Contracts: Smart contracts automate water management processes, triggering alerts or actions when water usage exceeds limits or quality parameters are not met. This automation ensures swift responses to irregularities (Sunny et al., 2020).

Billing and Payment: Blockchain facilitates transparent and automated billing systems. Consumers can view real-time usage data and receive accurate bills, raising awareness of consumption patterns and costs, thereby encouraging responsible water usage (Swan and De Filippi, 2017).

Supply Chain Management: In agriculture, blockchain tracks the water supply chain from source to distribution to irrigation. Farmers can optimize water usage, ensuring efficient resource allocation and minimizing wastage (Syme et al., 1999).

Water Trading: Blockchain enables secure peer-to-peer water trading platforms. Excess water resources can be transparently traded, incentivizing responsible usage and reducing waste (Tayal et al., 2021).

Verification of Sustainable Practices: Businesses can use blockchain to verify eco-friendly water management practices, enhancing their environmental credentials and encouraging responsible water usage (Thakur et al., 2021).

Regulatory Compliance: Blockchain helps ensure compliance with water regulations. Records of water quality tests, emissions data, and consumption reports are easily accessible for audits and reporting (Tian, 2017).

Community Engagement: Decentralized platforms allow residents to engage in discussions about water usage and conservation measures, fostering community involvement in water management decisions (Venkatesh et al., 2020).

Environmental Impact Assessment: Blockchain provides a comprehensive record of water usage and its environmental impact, aiding in the assessment of a region's sustainability efforts and identifying areas for improvement (Saleh et al., 2008).

By leveraging blockchain technology in water management, organizations and communities can enhance transparency, reduce waste, optimize resource allocation, and promote responsible water consumption. These practices benefit the environment and contribute to more sustainable and efficient water management (Swan and De Filippi, 2017).

10.4.3 Carbon Footprint Tracking

Blockchain technology can be a valuable tool for SMEs to accurately measure and track their carbon emissions, enabling them to make informed decisions for reducing their environmental impact.

Data Accuracy and Transparency: Blockchain's immutable ledger ensures that once carbon emissions data is recorded, it cannot be altered or tampered with. This guarantees data accuracy and reliability, a critical factor when calculating emissions.

Supply Chain Tracking: Blockchain can be used to trace emissions throughout the supply chain. SMEs can record and verify emissions data from their suppliers, transportation partners, and other stakeholders, providing a comprehensive view of their carbon footprint (Figure 10.4).

IoT Integration: IoT devices can collect real-time data on energy consumption, emissions from machinery, and other relevant factors. This data can be securely integrated with the blockchain, providing continuous monitoring and precise emissions calculations.

Smart Contracts: Smart contracts can automate emissions reporting and verification processes. These contracts can trigger emissions calculations based on data inputs and initiate notifications or actions when emissions thresholds are exceeded.

FIGURE 10.4 Role of blockchain in carbon footprint tracking (Authors' own).

Certification and Verification: Blockchain can store certifications related to carbon reduction efforts, such as carbon credits or offsets. These certifications can be easily accessed and verified, providing third-party validation of sustainability claims.

Data Sharing and Collaboration: SMEs can securely share emissions data with partners, regulators, and other stakeholders using blockchain. This facilitates collaboration on sustainability initiatives and ensures transparency in reporting.

Compliance with Regulations: Blockchain can help SMEs ensure compliance with carbon emissions regulations. By maintaining a transparent and auditable record of emissions data, SMEs can easily demonstrate adherence to environmental standards.

Real-Time Reporting: Blockchain allows for real-time emissions reporting and monitoring. SMEs can assess their emissions performance continuously, rather than relying on periodic assessments, enabling timely adjustments to reduce environmental impact.

Data-Driven Decision-Making: Accurate and up-to-date emissions data empowers SMEs to make data-driven decisions about sustainability strategies. They can identify areas of high emissions, set reduction targets, and implement efficient mitigation measures.

Green Marketing and Reputation Building: SMEs can use blockchain-verified emissions data to promote their environmentally friendly practices and build a green brand image. Transparency in emissions reporting can enhance their reputation among eco-conscious consumers.

By leveraging blockchain technology for carbon emissions measurement and tracking, SMEs can not only fulfill their environmental responsibilities but also identify cost-saving opportunities, improve operational efficiency, and enhance their competitiveness in a market increasingly focused on sustainability. This technology enables SMEs to be proactive in reducing their carbon footprint and contributing to a more sustainable future.

10.4.4 Renewable Energy Trading

Blockchain technology holds significant potential for facilitating peer-to-peer (P2P) renewable energy trading among SMEs, promoting the use of sustainable energy sources. Here's how blockchain can empower this transformation:

Decentralized Energy Exchange: Blockchain creates decentralized, trustless networks where SMEs can directly buy and sell excess renewable energy among themselves. This removes the need for intermediaries, such as traditional energy providers, reducing costs and promoting energy independence.

Transparency and Trust: Blockchain's transparency ensures that all energy transactions are recorded on an immutable ledger. SMEs can verify the source of the renewable energy and be confident that the energy they purchase is genuinely green. This transparency fosters trust among participants.

Smart Contracts: Smart contracts can automate energy trading agreements, executing transactions automatically when predefined conditions are met. For instance, when a surplus of renewable energy is generated, it can be sold to other SMEs in the network without manual intervention.

Grid Optimization: Blockchain-enabled P2P energy trading allows for efficient distribution of energy within local grids. Excess energy from one SME can be utilized by others in the same vicinity, reducing the strain on the central grid and minimizing transmission losses.

Fair and Transparent Pricing: Blockchain can facilitate dynamic pricing models where energy prices fluctuate based on supply and demand. This can result in fairer pricing for consumers and incentivize energy conservation during peak demand times.

Energy Traceability: SMEs can trace the source and production methods of the renewable energy they purchase. This information can be a valuable marketing tool for businesses committed to sustainability, demonstrating their use of green energy to customers and investors.

Incentivizing Renewable Energy Production: By providing SMEs with a platform to sell excess renewable energy, blockchain incentivizes the production of clean energy. This can encourage more SMEs to invest in renewable energy sources like solar panels or wind turbines.

Compliance and Reporting: Blockchain simplifies the process of verifying and reporting renewable energy generation and consumption for regulatory compliance, supporting SMEs in meeting environmental standards.

Resilience and Security: Blockchain networks are inherently secure and resilient, making them less susceptible to cyberattacks or disruptions. This ensures the stability of P2P energy trading platforms.

Community Engagement: P2P energy trading fosters a sense of community and cooperation among SMEs. Businesses can work together to ensure a stable and sustainable local energy supply.

Blockchain-driven P2P renewable energy trading holds immense promise for SMEs in promoting the adoption of sustainable energy sources. It not only offers economic benefits through cost savings but also aligns with environmental goals, enhances energy security, and contributes to a more resilient and sustainable energy infrastructure. This technology has the potential to revolutionize the way SMEs access and utilize renewable energy, driving positive environmental and economic outcomes.

10.4.5 Circular Economy Initiatives

Blockchain technology plays a pivotal role in supporting SMEs as they implement circular economy models by enabling secure sharing and tracking of resources. Here's how blockchain accomplishes this:

Resource Traceability: Blockchain records the entire lifecycle of resources, from extraction or production to consumption and recycling. This traceability ensures transparency and accountability in resource management, crucial for a circular economy.

Immutable Records: Once data about resources is recorded on the blockchain, it becomes tamper proof. SMEs can rely on the integrity of records, which is essential when sharing and tracking resources within a circular system.

Ownership and Provenance: Blockchain establishes ownership and provenance of resources. SMEs can verify the origin and history of resources, ensuring they meet sustainability criteria and ethical sourcing standards.

Smart Contracts: Smart contracts can automate resource-sharing agreements. For instance, when a company is finished with a resource, a smart contract can trigger its transfer to another SME within the circular network, streamlining resource allocation.

Inventory Management: Blockchain can optimize inventory management by tracking the availability of shared resources in real time. SMEs can reduce over-purchasing and under-utilization, resulting in cost savings and waste reduction.

Collaborative Ecosystems: Blockchain facilitates collaboration among SMEs within a circular ecosystem. These businesses can securely share data about available resources, demand, and utilization patterns, leading to more efficient resource allocation.

Waste Reduction: Circular economy models aim to minimize waste by reusing and recycling resources (Furze et al., 2022). Blockchain helps SMEs identify opportunities for reusing resources within their network and ensures that recyclable materials are efficiently processed and returned to the supply chain.

Data-Driven Decision-Making: Blockchain's data transparency enables SMEs to make informed decisions about resource allocation, usage, and recycling. They can identify trends and opportunities for further circularization of their operations.

Regulatory Compliance: By providing a transparent record of resource usage and disposal, blockchain helps SMEs demonstrate compliance with environmental regulations and sustainability standards.

Marketplace for Circular Resources: Blockchain can support the creation of digital marketplaces where SMEs can trade or exchange resources within the circular economy. This can stimulate economic growth and encourage sustainable practices.

Blockchain technology empowers SMEs to transition to circular economy models by ensuring transparency, traceability, and efficient resource sharing within collaborative ecosystems. By securely sharing and tracking resources, SMEs can reduce waste, minimize environmental impact, and unlock economic opportunities in a sustainable and environmentally responsible manner. This technology is a cornerstone in the journey toward a more circular and sustainable future for SMEs.

10.5 BENEFITS AND ADVANTAGES

Blockchain technology offers numerous advantages to SMEs striving to achieve sustainability goals. Firstly, its inherent transparency ensures that all transactions and data related to sustainability practices are visible and tamper proof, building trust with stakeholders and allowing for credible reporting. Secondly, blockchain's decentralized nature reduces the need for intermediaries, leading to cost savings and greater operational efficiency, which can be redirected toward eco-friendly initiatives. Furthermore, the immutability of blockchain records ensures the integrity of environmental data, vital for regulatory compliance and demonstrating adherence to sustainability standards. Through automation via smart contracts, SMEs can streamline sustainability processes, such as supply chain traceability or carbon emissions tracking, improving accuracy and reducing human errors. Finally, blockchain can foster collaboration within sustainability-focused ecosystems, enabling SMEs to engage in circular economy practices and promote responsible consumption. In summary, blockchain empowers SMEs to navigate the sustainability landscape effectively, offering transparency, efficiency, cost savings, and a competitive edge in the journey toward a more sustainable future.

10.5.1 Economic, Social, and Environmental Benefits of Integrating Blockchain

Integrating blockchain technology into their operations can bring a range of economic, social, and environmental benefits to SMEs:

Economic Benefits:

1. Cost Reduction: Blockchain streamlines processes by eliminating intermediaries, reducing transaction costs, and automating tasks. This leads to significant cost savings, enabling SMEs to allocate resources more efficiently.
2. Improved Efficiency: Automated processes and real-time data access provided by blockchain enhance operational efficiency. SMEs can optimize their supply chains, reduce waiting times, and make better-informed decisions, ultimately increasing productivity.
3. Access to New Markets: Blockchain can open doors to markets that prioritize sustainable and transparent practices. SMEs with blockchain-backed sustainability credentials may find it easier to access these markets, expanding their customer base and revenue streams.
4. Enhanced Financial Access: Blockchain can improve access to financing for SMEs through mechanisms like tokenization and peer-to-peer lending, enabling them to invest in sustainability initiatives and business growth.

Social Benefits:

1. Transparency and Trust: Blockchain fosters trust by providing transparent and immutable records. Consumers, partners, and stakeholders can trust the accuracy of data related to sustainability practices, fostering stronger relationships.
2. Consumer Empowerment: Through blockchain, consumers gain access to detailed information about products' origins and environmental impact. This empowers them to make informed, eco-conscious choices, encouraging sustainable consumption.
3. Community Engagement: SMEs can engage with local communities, demonstrating their commitment to sustainability through transparent and collaborative blockchain networks. This can foster goodwill and positive relationships.
4. Fair Trade and Ethical Practices: Blockchain can ensure that ethical and fair-trade practices are upheld throughout the supply chain, benefiting workers and communities involved in production.

Environmental Benefits:

1. Resource Efficiency: Blockchain supports efficient resource allocation and waste reduction. By optimizing processes and sharing resources, SMEs can minimize environmental impact, conserve resources, and reduce waste generation.
2. Emissions Reduction: Blockchain-enabled supply chain transparency helps identify opportunities to reduce carbon emissions and implement sustainable transportation practices, contributing to climate change mitigation.
3. Promotion of Sustainable Practices: Through transparent data recording, blockchain encourages businesses to adopt sustainable practices, such as responsible sourcing, circular economy models, and renewable energy adoption.
4. Ecosystem Preservation: Sustainable resource management and reduced environmental impact contribute to ecosystem preservation, safeguarding biodiversity and the long-term health of ecosystems.

Incorporating blockchain technology into their operations not only improves economic efficiency but also enhances social trust and responsibility, ultimately contributing to environmental sustainability. SMEs can position themselves as responsible corporate citizens, strengthen their market presence, and make significant strides toward achieving their sustainability objectives.

10.6 CASE STUDIES

There are several real-world examples of SMEs successfully incorporating blockchain technology to improve their sustainability practices in water and environmental systems. Please note that the landscape may have evolved since then, and it's advisable to check for more recent developments. Here are a few noteworthy examples:

SweetSense: SweetSense, an IoT technology provider, partnered with IBM to develop a blockchain-based system that monitors and manages groundwater usage in rural Kenya. This system helps SMEs and communities ensure the responsible use of water resources, particularly in areas where water scarcity is a significant concern.

WaterChain: WaterChain is a blockchain initiative that focuses on water resource management. It seeks to address water scarcity issues by incentivizing the reduction of water wastage through a blockchain-based platform. While it's still in its early stages, it has the potential to benefit SMEs involved in water-intensive industries.

These examples showcase how blockchain technology is being adopted across different sectors and regions to enhance sustainability practices related to water, energy, and environmental systems. As blockchain continues to evolve, it is likely that more SMEs will find innovative ways to incorporate this technology into their operations to promote responsible resource management and environmental stewardship.

10.7 CHALLENGES AND CONSIDERATIONS

Implementing blockchain solutions, while promising for SMEs in the pursuit of sustainability, can indeed present several challenges:

Technological Barriers: SMEs may face hurdles in terms of technological readiness. Blockchain implementation often requires advanced IT infrastructure, including high-speed internet, compatible hardware, and skilled personnel. Overcoming these barriers may require upfront investments and training, which some SMEs might find financially daunting.

Integration Issues: Integrating blockchain into existing systems and processes can be complex. Compatibility issues, data migration challenges, and the need for collaboration among multiple stakeholders can pose integration difficulties. SMEs might need to carefully plan and manage the integration process to minimize disruptions to their operations.

Regulatory Concerns: Depending on the region and industry, there might be regulatory uncertainties or barriers related to blockchain technology. Compliance with data protection laws, financial regulations, and environmental standards can be complex when blockchain is involved. SMEs need to navigate this regulatory landscape carefully and possibly engage legal expertise to ensure they remain compliant.

Costs: While blockchain can lead to long-term cost savings, the initial investment can be significant. Implementing blockchain solutions may require funding for software development, infrastructure upgrades, and ongoing maintenance. SMEs with limited budgets may need to carefully assess the return on investment.

Scalability: Blockchain networks can face scalability challenges, especially with growing data volumes. As SMEs expand their operations, they must consider whether their chosen blockchain solution can scale to accommodate increased transactions and data without sacrificing performance or efficiency.

Interoperability: Ensuring that different blockchain systems can communicate and work together seamlessly is a concern, particularly when dealing with partners or clients who use different blockchain platforms. Interoperability standards are still evolving, and SMEs must consider how to navigate this issue.

Security and Privacy: While blockchain is known for its security, it's not immune to vulnerabilities. SMEs need to implement robust cybersecurity measures to protect their blockchain systems from attacks and breaches. Additionally, privacy concerns may arise if sensitive data is recorded on a public blockchain, necessitating careful consideration of private or hybrid blockchain solutions.

While blockchain offers considerable advantages to SMEs in enhancing sustainability practices, these potential challenges should not be underestimated. A strategic and well-planned approach, combined with a thorough assessment of costs, benefits, and regulatory implications, is essential to the successful implementation of blockchain solutions in SME operations.

10.8 PROSPECTS AND TRENDS

Several emerging trends in blockchain technology hold the potential to further enhance its utility for SMEs in their sustainability efforts:

1. *Decentralized Finance (DeFi) and Tokenization*: DeFi platforms are leveraging blockchain to create decentralized financial ecosystems. SMEs can use DeFi for access to sustainable financing, making it easier to fund green initiatives. Tokenization of assets, including carbon credits, renewable energy certificates, and even company shares, offers new avenues for SMEs to raise capital for sustainability projects while ensuring transparency and liquidity.
2. *NFTs for Sustainability*: Non-fungible tokens (NFTs) have gained popularity in the art and entertainment world, but their application in sustainability is emerging. SMEs can create NFTs to represent eco-friendly products, carbon offsets, or contributions to environmental causes, allowing consumers to trade or display their support for sustainable practices.
3. *Supply Chain Oracles*: Oracles, which provide real-world data to blockchain smart contracts, are evolving to enhance supply chain transparency. SMEs can integrate supply chain oracles to automatically verify the authenticity and sustainability of products and materials, thereby ensuring that their supply chains adhere to eco-friendly practices.
4. *Cross-Chain Integration*: As blockchain networks proliferate, cross-chain integration solutions are emerging. These technologies facilitate interoperability between different blockchains, making it easier for SMEs to collaborate with partners on various blockchain platforms while maintaining data integrity. This fosters broader adoption of blockchain in sustainability initiatives.
5. *Energy-Efficient Blockchains*: Recognizing the energy consumption concerns associated with blockchain, emerging solutions focus on energy efficiency. Some blockchains are transitioning to consensus mechanisms that require significantly less energy, making them more sustainable and appealing for SMEs committed to reducing their carbon footprint.
6. *Sustainability-Focused Consortia*: Consortia and industry alliances are forming to promote sustainability-focused blockchain standards and best practices. SMEs can benefit from these initiatives by accessing shared resources, knowledge, and networks to advance their sustainability goals more efficiently.

7. *Carbon Accounting Blockchains*: New blockchain platforms are designed specifically for carbon accounting and offset trading. SMEs can use these platforms to accurately measure, report, and trade carbon credits, simplifying their participation in carbon markets and emissions reduction efforts.
8. *Decentralized Autonomous Organizations* (DAOs): DAOs are blockchain-based organizations governed by smart contracts and community members. SMEs can use DAOs to create decentralized decision-making structures for sustainability initiatives, enabling stakeholders to collectively steer the company's green efforts.

These emerging trends in blockchain technology are expanding the possibilities for SMEs to enhance their sustainability practices. By staying informed about these developments and exploring how they can be tailored to their specific sustainability goals, SMEs can take advantage of the latest blockchain innovations to make meaningful environmental and social impacts.

10.9 CONCLUSION

In the face of pressing environmental challenges, small and medium-sized enterprises (SMEs) are increasingly turning to blockchain technology as a powerful ally in their quest for sustainability. Blockchain's impact on SMEs spans economic, social, and environmental dimensions. Economically, it reduces costs, enhances efficiency, and opens new avenues for sustainable financing. Socially, it fosters transparency and trust, empowers consumers to make eco-conscious choices, and strengthens community engagement in sustainability initiatives. Environmentally, it enables responsible resource management, carbon emissions tracking, and the promotion of circular economy practices.

The potential of blockchain in SMEs' sustainability efforts extends beyond the present into an era of continued evolution. Emerging trends like DeFi, tokenization, NFTs for sustainability, and interoperability are paving the way for more accessible financing, innovative sustainability models, and seamless collaboration between SMEs. As blockchain technology becomes more energy efficient and adaptable to various industries, its role in addressing environmental challenges such as carbon neutrality, sustainable supply chains, renewable energy integration, and conservation is poised to expand.

The collaborative nature of blockchain aligns with the global imperative to tackle environmental crises collectively. SMEs, governments, organizations, and individuals can collaborate across borders, leveraging blockchain's secure and transparent infrastructure to create comprehensive solutions for shared planetary issues. Blockchain's role in enhancing environmental governance and promoting sustainable decision-making is becoming increasingly evident.

The integration of blockchain technology into SME operations is not merely a technological advancement but a transformative force for sustainability. It enables SMEs to align their economic pursuits with social and environmental responsibilities, fostering a harmonious relationship between business success and planetary well-being. As blockchain's potential unfolds, SMEs are not only contributing to their own sustainable growth but also becoming vital players in the broader mission to address the urgent environmental challenges that affect us all. The journey toward a more sustainable future, driven by blockchain technology and the vision of responsible SMEs, is a testament to human ingenuity and our capacity to adapt, innovate, and collaborate for the betterment of our planet.

REFERENCES

Aste T, Tasca P, Di Matteo T. (2017) Blockchain technologies: The foreseeable impact on society and industry. *Computer (Long Beach Calif)*;50(9):18–28.

Avevor E. (2016) *Challenges Faced by SMEs When Accessing Fund from Financial Institutions in Ghana.; Business Economics and Tourism*, University of Applied Sciences, 63.

Badhwar A, Islam S, Tan CSL. (2023) Exploring the potential of blockchain technology within the fashion and textile supply chain with a focus on traceability, transparency, and product authenticity: A systematic review. *Front Blockchain* 2023;6:1044723.

Bilal ZO, Al Mqbali NS. (2015) Challenges and constrains faced by small and medium enterprises (SMEs) in Al Batinah governorate of Oman. *World J Entrep Manag Sustain Dev*;11(2):120–130.

Bordel B, Martin D, Alcarria R, Robles T. (2019) A blockchain-based water control system for the automatic management of irrigation communities. In: *2019 IEEE International Conference on Consumer Electronics (ICCE)*. IEEE. pp. 1–2.

Cant M. (2012) Challenges faced by SMEs in South Africa: Are marketing skills needed? *Int Bus Econ Res J*;11(10):1107–1116.

Centobelli P, Cerchione R, Del Vecchio P, Oropallo E, Secundo G. (2022) Blockchain technology for bridging trust, traceability and transparency in circular supply chain. *Inf Manag*;59(7):103508.

Chapman DV. (2021) *Water Quality Assessments: A Guide to the Use of Biota, Sediments and Water in Environmental Monitoring*. CRC Press

Daghighi A, Shoushtari F. (2023) Toward sustainability of supply chain by applying blockchain technology. *Int J Ind Eng Oper Res.*;5(2):60–72.

Dasaklis TK, Voutsinas TG, Tsoulfas GT, Casino F. (2022) A systematic literature review of blockchain-enabled supply chain traceability implementations. *Sustainability*;14(4):2439.

Diegtiar OA, Hornyk VH, Kravchenko SO, Valentyna V, Shtal TV. (2020) Improving public water resources policy in Ukraine: Municipal and environmental issues. *J Environ Manag Tour.*;11(3):672–678.

Dogo EM, Salami AF, Nwulu NI, Aigbavboa CO. (2019) Blockchain and internet of things-based technologies for intelligent water management system. *Artif Intell IoT*;129–150.

Drăgulinescu A-M, Constantin F, Orza O, Bosoc S, Streche R, Negoita A, et al. (2021) Smart watering system security technologies using Blockchain. In *13th International Conference on Electronics, Computers and Artificial Intelligence (ECAI)*. IEEE; 2021. pp. 1–4.

Vernekar AG. (2020) Blockchain based water management system. *Int Res J Eng Technol.* 7:7505–7507.

Efanov D, Roschin P. (2018) The all-pervasiveness of the blockchain technology. *Procedia Comput Sci.*;123:116–121.

Erol I, Neuhofer IO, Dogru T, Oztel A, Searcy C, Yorulmaz AC. (2022) Improving sustainability in the tourism industry through blockchain technology: Challenges and opportunities. *Tour Manag.*;93:104628.

Friedman N, Ormiston J. (2022) Blockchain as a sustainability-oriented innovation?: Opportunities for and resistance to Blockchain technology as a driver of sustainability in global food supply chains. *Technol Forecast Soc Change.*;175:121403.

Frikha T, Ktari J, Ben AN, Chaabane F, Hamdi M, Denguir F, et al. (2023) Low power blockchain in Industry 4.0 case study: Water management in Tunisia. *J Signal Process Syst*;96:257–271.–15.

Furze JN, Eslamian S, Raafat S, Swing K. (2022) *Mathematical Advances Towards Earth Systems Protection and Sustainability* Vol. 1, Springer, 337

Garcia-Torres S, Rey-Garcia M, Sáenz J, Seuring S. (2022) Traceability and transparency for sustainable fashion-apparel supply chains. *J Fash Mark Manag An Int J.* 26(2):344–364.

Harper C, Snowden M. (2017) *Environment and Society: Human Perspectives on Environmental Issues*. Routledge.

Jena AK, Dash SP. (2021) Blockchain technology: Introduction, applications, challenges. In Sandeep Kumar Panda, Ajay Kumar Jena, Santosh Kumar Swain, Suresh Chandra Satapathy (eds.) *Blockchain Technology: Applications and Challenges*. Springer. pp. 1–11.

Khan FA, Asif M, Ahmad A, Alharbi M, Aljuaid H. (2020) Blockchain technology, improvement suggestions, security challenges on smart grid and its application in healthcare for sustainable development. *Sustain Cities Soc.*;55:102018.

Khan M, Parvaiz GS, Dedahanov AT, Abdurazzakov OS, Rakhmonov DA. (2022) The impact of technologies of traceability and transparency in supply chains. *Sustainability.*;14(24):16336.

Laurence T (2011) *Introduction to Blockchain Technology*. Van Haren; 2019.

Lee K, Neff J. (2011) *Produced Water: Environmental Risks and Advances in Mitigation Technologies*. Springer.

Leng J, Ruan G, Jiang P, Xu K, Liu Q, Zhou X, et al. (2020) Blockchain-empowered sustainable manufacturing and product lifecycle management in industry 4.0: A survey. *Renew Sustain Energy Rev.*;132:110112.

Maulekhi A, Patani KA, Sangani V, Pawar M V. (2020) The blockchain-based smart water harvesting system. In *2020 International Conference on Smart Electronics and Communication (ICOSEC)*. IEEE; pp. 771–776.

Mishra L, Kaushik V. (2023) Application of blockchain in dealing with sustainability issues and challenges of financial sector. *J Sustain Financ Invest.*;13(3):1318–1333.

Mora H, Mendoza-Tello JC, Varela-Guzmán EG, Szymanski J. (2021) Blockchain technologies to address smart city and society challenges. *Comput Human Behav.* 122:106854.

Namasudra S, Akkaya K. (2023) Introduction to blockchain technology. In *Blockchain and Its Applications in Industry 40*. Springer; pp. 1–28.

Naqash MT, Syed TA, Alqahtani SS, Siddiqui MS, Alzahrani A, Nauman MA. (2023) Blockchain based framework for efficient water management and leakage detection in Urban Areas. *Urban Sci.*;7(4):99.

Ogbokor CA, Ngeendepi EJ. (2012) Investigating the challenges faced by SMEs in Namibia; *International Bilingual Conference on Sustainable Entrepreneurship*. University of Mauritius, Mauritius.

Owen LA, Pickering KT. (2006) *An Introduction to Global Environmental Issues*. Routledge;2006.

Paiva Sobrinho R, Garcia JR, Maia AG, Romeiro AR. (2022) Blockchain technology and complex flow systems as opportunities for water governance innovation. *Rev Bras Inovação.*;18:157–176.

Pant RR, Prakash G, Farooquie JA. (2015) A framework for traceability and transparency in the dairy supply chain networks. *Procedia-Social Behav Sci.* 2015;189:385–394.

Pimentel D, Berger B, Filiberto D, Newton M, Wolfe B, Karabinakis E, et al. (2004) Water resources: Agricultural and environmental issues. *Bioscience.*;54(10):909–918.

Qin B, Xu P, Wu Q, Luo L, Zhang Y. (2007) Environmental issues of lake Taihu, China: Eutrophication shallow lakes with Spec Ref to Lake Taihu, China. *Hydrobiologia*; 581:3–14, DOI: 10.1007/s10750-006-0521-5

Saleh AS, Caputi P, Harvie C. (2008) Perceptions of business challenges facing Malaysian SMEs: some preliminary results 79–106. https://ro.uow.edu.au/articles/conference_contribution/Perceptions_of_business_challenges_facing_Malaysian_SMEs_some_preliminary_results/27692418/1/files/50430912.pdf

Saleh AS, Caputi P, Harvie C. (2008) Perceptions of business challenges facing Malaysian SMEs: some preliminary results. In M. Obayashi & N. Oguchi (Eds.), *5th SMEs in a Global Economy Conference 2008* (pp. 79–106). Senshu University Tokyo, Japan.

Shaikh AA, Kumar A, Syed AA, Shaikh MZ. (2021) A two-decade literature review on challenges faced by SMEs in technology adoption. *Acad Mark Stud J*;25(3), 13.

Sunny J, Undralla N, Pillai VM. (2020) Supply chain transparency through blockchain-based traceability: An overview with demonstration. *Comput Ind Eng.* 150:106895.

Swan M, De Filippi P. (2017) Toward a philosophy of blockchain: A symposium: introduction. *Metaphilosophy*;48(5):603–619.

Syme GJ, Nancarrow BE, McCreddin JA. (1999) Defining the components of fairness in the allocation of water to environmental and human uses. *J Environ Manage*;57(1):51–70.

Tayal A, Solanki A, Kondal R, Nayyar A, Tanwar S, Kumar N. (2021) Blockchain-based efficient communication for food supply chain industry: Transparency and traceability analysis for sustainable business. *Int J Commun Syst.*;34(4):e4696.

Thakur T, Mehra A, Hassija V, Chamola V, Srinivas R, Gupta KK, et al. (2021) Smart water conservation through a machine learning and blockchain-enabled decentralized edge computing network. *Appl Soft Comput.*;106:107274.

Tian F. (2017) A supply chain traceability system for food safety based on HACCP, blockchain & Internet of things. In *2017 International Conference on Service Systems and Service Management*. IEEE; pp. 1–6.

Venkatesh VG, Kang K, Wang B, Zhong RY, Zhang A. (2020) System architecture for blockchain based transparency of supply chain social sustainability. *Robot Comput Integr Manuf.* 63:101896.

11 Blockchain in Agriculture
Transforming Crop Management with Transparency and Efficiency

Rahul Ganpatrao Sonkamble, Swati Dattatray Shirke, Pathan Mohd. Shafi, Shraddha Phansalkar, and Saeid Eslamian

11.1 INTRODUCTION

Crop management encompasses a range of practices and techniques aimed at maximizing crop yield and quality while minimizing losses from factors such as pests, diseases, and environmental conditions (University of Nevada, Reno, Maninder, 2021). It plays a pivotal role in agriculture, involving activities like soil preparation, irrigation, nutrient management, pest and disease control, and crop rotation (Ogwu & Osawaru, 2015). Effective crop management requires strategic planning and decision-making based on insights from plant physiology, soil science, pest control, and environmental considerations (Dalezios et al., 2023). Additionally, it takes into account economic and market factors that influence crop production. Sustainable crop management systems strive to strike a balance in nutrient utilization to maintain high yields while minimizing ecological impacts (Bučienė et al., 2007).

Despite its significance, crop management faces numerous challenges. Pests and diseases pose a major threat to crop yield and quality (Tian et al., 2023), resulting in substantial losses – approximately 17% of global grain yield is lost to pests annually. Unpredictable weather and climate conditions further compound the difficulties in implementing effective crop management practices (Kumar & Prashant, 2023). Another prominent challenge is the complexity of the agricultural supply chain. Crop managers must navigate this intricate network involving various stakeholders and processes, from production to consumption. This includes coordination with seed suppliers, fertilizer manufacturers, equipment providers, transportation logistics, and market access (Duan et al., 2021; Prashanth et al., 2023).

Traceability is another pressing issue in crop management (Lin, 2022). It refers to the ability to track and trace products throughout the supply chain, ensuring food safety, quality, and the ability to identify the source of contamination or disease outbreaks. Achieving traceability can be complex and time consuming, necessitating the implementation of systems and technologies (Wang et al., 2021; Rana, 2020).

Data security is a paramount concern in crop management (Brogi et al., 2020). Crop managers handle vast amounts of sensitive data related to crop production, weather conditions, market trends, and more. Robust data security measures, including secure storage, transmission, and access protocols, as well as encryption and authentication measures, are crucial to protect this valuable information (Hua et al., 2018).

To address these challenges, technology and models play a crucial role. Remote sensing, satellite imagery, and precision agriculture techniques have revolutionized crop management, offering precise monitoring and input optimization (Wilson et al., 2014; Xu et al., 2018; Oliveira et al., 2020; Kang et al., 2021; Cortez et al., 2020; Nick, 2021). Collaboration and knowledge sharing among crop managers also contribute to overcoming these hurdles.

DOI: 10.1201/9781003599593-17

Blockchain technology emerges as a potential solution to enhance data security and traceability in crop management (Marfuah & Yuliasih, 2022). By leveraging blockchain's decentralized and transparent ledger system, crop managers can securely record and track crop production data, ensuring data authenticity and accuracy while preventing issues like fraud or tampering.

In this chapter, the following research questions are addressed.

1. What challenges can blockchain technology address in crop management?
2. What are the potential benefits of integrating blockchain technology into crop management?
3. How can blockchain technology be utilized to enhance data security and traceability insurance in crop management?
4. How can the integration of blockchain technology with other technologies improve efficiency and productivity in crop management?

This chapter is divided into six sections with respect to traceability, supply chain management, insurance, data security, and integration of blockchain with other technologies. In each case, a summary is provided concerning the recent history of developments and challenges.

11.2 TRACEABILITY USING BLOCKCHAIN

Blockchain technology offers a promising solution to address challenges in crop management, primarily by enhancing traceability. By leveraging blockchain, the entire agricultural product supply chain can be meticulously tracked and recorded, ensuring transparency and immutability (Hasan & Habib, 2023). This empowers seamless tracking of transactions and crop-related data, spanning from planting to distribution. Blockchain technology guarantees the authenticity and accuracy of this data through verification by multiple network participants (Kravenkit & So-In, 2022).

Blockchain's adoption in crop management brings several advantages, particularly in traceability. Firstly, it amplifies transparency by providing a decentralized ledger accessible to all supply chain stakeholders. This transparency nurtures trust among participants and offer a clear overview of crop movement throughout production (Hasan & Habib, 2023). Secondly, blockchain strengthens data security (Lei et al., 2022). Its decentralized architecture thwarts malicious attempts to alter or manipulate stored data, preserving information integrity and reducing fraud risks in crop management. Thirdly, blockchain technology boosts efficiency and accountability. Supply chain participants can effortlessly trace and authenticate the origin, quality, and journey of crops, streamlining processes like recalls and compliance with regulatory standards. Blockchain's potential in crop management is substantial, as it revolutionizes traceability and data security while optimizing efficiency and accountability across the agricultural supply chain. Works for the traceability are mentioned in Table 11.1.

11.3 SUPPLY CHAIN MANAGEMENT USING BLOCKCHAIN

Supply chain management is a multifaceted process encompassing the production, distribution, and coordination of goods and services through various stages. Traditionally, supply chains have grappled with inefficiencies, opacity, and trust issues among participants. Blockchain technology has emerged as a promising remedy for these challenges by delivering a decentralized, transparent, and immutable ledger for recording transactions and data across the supply chain (Gomber et al., 2017).

Benefits of Blockchain in Supply Chain Management:

1. **Transparency and Traceability**: Blockchain's distributed ledger offers real-time visibility into the entire supply chain, enabling stakeholders to trace products' journey from source to destination (Fan, 2022).

TABLE 11.1
Traceability in Crop Management Using Blockchain

Works	Blockchain Applications	Challenges and Considerations	Case Studies
Li et al. (2020)	Supply Chain Transparency	• Scalability	• Thai Agriculture Pilot Project (Rahman et al., 2020)
	Food Safety and Quality Assurance	• Interoperability	• IoT-Blockchain Framework for Food Safety (Bhagat & Rani, 2019)
	Smart Contracts	• Ensuring data privacy	Not specified
		• Compliance with regulations (e.g., GDPR)	
		• Ensuring interoperability	
		• Addressing scalability	
Bassi & Joshi (2019)	Supply Chain Transparency	Not specified	Not specified
	Food Safety and Quality Assurance		
	Smart Contracts	• Data privacy	
		• Compliance with regulations (e.g., GDPR)	
		• Ensuring interoperability	
Rahman et al. (2020)	Supply Chain Transparency	Ensuring data privacy	• Thai Agriculture Pilot Project (Rahman et al., 2020)
Bhagat & Rani (2019)	Supply Chain Transparency	Not specified	• IoT-Blockchain Framework for Food Safety (Bhagat & Rani, 2019)
	Food Safety and Quality Assurance		Not specified
	Smart Contracts		
Miah et al. (2019)	Supply Chain Transparency	• Ensuring data privacy	Not specified
	Food Safety and Quality Assurance	• Compliance with regulations (e.g., GDPR)	
	Smart Contracts	• Ensuring interoperability	
Vu (2023)	• Tracking product origins – Ensuring supply chain transparency	• Integration with existing systems • Standardization • Data accuracy • Data sharing	Not specified
Shen (2023)	• Supply chain traceability	• Interoperability	Not specified

2. **Quality Control and Provenance**: Blockchain bolsters quality control by providing a complete product history, essential for verifying authenticity and compliance with quality standards (Xu et al., 2018).
3. **Counterfeit Prevention**: Blockchain combats counterfeit products by allowing consumers and businesses to verify product authenticity via QR codes or RFID tags (Aguilar-Gallardo & Bonora-Centelles, 2023; Silva, 2023).

4. **Streamlined Documentation**: Smart contracts on blockchain automate supply chain processes like payments, customs clearance, and quality inspections, reducing manual paperwork (Kim & Laskowski, 2018).
5. **Supply Chain Finance**: Blockchain facilitates secure and transparent supply chain finance, reducing lending risks for suppliers and financiers (Beraldi et al., 2021).
6. **Provenance and Ethical Sourcing**: Blockchain tracks product origin and ethical sourcing, empowering informed choices, especially in industries like fashion and jewelry (Zheng et al., 2017).
7. **Fraud Reduction**: Blockchain's data immutability minimizes fraud risk in supply chains, making it challenging for malicious actors to manipulate records (Tapscott & Tapscott, 2016).

Challenges and Future Directions: Despite its potential, blockchain faces scalability, interoperability, regulatory compliance, and collaboration challenges (Nofer et al., 2017; Wang et al., 2019). Ongoing research and industry partnerships are crucial to overcoming these hurdles and fully realizing blockchain's potential in supply chain management. As the technology evolves and more use cases emerge, it is poised to become integral to modern supply chains (Bettín-Díaz et al., 2018).

Blockchain technology promises to revolutionize supply chain management, enhancing transparency, traceability, security, and efficiency. As blockchain matures and gains broader adoption, it will play a pivotal role in reshaping the global supply chain landscape (Fan, 2022). Blockchain's tamper-resistant nature simplifies compliance with industry regulations and standards, streamlining auditing and reporting processes (Xu et al., 2018).

In summary, blockchain technology offers a multitude of advantages for supply chain management, ranging from transparency and traceability to fraud prevention and automation. Its adoption is gaining momentum across industries as organizations seek to optimize their supply chains and establish more efficient, secure, and trustworthy operations (Beraldi et al., 2021). Comparative analysis is presented in Table 11.2.

11.4 CROP INSURANCE

Crop insurance serves as a vital financial tool, providing essential protection to farmers against various risks associated with their agricultural activities. It plays a strategic role in risk management, aiming to safeguard farmers' livelihoods by mitigating financial losses resulting from unforeseen circumstances. These events can range from crop failures and natural disasters to market fluctuations, and the primary goal is to offer financial security to farmers, thereby promoting agricultural production.

Importance of Crop Insurance:

1. **Risk Mitigation**: Crop insurance empowers farmers to manage the uncertainties posed by unpredictable factors, including adverse weather conditions, pests, diseases, and market price fluctuations.
2. **Investment Confidence**: With insurance coverage in place, farmers gain the confidence to make investments in their agricultural operations, such as acquiring modern equipment or expanding their farming activities.
3. **Financial Stability**: Crop insurance reinforces the financial stability of farmers, protecting them from the risk of falling into debt or poverty in case of crop failures or other disasters.
4. **Food Security**: It plays a crucial role in ensuring food security by enabling uninterrupted production even in the face of unforeseen setbacks.

TABLE 11.2
Supply Chain Management Using Blockchain

Works	Aspects	Blockchain Applications	Challenges and Considerations
Fan (2022)	Blockchain technology supply chain landscape	Transparency and Traceability	Reducing fraud and inefficiencies
Xu et al. (2018)	Blockchain in supply chain management	Quality Control and Provenance	• Enhances quality control by providing a complete history of a product's origin and production process • Simplifies auditing and reporting processes by providing a comprehensive and immutable record of transactions
		Reducing Counterfeits	Simplifies auditing and reporting processes by providing a comprehensive and immutable record of transactions
Beraldi et al. (2021)	Supply chain management using blockchain technology	Supply Chain Finance	Including enhanced transparency, traceability, quality control, fraud prevention, and automation.
Kim & Laskowski (2018)	Supply chain management using blockchain technology	Streamlining Documentation	Reduces manual paperwork and accelerates the flow of goods
Zheng et al. (2017)	Supply chain management using blockchain technology	Provenance and Ethical Sourcing	Track the origin and ethical sourcing of products
Tapscott & Tapscott (2016)	Supply chain management using blockchain technology	Reducing Fraud	Blockchain data reduces the risk of fraud

Types of Crop Insurance:

1. **Crop Disaster Insurance**: This category offers protection against losses caused by weather events, pests, or diseases affecting crops.
2. **Livestock Insurance**: It covers losses associated with livestock production, including diseases, accidents, or natural disasters.
3. **Income Protection Insurance**: This form of insurance provides coverage for income losses resulting from crop failures or price fluctuations.
4. **Multi-Peril Crop Insurance**: This comprehensive coverage option offers extensive indemnity in the event of various perils, ensuring robust protection against losses (Chatterjee, 2017).

The crop insurance market holds significant importance within the broader insurance industry and is segmented based on coverage, distribution channels, and geographical regions. Ongoing technological advancements and increasing awareness of its benefits, particularly in countries heavily reliant on agriculture, are driving the market's evolution (Agricultural Insurance Market Size, Share & Analysis). While crop insurance offers numerous advantages, challenges remain, including concerns about affordability, low market penetration in certain regions, and administrative complexities. Nevertheless, the integration of technology, data analytics, and innovative insurance products presents new opportunities to enhance coverage and reach a broader spectrum of farmers.

Blockchain technology finds valuable applications in smart index–based agricultural insurance solutions. These solutions leverage blockchain to collect data from sources like weather stations and IoT devices. When predefined conditions, such as adverse weather events, are met, insurance payouts are automatically triggered, streamlining the claims process and enhancing transparency (Xiong et al., 2020). In the realm of crop insurance, initiatives like Etheric, in collaboration with Aon and Oxfam, have successfully tested blockchain-based solutions in Sri Lanka. These efforts aim to enhance risk reduction and coverage for farmers by harnessing blockchain's trust and transparency features (Jha et al., 2021).

Blockchain also plays a pivotal role in ensuring supply chain traceability within agriculture. It enables the tracking of agricultural product origins, ensuring transparency throughout the supply chain. This not only improves food safety but also allows consumers to trace the journey of their food from the farm to their table (Shen et al., 2023). Global risk management companies like AON have developed blockchain-based agricultural insurance policies, offering heightened security and transparency in insurance processes, ultimately benefiting farmers (Kamilaris et al., 2019). Furthermore, blockchain technology automates various agricultural processes, including record-keeping, payment settlements, and contract enforcement, reducing the need for intermediaries and enhancing efficiency in the agricultural sector (Xiong et al., 2020).

Comparative analysis is presented in Table 11.3.

TABLE 11.3
Insurance Using Blockchain

Paper	Aspects	Blockchain Applications	Benefits	Challenges	Considerations	Case Studies
Xiong (2020)	Smart Agriculture Insurance, Crop Insurance	• Smart index–based insurance • Crop insurance	Quicker and transparent claims processing	• Data accuracy and integration • Regulatory challenges	• Privacy and data security • Integration with existing systems	Not specified
Jha (2021)	Crop Insurance	Decentralized crop insurance using smart contracts	• Enhanced risk reduction • Improved coverage for farmers	• Multiple claims • Data security	• Scalability and adoption • Trust among stakeholders	Sri Lanka (Etherisc, Aon, Oxfam)
Shen (2023)	Automation	Automation of agricultural processes	Efficiency gains	Data integration	• Regulatory compliance • Data privacy	Not specified
Kshetri (2021)	Crop Insurance	• Blockchain-based smart contracts for crop insurance	• Accessibility for smallholder farmers • Enhanced risk management	• Adoption barriers in developing countries • Trust issues	• Regulatory compliance • Scalability	Not specified
World Bank (2021)	Risk Management, Traceability	• Blockchain for risk management and supply chain traceability in agriculture	• Enhanced risk assessment • Supply chain transparency	• Data integration • Scalability	• Regulatory compliance • Data security	Not specified

11.5 DATA SECURITY

Crop management involves the collection and analysis of extensive data, encompassing aspects like weather patterns, soil conditions, crop growth rates, and occurrences of pests and diseases (Flint et al., 2023). This data is pivotal for informed decisions in crop production and yield optimization. However, data security in crop management faces risks from cyber threats and data breaches (Pyke et al., 2021).

To enhance data security, blockchain technology offers a robust solution. Blockchain ensures secure and tamper-proof data storage and sharing. Its decentralized nature eliminates centralized vulnerabilities, reducing the risk of hacking or unauthorized access. Additionally, cryptographic algorithms in blockchain encrypt data, limiting access to authorized parties.

Moreover, blockchain facilitates the implementation of smart contracts in crop management. These contracts automate processes and ensure that all parties fulfil their obligations. For instance, a smart contract can trigger actions or payments automatically when specific conditions are met, such as confirming the delivery of crops in a sale agreement.

Furthermore, blockchain, combined with IoT and big data analysis, improves crop and food production and encourages sustainable practices in agriculture (Jiang et al., 2022). It enhances product traceability across the supply chain, promoting transparency and accountability.

Blockchain also enhances data privacy in crop management by securely storing personal data on a decentralized network. This safeguards sensitive information from unauthorized access. Additionally, blockchain helps promote sustainable practices and trust among stakeholders. Farmers can securely record and verify practices, like organic fertilizers or eco-friendly pest control methods, with validation from other supply chain stakeholders. This instils confidence in consumers regarding the sustainability and quality of purchased crops (Chen et al., 2022).

Crop data validation and verification by various supply chain stakeholders, including buyers and certification bodies, instill consumer confidence in the sustainability and quality of purchased crops. Data privacy is a significant concern in crop management due to the sensitive information involved (Kairaldeen et al., 2023; Kaur et al., 2022). However, blockchain technology offers a secure decentralized network for storing and sharing personal data, granting access solely to authorized parties. Comparative analysis is given in Table 11.4.

Furthermore, the integration of smart contracts with blockchain technology can revolutionize crop management. Smart contracts automate and enforce agreements among different participants in crop management, such as farmers, distributors, and retailers. Through smart contracts, farmers can ensure that their crops adhere to agreed-upon terms during distribution, with payments triggered automatically when contract conditions are met. This combination of blockchain and smart contracts enhances transparency and efficiency in crop management practices. By using these technologies, the agricultural industry can address data privacy concerns, streamline processes, and improve the overall quality and sustainability of crops.

11.6 BLOCKCHAIN INTEGRATION WITH OTHER TECHNOLOGIES

The integration of blockchain and IoT technologies offers significant advantages to farmers (IvyPanda, 2023). It allows for the real-time monitoring of environmental factors crucial for crop growth and quality. For instance, field sensors can measure parameters like soil moisture, temperature, and humidity (Qi et al., 2022). This data is then automatically recorded on the blockchain, establishing a transparent and unchangeable record of environmental conditions throughout the crop's growth cycle. This valuable information empowers farmers to make data-driven decisions, optimizing their farming practices to enhance crop productivity while minimizing resource wastage (Marfuah & Yuliasih, 2022).

Furthermore, blockchain combined with IoT plays a crucial role in ensuring the traceability of crops within the supply chain (Marchesi et al., 2022). By integrating sensors and tracking devices

TABLE 11.4
Data Security and Privacy Using Blockchain

Paper	Aspect	Blockchain Application	Benefits	Challenges and Considerations
Aliyu, & Liu (2023)	Data Security	Blockchain secures crop data through encryption and decentralized storage.	• Protection against data tampering • Secure data sharing	• Data access control and permission • Scalability and network congestion • Integration with existing systems
Ronaghi (2021)	Data Privacy	Blockchain ensures data privacy by allowing selective data sharing with cryptographic keys.	• Privacy preservation • User consent control • Immutable audit trails	• Identity management and authentication • Regulatory compliance • Data anonymization
Zhao, et al. (2023)	Data Security	Blockchain optimizes IoT data storage for efficiency security in agriculture.	• Efficient data collection • Enhanced data security	• Ensuring data integrity • Scalability and resources constraints • Integration with legacy systems
Jaber et al. (2022)	Data Privacy	Blockchain maintains user access control and data security in green IoT–based systems.	• Enhanced data privacy • Auditability and accountability	• User management and permissions • Privacy versus transparency
Khan, et al. (2022)	Data Security	Blockchain technology enhances data security in agricultural supply chains.	• Data immutability and security • Traceability and transparency	• Interoperability and data standardization • Adoption and integration challenges • Governance and regulatory compliance
	Data Privacy	Blockchain ensures data privacy and access control in crop insurance solutions.	• Secure and private data sharing • Audit trails for regulatory compliance	• Identity and access management • Smart contract vulnerabilities • Integration with existing insurance systems
Xiong, (2020)	Data Accuracy, Transparency	• Blockchain to ensure data accuracy and transparency in agriculture	• Improved decision-making • Reduced fraud	• Integration challenges • Data privacy • Interoperability • Regulatory compliance

with blockchain technology, the movement of crops from the farm to the consumer can be meticulously monitored and recorded. This fosters transparency and mitigates the risks associated with fraud or misrepresentation in the supply chain, as stakeholders can readily access details concerning crop origin, handling, and transportation (Huang et al., 2022).

A notable example of the integration of blockchain and IoT for crop traceability is the "Harvest Network" developed by Kim et al. (2018). This system harnesses the Ethereum platform and IoT devices to meticulously track and document the safety measures applied to agricultural products on the blockchain. Consequently, consumers can access a transparent and secure system to trace the journey of crops from the farm to their plate. The blockchain's transparency and immutability serve as safeguards against fraudulent or unethical practices, preserving the integrity of the supply chain.

Furthermore, the synergy of blockchain and IoT in crop management significantly enhances traceability across the entire agri-food supply chain, as demonstrated by (Krishna et al., 2021). IoT devices capture and store data on the blockchain, allowing for accurate tracking of agricultural product origins and journeys from the farm to the consumer. This heightened traceability plays a critical role in ensuring food safety, as it enables swift detection and containment of potential contamination or adulteration.

Additionally, the implementation of blockchain and IoT technologies in crop management effectively addresses concerns about data authenticity and trustworthiness, as highlighted by Marfuah and Yuliasih (2022). The decentralized and transparent nature of blockchain assures the integrity of recorded data. It also offers unique identification and tracking capabilities, swiftly identifying and rectifying fraudulent or erroneous data entries, as emphasized by Zhong et al. (2023).

Blockchain and IoT integration in crop management offers enhanced security and data integrity. Data generated by IoT devices is securely stored and encrypted on the blockchain, ensuring its tamper-proof nature, preventing unauthorized alterations or deletions. Additionally, smart contracts in blockchain and IoT systems automate actions based on predefined conditions, such as triggering irrigation in response to low soil moisture or initiating pest control measures upon exceeding specific thresholds. This automation not only boosts efficiency but also mitigates the risk of human error in crop management practices.

Furthermore, this integration contributes significantly to sustainability efforts (Sun et al., 2022; Hasan et al., 2020). Blockchain establishes a decentralized and transparent network, allowing for the verification of sustainability claims within the agricultural industry. IoT devices collect real-time data on factors like soil conditions, water usage, and energy consumption throughout the crop production process. This data is then stored and verified on the blockchain, creating an auditable record of sustainable practices.

Moreover, blockchain and IoT integration aids resource optimization and waste reduction (Jiang et al., 2022). Analysis of IoT data empowers farmers with insights into resource consumption patterns, enabling informed decisions to minimize waste and maximize efficiency. For instance, monitoring water usage via IoT devices helps identify areas of wastage, facilitating the implementation of efficient irrigation systems. This not only conserves water resources but also reduces costs for farmers.

Furthermore, this combination enhances transparency and accountability in the agricultural supply chain (Alissa et al., 2023). IoT sensors track the movement and conditions of agricultural products in real time, with data recorded on the blockchain. This establishes a transparent and immutable record of the product's journey from farm to consumer, instilling confidence in consumers regarding the origin and quality of their purchases. The works using blockchain and IoT are presented in Table 11.5.

The integration of blockchain and IoT in crop management offers an additional realm of benefits through the application of machine learning algorithms (Jebamikyous et al., 2023). These algorithms harness the extensive data gathered by IoT devices and stored on the blockchain to offer predictions and recommendations for crop management (Alissa et al., 2023). Through the analysis of historical data encompassing climate conditions, soil quality, and crop yields, machine learning

TABLE 11.5
Crop Management Using Blockchain & IoT

Paper	Aspect	Blockchain Application	Benefits	Challenges and Considerations
Krishna et al. (2021), Ferrández-Pastor et al. (2022)	Data Security	Use of blockchain ensures secure data storage and access in agricultural IoT systems.	• Protection against data tampering • Enhanced data security	• Identity management and access control • Privacy concerns and data anonymization
Adow et al. (2022), Ronaghi (2021)	Transparency	Blockchain provides a transparent and tamper-proof ledger for crop-related data.	• Improved traceability throughout the supply chain	• Scalability and network congestion • Integration with existing systems
van Hilten (2020)	Supply Chain Traceability	Blockchain enables end-to-end traceability of agricultural products in the supply chain.	• Enhanced food safety • Consumer confidence in product origins	• Standardization of data formats and protocols • Interoperability with various IoT devices
Xiong et al. (2020)	Quality Assurance	Monitoring and ensuring crop quality using IoT sensors and blockchain technology.	• Improved crop quality assessment • Enhanced decision-making for farmers	• Sensor data accuracy and calibration • Maintenance and scalability of IoT sensor networks
Zhao et al. (2023)	Traceability	Leveraging blockchain for tracing the origin and journey of agricultural products.	• Enhanced traceability of agricultural products • Reduced food fraud and counterfeit products	• Data privacy and GDPR compliance • Cross-border regulatory challenges
Zhao et al. (2023)	Sustainability	Using blockchain and IoT for sustainable agriculture practices, such as precision farming.	• Improved resource management and sustainability • Reduced environmental impact	• Integration with existing agricultural practices • Education and training for farmers and stakeholders

algorithms provide valuable insights for optimizing farming practices (Merenda et al., 2020). These insights span a spectrum, ranging from determining the optimal time for crop planting to refining irrigation schedules and executing pest management strategies. The amalgamation of blockchain and IoT furnishes a dependable and secure platform for data collection and storage, ensuring data accuracy and integrity, thereby enabling effective machine learning analysis.

Furthermore, the amalgamation of machine learning with blockchain and IoT in crop management can pave the way for the development of intelligent farming systems. These systems possess the capability to autonomously monitor and regulate various facets of crop production. For instance, they can dynamically adjust watering schedules based on real-time weather conditions or employ AI–powered image recognition to detect and respond to pest infestations.

Incorporating machine learning into blockchain and IoT–based crop management goes beyond optimizing farming practices; it unlocks new horizons for predictive analytics and decision-making in agriculture. Machine learning algorithms, by scrutinizing vast datasets from diverse sources, can unveil hidden patterns and trends imperceptible to humans. This empowers farmers to make informed choices regarding planting, harvesting, and resource allocation. For instance, historical data on crop yields, weather conditions, and market demand can be analyzed by machine learning algorithms to offer precise forecasts about future crop production and market dynamics. Consequently, farmers can refine planting schedules and production levels to align with market needs, ultimately enhancing profitability.

Furthermore, integrating machine learning into blockchain and IoT–based crop management aids in tackling agricultural challenges, notably effective supply chain management. By merging blockchain and IoT systems, a transparent, adaptable, and secure data transmission infrastructure can be established for the entire supply chain. Machine learning algorithms can scrutinize blockchain-stored data to detect anomalies or suspicious activities, mitigating fraud or quality issues. This fortifies the authenticity and quality of agricultural products, bolstering consumer trust throughout the supply chain.

Another agricultural challenge, resource optimization (e.g., water and fertilizers), can be addressed through machine learning. By analyzing real-time IoT data encompassing soil moisture, weather conditions, and crop growth, machine-learning algorithms furnish accurate predictions and recommendations for smart irrigation practices. Comparative works are presented in Table 11.6.

In addition to machine learning, the fusion of blockchain technology with cloud computing holds promise for advancing crop management practices (Vitali et al., 2021). Blockchain offers a secure and decentralized platform for storing, sharing, and accessing agricultural data, ensuring data integrity and enabling transparency throughout the agricultural process. On the other hand, cloud computing facilitates efficient storage and processing of vast data volumes. By integrating blockchain

TABLE 11.6
Crop Management Using Blockchain & Machine Learning

Feature	Blockchain-Based Crop Management	Machine Learning-Based Crop Management
Data Transparency	Offers transparency through immutable ledgers, allowing stakeholders to track every stage of crop production and supply chain. It ensures data integrity and reduces fraud risk (Durai & Shamili, 2022).	Provides data-driven insights into crop conditions, yield predictions, and disease detection through algorithms and models (Benos et al., 2021).
Security and Traceability	Ensures data security with decentralized architecture and cryptographic techniques, reducing the risk of fraud and ensuring traceability (Alahmadi et al., 2022).	Focuses on analyzing historical crop data and real-time sensor data to detect anomalies, pests, and diseases, enhancing traceability (Khan et al., 2022).
Yield Forecasting and Optimization	Utilizes historical data and smart contracts to optimize resource allocation, predicting crop yields and guiding efficient cultivation practices (Durai & Shamili, 2022).	Leverages predictive models and data analytics to forecast crop yields, helping farmers make informed decisions about planting and harvesting Benos et al., 2021).
Decision Support System	Offers decision support by providing real-time data on crop conditions, enabling timely actions for irrigation, pest control, and nutrient management (Jebamikyous et al., 2023).	Employs machine learning algorithms to assist farmers in making decisions regarding planting, harvesting, and crop rotation based on data-driven insights (Jebamikyous et al., 2023).
Cost Reduction and Resource Optimization	Reduces costs by eliminating intermediaries in supply chains, ensuring fair payments through smart contracts, and optimizing resource usage (Durai, & Shamili, 2022).	Optimizes resource usage by predicting crop health, reducing waste, and minimizing resource input based on data analysis (Benos et al., 2021).

and cloud computing, stakeholders in the agriculture industry, including farmers, gain access to real-time data from diverse sources like weather forecasts (Parwada & Marufu, 2023).

The inclusion of cloud computing provides farmers with increased scalability and flexibility in managing their agricultural data (Debauche et al., 2022). Cloud computing enables cost-effective and efficient handling of extensive data, allowing farmers to access their data from anywhere at any time and adapt computing resources to their specific requirements. The briefing of such works is represented in Table 11.7.

Moreover, the amalgamation of blockchain and cloud computing fosters collaboration and information sharing among various agricultural stakeholders (Marchesi et al., 2022). Farmers, suppliers, distributors, and consumers can securely and transparently share and verify data concerning crop production, quality control, and logistics. This collaboration enhances coordination, efficiency, and traceability within the agricultural supply chain.

In summary, the integration of technologies such as machine learning, IoT, blockchain, and cloud computing has the potential to revolutionize crop management practices in agriculture.

TABLE 11.7
Crop Management Using Blockchain & Cloud Computing

Feature	Blockchain-Based Crop Management	Cloud Computing-Based Crop Management
Data Security and Immutability	Utilizes blockchain's decentralized ledger to ensure data security and immutability, preventing unauthorized access and tampering (Khanna et al., 2022).	Relies on cloud-based security measures to protect data, providing secure storage and access controls (Debauche et al., 2022).
Transparency and Traceability	Offers transparent tracking of crop-related data through blockchain, enabling stakeholders to trace the entire supply chain (Ronaghi, 2021).	Provides traceability through cloud-based data storage, allowing authorized parties to access information at various stages of crop management (Debauche et al., 2022).
Smart Contracts and Automation	Deploys smart contracts on the blockchain to automate processes such as payments, reducing manual intervention and ensuring fair transactions (Fan, 2022).	Utilizes cloud-based automation tools and IoT devices to optimize tasks like irrigation and monitoring, enhancing efficiency (Debauche et al., 2022).
Scalability and Resource Allocation	Offers scalability through distributed blockchain nodes, ensuring efficient resource allocation and reduced overhead (Ronaghi, 2021).	Leverages cloud computing's scalability to handle varying workloads and allocate resources dynamically based on demand (Debauche et al., 2022).
Accessibility and Collaboration	Provides real-time data accessibility to authorized users through blockchain, enabling collaborative decision-making among stakeholders (Cambra Baseca et al., 2019).	Allows remote access and collaboration through cloud-based platforms, facilitating data sharing and cooperative farming practices (Cambra Baseca et al., 2019).
Cost-Efficiency and Maintenance	Minimizes costs by eliminating intermediaries in transactions and reducing the need for extensive maintenance through blockchain's self-regulating nature (Fan, 2022).	Offers cost savings by reducing the need for on-premises hardware and maintenance while providing scalable solutions (Debauche et al., 2022).

11.7 CONCLUSIONS

Effective management of crops is essential for improving agricultural productivity and sustainability. By implementing best practices in crop management, such as optimizing resource use and reducing environmental impacts, farmers can enhance crop yields. Accurate forecasting of crop yield is crucial not only for advanced planning and policy formulation but also for empowering farmers to make informed decisions on planting schedules, irrigation needs, fertilization strategies, and pest control. In countries where agriculture is a critical sector of the economy and plays a significant role in people's livelihoods, reliable forecasts of crop yield are vital for ensuring food security, minimizing waste, and promoting sustainable agricultural practices. Crop management plays a pivotal role in enhancing agricultural productivity and ensuring sustainability. By employing effective crop management practices, farmers can achieve higher crop yields while optimizing resource utilization and minimizing environmental impact. Accurate forecasting of crop yield is indispensable for informed decision-making related to food procurement, distribution, pricing structures, import export policies, and the formulation of sustainable farming strategy. In countries where agriculture holds significant economic importance and supports livelihood, reliable forecasting of crop yield becomes crucial for maintaining food security, reducing waste, and promoting sustainable agricultural practices.

Crop management is effective if it is combined with blockchain technology. If we use blockchain technology then all aspects of traceability, supply chain management, insurance, data security, and integration of blockchain with other technologies will be handled. We have briefed every aspect in this chapter. Blockchain technology should be used for proper management of the crops.

REFERENCES

Adow, A. H., Shrivas, M. K., Mahdi, H. F., Zahra, M. M. A., Verma, D., Doohan, N. V., & Jalali, A. (2022). Analysis of Agriculture and Food Supply Chain through Blockchain and IoT with Light Weight Cluster Head. *Computational Intelligence and Neuroscience*, 2022. https://doi.org/10.1155/2022/1296993

Aguilar-Gallardo, C., & Bonora-Centelles, A. (2023). Blockchain technology for advanced therapy medicinal products: Applications in tracking, data sharing, and supply chain automation. *Journal of Pharmaceutical and Biopharmaceutical Research*, *5*(2), 430–443.

Alahmadi, A. N., Rehman, S. U., Alhazmi, H. S., Glynn, D. G., Shoaib, H., & Solé, P. (2022). Cyber-security threats and side-channel attacks for digital agriculture. *Sensors*, *22*(9), 3520.

Alissa, K., Maray, M., Malibari, A., Alazwari, S., Alqahtani, H., Nour, M. K., Obbaya, M., Shamseldin, M. A., & Duhayyim, M. A. (2023). Optimal Deep Learning Model Enabled Secure UAV Classification for Industry 4.0, Computers. *Materials & Continua*, 74(3), 5349–5367, https://doi.org/10.32604/cmc.2023.033532

Aliyu, A. A., & Liu, J. (2023). Blockchain-Based Smart Farm Security Framework for the Internet of Things. *Sensors*, 23(18), 7992.

Bassi, A., & Joshi, G. (2019). Blockchain Technology for Enhancing Supply Chain Transparency in the Agri-Food Sector. *Computers in Industry*, 110, 103–113.

Benos, L., Tagarakis, A. C., Dolias, G., Berruto, R., Kateris, D., & Bochtis, D. (2021). Machine learning in agriculture: A comprehensive updated review. *Sensors*, *21*(11), 3758.

Beraldi, P., Violi, A., Ferrara, M., Ciancio, C., & Pansera, B. A. (2021). Dealing with complex transaction costs in portfolio management. *Annals of Operations Research*, *299*(1), 7–22.

Bettín-Díaz, R., Rojas, A. E., & Mejía-Moncayo, C. (2018, May). Methodological approach to the definition of a blockchain system for the food industry supply chain traceability. In *International conference on computational science and its applications* (pp. 19–33). Cham Germany Springer International Publishing.

Bhagat, K., & Rani, R. (2019). Blockchain and IoT-Based Traceability Framework for Food Quality and Safety. *Computers in Industry*, 113, 103130.

Brogi, C., Huisman, J. A., Herbst, M., Weihermüller, L., Klosterhalfen, A., Montzka, C., Reichenau, T. G., & Vereecken, H. (2020). Simulation of Spatial Variability in Crop Leaf Area Index and Yield Using Agroecosystem Modeling and Geophysics-Based Quantitative Soil Information. *Vadose Zone Journal*, *19*(1), e20009.

Bučienė, A., Antanaitis, Š., Mašauskienė, A., & Simanskaite, D. (2007). Nitrogen and Phosphorus Losses with Drainage Runoff and Field Balance as a Result of Crop Management. *Communications in Soil Science and Plant Analysis*, 38, 2177–2195.

Cambra Baseca, C., Sendra, S., Lloret, J., & Tomas, J. (2019). A Smart Decision System for Digital Farming. *Agronomy*, 9(5), 216.

Chatterjee, A. (2017). Agriculture Insurance. *ADB Brief.* https://www.adb.org/sites/default/files/publication/239181/adb-brief-agriculture-insurance-rt-4th_1.pdf

Chen, J., Chen, X., & Chen, C. (2022, March 9). A Traceable Blockchain-Based Vaccination Record Storage and Sharing System. *Journal of Healthcare Engineering*, https://doi.org/10.1155/2022/2211065

Cortez, M. L. J., Alves, M. D. C., Carvalho, G. R., & Pozza, E. A. (2020). Relationship Between Sentinel-2 Orbital Data and *In Situ* Monitoring of Coffee Rust. *SN Applied Sciences*, 2, 1460. https://doi.org/10.1007/s42452-020-03257-1

Dalezios, N. R., Dercas, N., Faraslis, I. N., Spiliotopoulos, M., Sidiropoulos, P., Sakellariou, S., & Eslamian, S. (2023). Irrigation and Agrometeorology: Innovative Remote Sensing Applications in Crop Monitoring, Ch. 11. In Eslamian, S. & F. Eslamian, *Handbook of Irrigation Hydrology and Management, Vol. 1: Irrigation Fundamentals*, Taylor & Francis Group, CRC Group, USA.

Debauche, O., Mahmoudi, S., Manneback, P., & Lebeau, F. (2022). Cloud and distributed architectures for data management in agriculture 4.0: Review and future trends. *Journal of King Saud University-Computer and Information Sciences*, *34*(9), 7494–7514.

Duan, S. X., Wibowo, S., & Chong, J. (2021). A Multicriteria Analysis Approach for Evaluating the Performance of Agriculture Decision Support Systems for Sustainable Agribusiness. *Mathematics*, 9(8), 884. https://doi.org/10.3390/math9080884

Durai, S. K. S., & Shamili, M. D. (2022). Smart Farming Using Machine Learning and Deep Learning Techniques. *Decision Analytics Journal*, 3, 100041.

Fan, M. (2022). International Logistics Management System Based on Cloud Computing Technology. *Wireless Communications and Mobile Computing, Wireless Communications and Mobile Computing*, (4), 1–8. https://doi.org/10.1155/2022/4317578

Ferrández-Pastor, Francisco-Javier, Mora-Pascual, Jerónimo, and Díaz-Lajara, Daniel. *Journal of Industrial Information Integration*, *29*(2022): 100381.

Flint, E. A., Hopkins, B. G., Svedin, J. D., Kerry, R., Heaton, M. J., Jensen, R. R., Campbell, C. S., Yost, M. A., & Hansen, N. C. (2023). Irrigation Zone Delineation and Management with a Field-Scale Variable Rate Irrigation System in Winter Wheat. *Agronomy*, 13(4), 1125. https://doi.org/10.3390/agronomy13041125

Gomber, P., Hinz, O., & Schiereck, D. (2017). Blockchain. *Business & Information Systems Engineering*, *59*(3), 183–187.

Hasan, A., Girona, M. M., Grosbois, G., Saha, N., & Halim, A. (2020). Land Sparing Can Maintain Bird Diversity in Northeastern Bangladesh. *Sustainability*, 12(16), 6472. https://doi.org/10.3390/su12166472

Hasan, I., & Habib, M. M. (2023). Revolutionizing Agriculture Supply Chain Management: Latest Technologies Making a Difference. *International Supply Chain Technology Journal*, 9(4). https://doi.org/10.1155/2022/3803566

Hua, P., Wang, J., Ma, Z., & Du, Y. (2018). Mini-Review of Application of IoT technology in Monitoring Agricultural Products Quality and Safety. *International Journal of Agricultural and Biological Engineering*, 11(5), 35–45.

Huang, Y., Liu, H., Guo, X., & Jiao, W. (2022). The Perception of the National Traceability Platform among Small-Scale Tea Farmers in Typical Agricultural Areas in Central China. *International Journal of Environmental Research and Public Health*, 19(23), 16280. https://doi.org/10.3390/ijerph192316280

IvyPanda. (2023). Blockchain and Internet-of-Things in Agriculture. https://ivypanda.com/essays/blockchain-and-internet-of-things-in-agriculture/

Jebamikyous, H., Li, M., Suhas, Y., & Kashef, R. (2023). Leveraging Machine Learning and Blockchain in E-commerce and Beyond: Benefits, Models, and Application. *Discover Artificial Intelligence*, 3(1), 3.

Jha, N., Prashar, D., Khalaf, O. I., Alotaibi, Y., Alsufyani, A., & Alghamdi, S. (2021). Blockchain Based Crop Insurance: A Decentralized Insurance System for Modernization of Indian Farmers. *Sustainability*, 13(16), 8921.

Jiang, S., Jakobsen, K., Bueie, J., Li, J., & Haro, P. H. (2022, January 1). A Tertiary Review on Blockchain and Sustainability with Focus on Sustainable Development Goals. *IEEE Access*, *10*, 114975–115006.

Kairaldeen, A. R., Abdullah, N. F., Abu-Samah, A., & Nordin, R. (2023). Peer-to-Peer User Identity Verification Time Optimization in IoT Blockchain Network. *Sensors*, 2023, 23(4). https://doi.org/10.3390/s23042106

Kamilaris, A., Fonts, A., & Prenafeta-Boldú, F. X. (2019). The rise of blockchain technology in agriculture and food supply chains. *Trends in Food Science & Technology*, *91*, 640–652.

Kang, Y., Meng, Q., Liu, M., You-Feng, Z., & Wang, X. (2021). Crop Classification Based on Red Edge Features Analysis of GF-6 WFV Data. *Sensors*, 21(13), 4328. https://doi.org/10.3390/s21134328

Kaur, J., Hazrati Fard, S. M., Amiri-Zarandi, M., & Dara, R. (2022). Protecting Farmers' Data Privacy and Confidentiality: Recommendations and Considerations. *Frontiers in Sustainable Food Systems*, 6, 903230. https://www.frontiersin.org/articles/10.3389/fsufs.2022.903230/full

Khan, H. H., Malik, M. N., Konečná, Z., Chofreh, A. G., Goni, F. A., & Klemeš, J. J. (2022). Blockchain Technology for Agricultural Supply Chains During the COVID-19 Pandemic: Benefits and Cleaner Solutions. *Journal of Cleaner Production*, 347, 131268.

Khanna, A., Sah, A., Bolshev, V., Burgio, A., Panchenko, V., & Jasiński, M. (2022). Blockchain–Cloud Integration: A Survey. *Sensors*, 22(14), 5238.

Kim, H. M., & Laskowski, M. (2018). Toward an Ontology-Driven Blockchain Design for Supply-Chain Provenance. *Intelligent Systems in Accounting, Finance and Management*, 25(1), 18–27.

Kim, M., Hilton, B., Burks, Z., & Reyes, J. (2018, November). Integrating Blockchain, Smart Contract-tokens, and IoT to Design a Food Traceability Solution. In *2018 IEEE 9th Annual Information Technology, Electronics and Mobile Communication Conference, IEMCON*, pp. 335–340, IEEE.

Kravenkit, S., & So-In, C. (2022, January 1). Blockchain-Based Traceability System for Product Recall, 95132–95150. https://doi.org/10.1109/ACCESS.2022.3204750

Krishna, K. M., Borole, Y. D., Rout, S., Negi, P., Deivakani, M., & Dilip, R. (2021, September). Inclusion of Cloud, Blockchain and IoT Based Technologies in Agriculture Sector. In *2021 9th International Conference on Cyber and IT Service Management*, IEEE. 22–23 September, pp. 1–8, Bengkulu, Indonesia

Kshetri, N. (2021). Blockchain-Based Smart Contracts to Provide Crop Insurance for Smallholder Farmers in Developing Countries. *IT Professional*, 23(6), 58–61.

Kumar, A., & Prashant, K. (2023). A Review on High Throughput Phenotyping for Vegetable Crops. *Journal of Botany Research*, 6(1), 170–175.

Lei, M., Xu, L., Liu, T., Liu, S., & Sun, C. (2022). Integration of Privacy Protection and Blockchain-Based Food Safety Traceability: Potential and Challenges. *Foods*, 11(15), 2262.

Li, Z., Guo, H., Barenji, A. V., Wang, W. M., Guan, Y., & Huang, G. Q. (2020). A sustainable production capability evaluation mechanism based on blockchain, LSTM, analytic hierarchy process for supply chain network. *International Journal of Production Research*, *58*(24), 7399–7419.

Lin, X. (2022). Analysis of Agricultural Product Information Traceability and Customer Preference Based on Blockchain. *Wireless Communications and Mobile Computing*, Special Issue: Advanced Information Security in Next Generation Wireless Communication-Enabled Internet of Things, Wiley, https://doi.org/10.1155/2022/1935233

Maninder, K. W., (2021), *Basics of Crop Management*. University of Nevada, Reno. https://naes.agnt.unr.edu/PMS/Pubs/2021-4103.pdf

Marchesi, L., Mannaro, K., Marchesi, M., & Tonelli, R. (2022). Automatic Generation of Ethereum-Based Smart Contracts for Agri-Food Traceability System. *IEEE Access*, 10, 50363–50383. https://doi.org/10.1109/ACCESS.2022.3171045

Marfuah, U., & Yuliasih, I. (2022). Blockchain Traceability for Agroindustry: A Literature Review and Future Agenda. https://doi.org/10.1088/1755-1315/1063/1/012056

Merenda, M., Porcaro, C., & Iero, D. (2020). Edge Machine Learning for Ai-Enabled IOT Devices: A Review. *Sensors*, 20(9), 2533. https://www.ncbi.nlm.nih.gov/pmc/articles/PMC7273223/

Miah, S. J., Gammack, J., Hasan, N., & Hasan, R. (2019). Blockchain Technology for Sustainable Agriculture: Applications and Challenges. *Sustainability*, 11(14), 3632.

Nick, M., (2021), How Satellite Imagery is Helping Precision Agriculture Grow to New Heights. https://www.eco-business.com/news/how-satellite-imagery-is-helping-precision-agriculture-grow-to-new-heights/

Nofer, M., Gomber, P., Hinz, O., & Schiereck, D. (2017). Blockchain. *Business & Information Systems Engineering*, *59*(3), 183–187.

Ogwu, M., & Osawaru, M. (2015). Soil Characteristics, Microbial Composition of Plot, Leaf Count and Sprout Studies of Cocoyam Collected in Edo State, Southern Nigeria. *Science, Technology and Arts Research Journal*, 4(1), 34–44.

Oliveira, J. T. D., Oliveira, R. A. D., Oliveira, L. A. A., Teodoro, P. E., & Montanari, R. (2020). Spatial Variability of Irrigated Garlic (Allium sativum L.) Production Components. 55(3), 300–303. https://doi.org/10.21273/HORTSCI14409-19

Parwada, C., & Marufu, H. (2023). Digitalisation of Agriculture in Zimbabwe: Challenges and Opportunities. *International Journal of Sustainable Agricultural Research, Conscientia Beam*, 10(1), 32–41.

Prashanth, D. S., Vk, T., Chinmye, K. N., Rishab, A., & Rashmi, K. S. (2023). A Model for Predicting Demand-Supply Information of Top Crops. *International Research Journal of Modernization in Engineering Technology and Science*, 04(03), 2093–2100.

Pyke, A., Rovira, E., Murray, S., Pritts, J., Carp, C. L., & Thomson, R. (2021). Predicting Individual Differences to Cyber-Attacks: Knowledge, Arousal, Emotional and Trust Responses. *Journal of Psychosocial Research on Cyberspace*, 15(4), https://doi.org/10.5817/CP2021-4-9

Qi, T., Cai, Z., & Cui, X. (2022). A Technological Quality Control System for Rice Supply Chain. *Food and Energy Equation*, 12(2), Special Issue: Food Security, https://doi.org/10.1002/fes3.382

Rahman, M. H., Majumder, S. C., & Debbarman, S. (2020). Examine the role of agriculture to mitigate the Co2 emission in Bangladesh. *Asian Journal of Agriculture and Rural Development, 10*(1), 392–405.

Rana, S. (2020). Blockchain-based Traceability and Transparency in Agricultural Supply Chains: Challenges and Opportunities. *Turkish Journal of Computer and Mathematics Education*, 11(3), 1948–1956. https://doi.org/10.17762/turcomat.v11i3.13591

Ronaghi, M. H. (2021). A Blockchain Maturity Model in Agricultural Supply Chain. *Information Processing in Agriculture, Information Processing in Agriculture*, 8(3), 398–408.

Shen, L., Zhang, Z., Zhou, Y., & Xu, Y. (2023). Applying Blockchain Technology and the Internet of Things to Improve the Data Reliability for Livestock Insurance. *Sensors*, 23(14), 6290.

Silva, P. A. D. (2023). Blockchain for Logistic and Supply Chain for Animal Protein and Genetic Improvement: A Systematic Literature Review. *Revista Científica Multidisciplinar Núcleo do Conhecimento*, https://doi.org/10.32749/nucleodoconhecimento.com.br/business-administration/blockchain-for-logistic

Sun, N., Tang, S. H., Zhang, J., Wu, J., & Wang, H. (2022). Food Security: 3D Dynamic Display and Early Warning Platform Construction and Security Strategy. *International Journal of Environmental Research and Public Health*. September 6;19(18):11169. https://doi.org/10.3390/ijerph191811169

Tapscott, D., & Tapscott, A. (2016). Blockchain Revolution: How the Technology Behind Bitcoin is Changing Money, Business, and the World. *Penguin*, Portfolio, 348.

Tian, L., Hu, Y., Shi, B., & Wu, W. (2023). Natural Product-Based Drugs Discovery: Semisynthesis and Biological Evaluation of Ester Derivatives from Pregn-5-ene-3β, 17α, 20(S)-triol. *Chemical and Pharmaceutical Bulletin*, 71(3), 220–228. https://doi.org/10.1248/cpb.c22-00698

van Hilten, M., Ongena, G., & Ravesteijn, P. (2020). Blockchain for organic food traceability: Case studies on drivers and challenges. *Frontiers in Blockchain, 3*, 567175.

Vitali, G., Francia, M., Golfarelli, M., & Canavari, M. (2021). Crop Management with the IoT: An Interdisciplinary Survey. *Agronomy*, 11, 181

Vu, N., Ghadge, A., & Bourlakis, M. (2023). Blockchain Adoption in Food Supply Chains: A Review and Implementation Framework. *Production Planning & Control*, 34(6), 506–523.

Wang, Y., Han, J. H., & Beynon-Davies, P. (2019). Understanding Blockchain Technology for Future Supply Chains: A Systematic Literature Review and Research Agenda. *Supply Chain Management: An International Journal*, 24(1), 62–84.

Wang, Z., Wang, L., Xiao, F., Chen, Q., Lu, L., & Hong, J. (2021). A Traditional Chinese Medicine Traceability System Based on Lightweight Blockchain. *Journal of Medical Internet Research*, 21;23(6): e25946. https://doi.org/10.2196/25946

Wilson, J. H., Zhang, C., & Kovacs, J. M. (2014). Separating Crop Species in Northeastern Ontario Using Hyperspectral Data. *Remote Sensing*. https://doi.org/10.3390/rs6020925

World Bank. (2021). Blockchain Application in Agriculture Insurance. Retrieved from https://www.indexinsuranceforum.org/blog/blockchain-application-agriculture-insurance

Xiong, H., Dalhaus, T., Wang, P., & Huang, J. (2020). Blockchain technology for agriculture: applications and rationale. *Frontiers in Blockchain, 3*, 7.

Xu, Y., Gao, Z., Khot, L. R., Meng, X., & Zhang, Q. (2018). A Real-Time Weed Mapping and Precision Herbicide Spraying System for Row Crops. *Sensors*, 18(12), 4245. https://doi.org/10.3390/s18124245

Zhao, Y., Li, Q., Yi, W., & Xiong, H. (2023). Agricultural IoT Data Storage Optimization and Information Security Method Based on Blockchain. *Agriculture*, 13(2), 274.

Zheng, Z., Xie, S., Dai, H., Chen, X., & Wang, H. (2017). An Overview of Blockchain Technology: Architecture, Consensus, and future Trends. In *2017 IEEE International Congress on Big Data (BigData congress)*, pp. 557–564, IEEE.

Zhong, M., Wang, Y., Yan, J., Cheng, Y., & Sun, P. (2023). Transformer-Based Comparative Multi-View Illegal Transaction Detection. *PLoS ONE*, 18(1), e0276495. https://doi.org/10.1371/journal.pone.0276495

Part VII

Blockchain Application in Disaster Management

12 Blockchain-Driven Resilience
Revolutionizing Flood Risk Management and Community Engagement

Saeid Eslamian, Yaser Sabzevari, and Faezeh Eslamian

12.1 INTRODUCTION

According to the index of the United Nations Sustainable Development Commission, if the total annual water consumption of any country exceeds 40% of the total annual renewable resources, the country will experience a severe water crisis (Igbatayo et al., 2022; Mohammadjani and Yazdanian, 2013). The competition for water between different agricultural, industrial and drinking uses is also increasing, and the production of more food for the growing world population will also cause more water to be harvested (Schultz, 2017). Rivers are considered as the main source of water supply for humans and other creatures, and sometimes this source of life causes destruction and irreparable damages (Eslamian and Maleki, 2021; Gregory, 2006; Nasim and Selajgeh, 2006). Every year, floods cause a lot of damage to the agricultural sector, roads, water structures, bridges and roads, and in some cases, it causes the death of many people and other creatures, which results in the destruction of the social structure and financial losses. Many lives are lost (Eslamian et al., 2021; Nazif et al., 2021). Therefore, it is necessary to conduct studies and research and use the appropriate methods in order to identify the causes, pathology and solutions to deal with floods.

In recent years, severe flooding has caused a lot of damage around the world. Due to the variability of climate and heavy precipitation, flood events occur regularly in several locations around the world (Hirabayashi et al., 2013; Khosravi et al., 2019; Ruidas et al., 2022). Population growth, urbanization, and agricultural development all increase the risk of flooding (Rahman et al., 2021; Zhang et al., 2018). Furthermore, flooding poses a significant risk to human life and can be caused by natural disasters, which can result in significant human casualties and economic destruction (Dottori et al., 2016). According to the Center for Research on the Epidemiology of Natural Disasters (Hoque et al., 2019), flooding has caused more direct damage to the locals than any other natural disaster during the last century. Implementing flood hazard, vulnerability, and risk assessments is essential to reducing the future threats for the aforementioned reasons (Fernandez et al., 2016; Wang et al., 2011).

Flooding is one of the natural phenomena that is likely to occur in different climatic conditions in all regions with different degrees of intensity and duration of rainfall.

One of the phenomena that has always threatened human lives and natural and economic resources is the huge flow of water along with alluvial sedimentary materials of the catchment basin, or in other words, flood with sedimentary materials. According to the definition of irrigation and drainage culture, flood is caused by strong current, accompanied by a relatively high rise of the water level in a river, which usually takes on the water of the surrounding lands.

A mathematical approach to flood definition can also be as follows: flood is the output of a random event in the form of a hydrological cycle, which, during a relatively continuous time interval,

DOI: 10.1201/9781003599593-19

the input function has triggered the operating system. The input function is the amount of precipitation and its temporal and spatial distribution patterns, and the operating system is the geographical area or watershed, which is the factor of converting the input function into the output function. The output or flood function is also the time trend, flow rate or volume changes during the flood. According to the set of definitions: an increase in the flow of a river to the extent that it exceeds its capacity and covers the surrounding lands and causes damage is called a flood. Long rains, heavy rains, rapid melting of snow and unnatural factors such as the failure of dams are the main causes of floods.

The issues related to floods are diverse and have a very complex nature. Flooding causes the destruction of facilities, human and financial losses and disruptions in the use of highways and railways.

What has placed the category of flood in the ranks of damage-causing risks is affected by two issues; first, natural factors, and on top of that, the intensification of extreme events of rainfall and flooding due to climate change; and secondly, human factors due to the destruction of natural resources, land use change, nonrecognition of flood zones, unauthorized encroachment and construction in flood channels. The factor usually considered to most influence aggravating flood damage is the human factor. Therefore, the most important solution in flood control is to change the approach of flood management from crisis management to risk management with coherent planning and directing budget resources from compensation to preventive measures, including watershed management activities, with an emphasis on flood prevention and control. The source is at the level of the upstream watersheds. With this change of attitude and modification of the flood threat approach, it will become a favorable opportunity to solve the problems of drought and water scarcity and, as a result, the stability of the land's ecosystem. Risk management keeps the system flexible, efficient and robust. Therefore, the use of efficient methods in this field helps in flood resilience.

The purpose of flood control is to eliminate or reduce the negative effects of flood disasters. Although it is not always possible to be successful in controlling floods, and it is not possible to solve the problem of flooding definitively, efforts should be made to reduce the harmful and destructive consequences of floods. There are various methods for flood control, and the choice of any method or combination of methods depends on the local conditions of the project, the importance of the areas under project protection and the economic conditions of the country.

The aim of this chapter is introducing the blockchain technology for flood resilience and management.

12.2 FLOOD CONTROL

12.2.1 Nonstructural Methods

Nonconstruction methods include the establishment and implementation of laws and regulations on how to use flood plains and various preventive measures to reduce the flood damage. Surveys and studies show that the reason for the increase in flood damages is not the increase in the amount of floods, but the intensification of the use of floodplain lands, that is, the increase in the use of floodplain lands adjacent to rivers, should be considered the main factor.

In general, nonstructural measures include the following: land zoning, anti-flooding of buildings and facilities, flood insurance, land segmentation and segregation rules, building regulations, and permanent or temporary evacuation of people and property and equipment from the flood crossing and flooded lands.

12.2.2 Flood Control Structural Methods

In structural methods, containment and control of damages caused by floods is done by means of facilities and structures that are built for the same purpose. This method has been used for centuries and refers to a variety of measures that are used to reduce the volume and intensity of floods, to prevent the floods from entering the target area, or to divert and dispose of flood flows. It is divided into the following types:

- Limiting floods by using the nearby gores
- Improvement of the course and sections of the river
- Flood diversion
- Watershed management methods
- Enclosing protected areas or flood retaining walls
- Storage tanks
- Delayed dams

12.3 FLOOD RESILIENCE AND RISK

Hochrainer-Stigler et al. (2020) graded the community resilience against floods in the form of an empirical analysis on a global scale. This study presented and analyzed a standard framework for measuring community resilience to floods. The relevant measurement tool is modeled and adapted based on a so-called technical risk classification approach used in the insurance sector. The method for ranking the indicators is based on a two-step procedure: (1) raw data is gathered, and (2) on the basis of this data, experts rank indicators, also known as resilience resources. Over 118 communities in nine countries, totaling about 1.25 million data points, were used to test this strategy. To support the findings from a qualitative viewpoint, content analysis was combined with the quantitative analysis. This study's key finding is that expert judgment matters much more than the actual raw data gathered when determining resilience scores related to subjective attributes like ability, emotion and trust. The study's conclusions regarding the importance of data and grade characteristics can help to determine how to evaluate the flood resilience standards in a better, more effective, and more robust manner.

Fraser et al. (2020) looked into three urban floods to determine the primary causes of small-scale disasters. Understanding how spatial factors and remote sensing contribute to meeting the local risk conditions remains a difficult challenge in our understanding of the causes of natural disasters. This is particularly true in the case of small-scale, high-impact disasters. This book chapter takes a thorough approach to discussing flood risk analysis, shown by three case studies of the management of coastal flooding in urban areas in the European Union. This chapter reveals the significance of local institutions in reducing the local risks as a result of higher-level political and economic changes. With regard to the institutional changes, local disaster risk reduction and the connection between flooding and austerity, it offers fresh empirical evidence.

12.4 WATER MANAGEMENT STRATEGIES

In general, water management is possible through three public sectors (government), the private sector and private organizations. The third method has now become the mainstream approach to water management because the government management is inefficient and the management of the private sector does not fit with the current structure of water management, which includes a large number of small farmers. On the other hand, participatory water management emphasizes that water users should be involved in all stages and levels of water management (Porzand, 2004). Some issues are irrigation, improper maintenance of water systems, inability of the government to collect enough capital for water users and poor performance of water systems (Munoz, 2007).

Many efforts have been made in different countries to transfer management rights and responsibilities from the government to the private sector or local organizations (Geijer et al., 1996; Vermillion, 1997). Governments' financial resources are considered to increase resilience against risks such as floods, improve the performance of hydroponics and ensure the long-term sustainability of water systems (Geijer et al., 1996; Munoz, 2007; Vermillion, 1997).

Transferring water management leads to achievements such as reducing water shortage problems, providing water during plant growth stages, increasing crop yield and excess water supply, and as a result, optimal use of water resources and increasing income and performance. Therefore, it will reduce the operating, maintenance and personnel costs, increase the spirit of cooperation between farmers and the government, increase the sense of responsibility and ownership of the farmers toward the use of water structures and their protection, and improve distribution of water (Mahboubi Ardakani and Razavi, 2004).

The success and efficiency of consumer management depends on their successful formation based on researching and studying them. The evidence has shown that if the field studies are not properly considered, the establishment, formation, continuation and stabilization of user management in rural and agricultural communities will fail.

The experience of the 1950s and 1960s has shown that although the government's entry into the field of construction of large water supply and distribution buildings promised the rapid growth of agriculture in developing countries, this approach, social issues and problems were imposed on the rural community. One of the negative results of this change of approach has been the imposition of issues such as poor maintenance of the irrigation system, water wastage and lack of motivation to save water. These problems started with neglecting the necessity of farmers' presence in the management and planning of irrigation networks, and the basic solution to solve this problem was to return the people to the field of management.

Today, it is believed that in order to achieve development, countries should use the presence of people and their participation in different stages, and to consolidate participation there should be planning from the bottom up. Transferring management responsibilities from government institutions to the people is now an important policy in many countries. So that the conducted studies show that almost all countries have accepted the participatory approach in water management with different degrees. Today, handing over water management to water users has become a global movement; in the process of its development, the role of governments in the operation and maintenance of structures has been reduced, and instead, local organizations of water users have the authority to manage the operation and maintenance of structures (Shahrodi et al., 2007). Accordingly, community participation is very important and necessary for the successful and sustainable management of water resources (Dungumaro and Madulu, 2003).

Participatory water management is actually handing over social privileges to the community. By using these social rights, people get opportunities and possibilities by taking control of management, as large socio-economic arenas, and engaging in managerial-economic activities such as planning, providing and allocating financial resources, attracting credits and monitoring and evaluation, and even political communication, to increase their skills and capabilities in the fields of planning, organizing, budgeting, and financial and accounting affairs (Zarei Dastgerdi et al., 2008).

12.5 INFORMATION TECHNOLOGY AND HYDROINFORMATICS

The adaptation of intricate man-made and natural systems, such as the hydrological cycle and urban water systems, is essential for modern societies to survive in the 21st century (Beck et al., 2010). Natural energy sources and renewable energy systems, extreme weather events and disasters (Ewing and Demir, 2021), the agricultural supply chain, and resilient infrastructure are all dependent on them (Alabbad et al., 2021). By measuring more natural systems and gathering more data, the integration of natural and artificial systems is made more compatible (Muste et al., 2017) and creating artificial systems using these data (Omatu et al., 2022; Sermet and Demir, 2020) and digital twin environments are both doable. This demand has increased the significance of trustworthy data and data processing technology (Hu and Demir, 2021). Due to this, over the past 20 years, information and communication technology approaches have been rapidly evolving, and digital transformation has accelerated in almost all sectors (Knell, 2021). The perspectives of different societies have developed concurrently on topics like production, planning and implementation (Savić, 2021). With advancements in hydroinformatics, the fields of hydrology and water resources followed a similar pattern (Demir et al., 2022). The most notable advancements are new deep learning (DL) applications in image synthesis and communication (Sermet and Demir, 2021), extensive modeling and analysis in side systems, Pluto (Ewing and Demir, 2021), virtual and augmented reality for hydrological modeling and education (Sermet and Demir, 2022), and new programming libraries and data standards (Nasrolahi et al., 2025). Xiang and Demir (2022) are mentioned. The water sector has started to generate, process and store more data as a natural result of this quick digital transformation (Haltas et al., 2021). However, in order for hydrological predictions to be more accurate, this process necessitates the reliability, security and standardization of the underlying data. Similar to this, data security and standardization are essential for stakeholders in water systems to communicate in a trustworthy and secure way (Voogd et al., 2022).

12.6 BLOCKCHAIN TECHNOLOGY

As the potential of blockchain technology is realized, trust issues are being addressed more quickly.

Blockchain technology is an advanced database mechanism that enables transparent information sharing in a network (Nakamoto, 2008). Data is stored in blocks that make up a chain in a blockchain database. Data is uniform in time because it cannot be altered or deleted on-chain without network consensus. As a result, using blockchain technology to track orders, payments, accounts and other transactions will result in an immutable ledger. Internal controls in the system guard against unauthorized transactions and ensure that everyone is observing authorized transactions in a fair and accurate manner.

One of the cutting-edge technologies that has been the focus of research and development over the past ten years is blockchain, which is an emerging technology that has been embraced by a number of industries. Energy companies utilizing blockchain technology may develop a marketplace for the exchange of electricity, or solar-paneled homeowners may use a similar marketplace to sell any excess electricity to their neighbors. The process is mostly automated because transactions are generated by smart meters and recorded on the blockchain. To manage online payments, accounts and stock market transactions, traditional financial systems like banks and stock exchanges can also use blockchain services. Additionally, blockchain technology can be used by media and entertainment companies to manage copyrighted data. Similar to this, retail businesses can utilize blockchain to monitor the flow of goods between suppliers and customers (Hu et al., 2022; Jing et al., 2021; Li et al., 2022). As water becomes a sensitive topic around the world, blockchain technology has the potential to expand the long-term vision of managing water resources, particularly in transferring management to consumers.

12.6.1 History of Blockchain Technology

Blockchain has the ability to track recorded data transparently, theoretically, in financial applications, as well as the ability to store data in a distributed ledger, choose which data to store using a consensus algorithm, define preconditions with a smart contract, and so forth. Discussions about it date back to the 1990s to solve issues with conventional data storage methods and the necessity of relying on third parties in financial transactions (Szabo, 1998). The first real-time application for storing financial transaction data was Bitcoin. Although the terms "block" and "chain" were always used separately in the original Bitcoin white paper, the name of the technology eventually changed to "blockchain" (Nakamoto, 2008). The Ethereum white paper (Buterin, 2014) announced the second major step in the development of blockchain technology as well as the first application of the use of recording transaction data in a database only if both parties meet specific requirements.

12.6.2 Blockchain Networks

Public blockchain networks, private blockchain networks, hybrid blockchains and consortium blockchain networks are the four main categories of decentralized or distributed blockchain networks. Anyone can join public blockchain networks, which lack access restrictions. Everyone who is a part of the blockchain has equal access to read, edit and verify the blockchain. In order to trade and mine cryptocurrencies like Bitcoin, Ethereum and Litecoin, public blockchain networks are primarily utilized (Wang et al., 2021). A single organization is in charge of private blockchains. Who can join the network and what privileges members have are determined by this authority. Due to access limitations, private blockchains are only partially decentralized. Some characteristics of private and public networks are combined in hybrid blockchains. Companies can create both a shared system and a private, license-based system. As a result, they keep some data public while controlling access to other data that is stored on the blockchain. To enable partners in the system to confirm whether or not private transactions have taken place, they use smart contracts. For instance, hybrid blockchains can offer a shared access to digital currency while maintaining the confidentiality of bank currencies. A collection of businesses manage consortium blockchain networks. Blockchain continuity and data access rights are jointly managed by preselected organizations. In industries where many organizations work toward a common objective and can gain from shared responsibility, consortium blockchain networks are typically preferred (Arooj et al., 2022).

12.6.3 Blockchain Protocols

The term "blockchain protocol" refers to various blockchain platforms that can be used to build applications. To fit particular industries or applications, each blockchain protocol modifies the fundamentals of the blockchain. Hyperledger, Ethereum, Corda and Quorum are the most well-known blockchain protocols. An open-source project with a number of tools and libraries is the Hyperledger protocol. Blockchain applications can be altered in this way. With special identity-management and access-control features, Hyperledger is an open-source, modular framework. These characteristics make it suitable for a range of applications, including supply chain tracking and monitoring, trade financing, incentive schemes and financial asset management (Hang and Kim, 2021). An alternative platform is Ethereum protocol. Because it is centralized, public blockchain applications can be built with it (Estevam et al., 2021). An open-source blockchain project called Corda is geared toward business. Ethereum is the basis for Quorum, an open source blockchain protocol. This application is specifically intended for use in private blockchain networks, where all nodes are owned by only one member, or consortium blockchain networks, where different members each own a portion of the network (Singh et al., 2022).

12.6.4 Blockchain Architecture

As a result, the following are the key characteristics of blockchain technology that distinguish it. Decentralization in blockchain refers to the transfer of control and decision-making authority from a central legal entity (for example, an individual, organization or group) to a distributed network. Transparency is a key tenet of decentralized blockchain networks that helps the participants build less trust with one another. These networks forbid users from exerting power or control over one another because doing so degrades network efficiency. Immutability is another crucial component. As a result, once the data is stored in the blockchain, it cannot be altered. If a report has an error, the error must be fixed by adding a new transaction, and both transactions must be made public on the network. The blockchain system's consensus mechanism, which establishes guidelines for participants to agree on recording transactions, is a prominent feature. Only after the majority of network users have approved them are new transactions allowed.

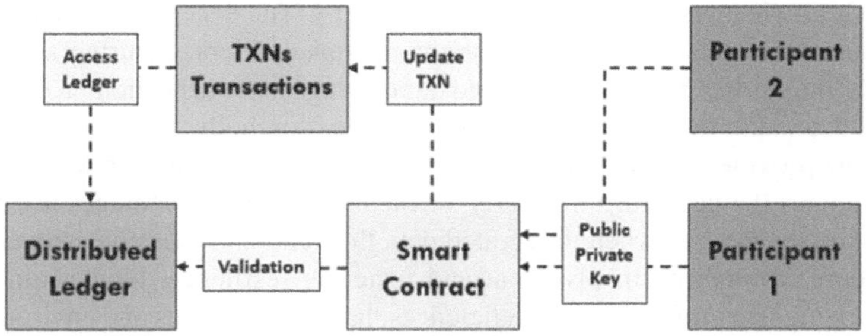

FIGURE 12.1 An image of the basic architecture of the blockchain.

On the other hand, blockchain architecture consists of the following technical elements: (a) distributed ledger, a shared database that is used to log transactions within the blockchain network; (b) smart contracts, computer programs that run instantly and are saved in the blockchain system when certain criteria are met. To ensure that transactions proceed securely, they perform "if-then" checks; and (c) a security feature known as public key encryption enables participants in the blockchain network to be identified only by their public key. For network users, two sets of keys are generated by this mechanism. A public key is one of the keys that the entire network uses at once. One private key belonging to each member is the other key. Data in the ledger is unlocked using a combination of private and public keys (Figure 12.1). Digital currencies like Bitcoin are the first generation of blockchain. Blockchain, also known as chain code, is the second generation of smart contracts. Additionally, as businesses identify and implement new application domains, the third or future generation of blockchain will continue to develop and grow.

12.7 BLOCKCHAIN APPLICATION IN FLOOD HYDROLOGY

The advantages of storing data with the highest level of reliability are the main focus of the initial ideas for blockchain implementation in hydrology and water resources management, as is the case in many other industries. Water management in all its forms depends heavily on the accuracy of data related to water. Since water is a multifaceted natural resource, it can best be managed by maximizing the interests of stakeholders with various priorities. Water has a social component in addition to different priorities for quantity and quality. Building trust requires accurate information about shared water resources (Pahl-Wostl, 2015). A straightforward distributed digital architecture based on blockchain and smart contracts makes it possible to directly guarantee the social trust in the hydrological community and the reliability of the data.

The use of unchangeable recorded data can help reduce uncertainty in hydrological systems for various stakeholders. Blockchain technology is also ideal for applications related to water finance. Water markets, where water rights are traded, or any investment in water management, financial support to improve water infrastructure and flood control structures, or enhancing sanitation in developing countries are possible application areas for this technology. From the viewpoint of investors or market participants, this technology enables transparent tracking of all financial transactions (Eslamian et al., 2025). The blockchain may also provide useful solutions when it comes to tracking water. Water-related traceability will be crucial for agricultural water management (Yildirim and Demir, 2022), water quality and control (Demir et al., 2009), urban water systems and the water footprint of a particular product.

12.8 CONCLUSIONS

Despite the challenges in standardizing blockchain applications within the diverse field of hydrology, numerous studies highlight four significant benefits: enhanced water trust, robust data security, immutable transaction ledgers and improved accountability. The decentralized nature of blockchain, facilitated by smart contracts, establishes trust among stakeholders, ensuring adherence to regulations even among unfamiliar parties. Distributed data storage and encryption protect water data from cyber threats and unauthorized access, offering security for both individuals and organizations. Immutable ledgers provide a unified, transparent record of transactions, while verifiable data on the blockchain eliminates the need for third-party verification. These advantages are particularly relevant in flood risk management, where integrated data flow and collaborative networks are essential. By enabling secure community involvement and fostering resilience, blockchain can address the vulnerabilities arising from infrastructure deficiences, land-use changes and environmental mismanagement. To fully realize this potential, future research should focus on developing adaptable blockchain frameworks for varied hydrological applications.

REFERENCES

Alabbad, Y., Mount, J., Campbell, A.M. and Demir, I. (2021). Assessment of transportation system disruption and accessibility to critical amenities during flooding: Lowa case study. *Science of the Total Environment*, 793, 148476.

Arooj, A., Farooq, M.S. and Umer, T. (2022). Unfolding the blockchain era: Timeline, evolution, types and real-world applications. *Journal of Network and Computer Applications*, 103511, Accessed November 2022, doi: 10.1016/j.jnca.2022.103511

Beck, M.B., Jiang, F., Shi, F., Walker, R.V., Osidele, O.O., Lin, Z., Demir, I. and Hall, J.W. (2010). March. Re-engineering cities as forces for good in the environment. In *Proceedings of the Institution of Civil Engineers-Engineering Sustainability*, Vol. 163, No. 1, pp. 31–46, Thomas Telford Ltd.

Buterin, V. (2014). Ethereum: A next-generation smart contract and decentralized application platform, ethereum.org, Accessed 21 November 2022, Available from: https://cryptorating.eu/whitepapers/Ethereum/Ethereum_white_paper.pdf

Demir, I., Jiang, F., Walker, R.V., Parker, A.K. and Beck, M.B. (2009). Information systems and social legitimacy scientific visualization of water quality. In *2009 IEEE International Conference on Systems, Man and Cybernetics*, pp. 1067–1072.

Demir, I., Sermet, Y. and Rink, K. (2022). Next generation visualization and communication systems for earth science using immersive reality and serious gaming. *Frontiers in Earth Science*, 10, 1101538.

Dottori, F., Salamon, P., Bianchi, A., Alfieri, L., Hirpa, F.A. and Feyen, L. (2016). Development and evaluation of a framework for global flood hazard mapping. *Advanced Water Resources*, 94, 87–102.

Dungumaro, E.W. and Madulu, N.F. (2003). Public participation in integrated water resources management: The case of Tanzania. *Physics and Chemistry of the Earth, Parts A/B/C*, 28, 1009–1014.

Eslamian, S., Fallah, A. E. & Sabzevari, Y. (2025). Optimizing Rain Gauges with the cuckoo Algorithm and Entropy of Information Transfer: a case Study on the Gavkhouni Basin in Iran. *Environmental Earth Sciences*, *84*(436), https://doi.org/10.1007/s12665-025-12433-8

Eslamian, S. and Maleki, M. (2021). Disaster resilience and computational methods for Urban infrastructures. In *Handbook of Disaster Risk Reduction for Resilience*, by Eslamian, S. and Eslamian, F., 279–298, Springer, Cham, Switzerland.

Eslamian, S., Parvizi, S. and Behnassi, M. (2021). New Frameworks for Building Resilience in Hazard Management, Chapter 5. In *Handbook of Disaster Risk Reduction for Resilience: New Frameworks for Building Resilience to Disasters*, by Eslamian, S. and Eslamian, F., 107–130, Springer, Cham, Switzerland.

Estevam, G., Palma, L.M., Silva, L.R., Martina, J.E. and Vigil, M. (2021). Accurate and decentralized timestamping using smart contracts on the Ethereum blockchain. *Information Processing & Management*, 58(3), 102471, Accessed 21 November 2022, Available from: doi: 10.1016/j.ipm.2020.102471

Ewing, G. and Demir, I. (2021). An ethical decision-making framework with serious gaming: A smart water case study on flooding. *Journal of Hydroinformatics*, 23(3), 466–482.

Fernandez, P., Mourato, S. and Moreira, M. (2016). Social vulnerability assessment of flood risk using GIS-based multicriteria decision analysis: A case study of Vila Nova de Gaia (Portugal). *Geomatical Natural Hazzard Risk*, 7(4), 1367–1389.

Fraser, A., Pelling, M., Scolobig, A. and Mavrogenis, S. (2020). Relating root causes to local risk conditions: A comparative study of the institutional pathways to small-scale disasters in three urban flood contexts. *Global Environmental Change*, 63, 102102.

Geijer, J.C., Svendsen, M. and Vermillion, D.L. (1996). *Transferring Irrigation Management Responsibility in Asia: Results of a Workshop*. International Irrigation Management Institute, Colombo, Sri Lanka.

Gregory, K.J. (2006). The human role in changing river channels. *Geomorphology*, 76, 172–191.

Haltas, I., Yildirim, E., Oztas, F. and Demir, I. (2021). A comprehensive flood event specification and inventory: 1930–2020 Turkey case study. *International Journal of Disaster Risk Reduction*, 56, 102086.

Hang, L. and Kim, D.-H. (2021). Optimal blockchain network construction methodology based on analysis of configurable components for enhancing Hyperledger Fabric performance. *Blockchain: Research and Applications*, 2(1), 100009, Accessed 21 November 2022, doi: 10.1016/j.bcra.2021.100009

Hirabayashi, Y., Mahendran, R., Koirala, S., Konoshima, L., Yamazaki, D., Watanabe, S., Kim, H. and Kanae, S. (2013). Global flood risk under climate change. *Nature Climate Change*, 3(9), 816–821.

Hochrainer-Stigler, S., Laurien, F., Velev, S., Keating, A. and Mechler, R. (2020). Standardized disaster and climate resilience grading: A global scale empirical analysis of community flood resilience. *Journal of Environmental Management*, 276, 111332.

Hoque, M., Tasfia, S., Ahmed, N. and Pradhan, B. (2019). Assessing spatial flood vulnerability at Kalapara Upazila in Bangladesh using an analytic hierarchy process. *Sensors*, 19(6), 1302.

Hu, A. and Demir, I. (2021). Real-time flood mapping on client-side web systems using hand model. *Journal of Hydrology*, 8(2), 65.

Hu, S., Huang, S. and Qin, X. (2022). Exploring blockchain-supported authentication based on online and offline business in organic agricultural supply chain. *Computers & Industrial Engineering*, 108738, Accessed 21 November 2022, doi: 10.1016/j.cie.2022.108738

Igbatayo, S.A., Eslamian, S., Babalola, O.O. and Makanju, A.A. (2022). Strengthening climate resilience and disaster risk reduction: Case study of the sahel adaptive social protection, Chapter 10. In *Disaster Risk Reduction for Resilience: Disaster and Social Aspects*, by Eslamian, S, and Eslamian, F., 221–243, Springer, Cham, Switzerland.

Jing, N., Liu, Q. and Sugumaran, V. (2021). A blockchain-based code copyright management system. *Information Processing & Management*, 58(3), 102518, doi: 10.1016/j.ipm.2021.102518

Khosravi, K., Shahabi, H., Pham, B.T., Adamowski, J., Shirzadi, A., Pradhan, B., Dou, J., Ly, H.-B., Grof, G., Ho, H.L., et al. (2019). A comparative assessment of flood susceptibility modeling using multi-criteria decision making analysis and machine learning methods. *Journal of Hydrology*, 573, 311–323.

Knell, M. (2021). The digital revolution and digitalized network society. *Review of Evolutionary Political Economy*, 2(1), 9–25, doi: 10.1007/s43253-021-00037-4

Li, Z., Mount, J. and Demir, I. (2022). Accounting for uncertainty in real-time flood inundation mapping using HAND model: Iowa case study. *Natural Hazards*, 112(1), 977–1004.

Mahboubi Ardakani, A.R. and Razavi, A. (2004). The role of participatory management in the operation and maintenance of irrigation networks and the review of Turkey's experience in handing over operation and maintenance management to users. Gohran Kavir, *The First Conference on Problems of Irrigation, Drainage and Optimal Use of Agricultural Water*, June, Tehran, Iran.

Mohammadjani, A. and Yazdanian, N. (2013). Analysis of the water crisis situation in the country and its management requirements. *Trend Quarterly*, 21(65–66), 117–144. (In Farsi)

Munoz, G. (2007). Irrigation management transfer: Worldwide efforts and results. *The 4th Regional Conference and 10th International Seminar on Participatory Irrigation Management*, 2–5 May, Tehran, Iran.

Muste, M., Lyn, D.A., Admiraal, D., Ettema, R., Nikora, V. and García, M.H. (Eds.). (2017). *Experimental Hydraulics: Methods, Instrumentation, Data Processing and Management: Volume I: Fundamentals and Methods*. CRC Press, USA.

Nakamoto, S. (2008). Bitcoin: A peer-to-peer electronic cash system, November 2022, https://bitcoin.org/bitcoin.pdf

Nasim, A. and Selajgeh, A. (2006). Determination of river border and bed using AUTOCAD, HEC-RAS software, case study: A section of the Karaj River (Sira-Pul Khob). *7th International Conference on River Engineering*, Shahid Chamran University, Ahvaz, Iran. (In Farsi)

Nasrolahi, A., Nazeri Tahroudi, M., & Sabzevari, Y. (2025). Investigating the probabilistic behavior of reference evapotranspiration using Vine tree sequence. *Applied Water Science*, *15*(100), https://doi.org/10.1007/s13201-025-02429-x

Nazif, S., Mohammadpour Khoie, M. M. and Eslamian, S. (2021). Urban Disaster Management and Resilience. Chapter 7. In *Handbook of Disaster Risk Reduction for Resilience: New Frameworks for Building Resilience to Disasters*, by Eslamian, S. and Eslmian, F., 157–185, Springer, Cham, Switzerland.

Omatu, S., Malluhi, Q.M., Gonzalez, S.R., Bocewicz, G., Bucciarelli, E., Giulioni, G. and Iqba, F., eds. (2022) Distributed computing and artificial intelligence, *12th International Conference[online]*, Cham Germany Springer, International Publishing, 18 December 2022, doi: 10.1007/978-3-319-19638-1

Pahl-Wostl, C. (2015). *Water Governance in the Face of Global Change*, Cham: Springer International Publishing, Accessed 21 November 2022. Available from: doi: 10.1007/978-3-319-21855-7

Porzand, A. (2004). Recent measures to make the management of exploitation and maintenance of irrigation and drainage networks private and competitive. *The First Conference on Investigating the Problems of Irrigation and Drainage Networks and Optimal Use of Agricultural Water*. Tehran, Iran, June. (in Persian).

Rahman, M., Chen, N., Elbeltagi, A., Islam, M.M., Alam, M., Pourghasemi, H.R., Tao, W., Zhang, J., Shufeng, T., Faiz, H., et al. (2021). Application of stacking hybrid machine learning algorithms in delineating multitype flooding in Bangladesh. *Journal Environmental Management*, 295, 113086.

Ruidas, D., Chakrabortty, R., Islam, A.R.M.T., Saha, A. and Pal, S.C. (2022). A novel hybrid of meta-optimization approach for flash flood-susceptibility assessment in a monsoon-dominated watershed: Eastern India. *Environmental Earth Science*, 81(5), 1–22.

Savić, D. (2021). Digital water developments and lessons learned from automation in the car and aircraft industries. *Engineering*, Accessed 21 November 2022, doi: 10.1016/j.eng.2021.05.013

Schultz, B. (2017). Agricultural Water Management and Food Security in a Sustainable Environment. *13th International Drainage, Workshop of ICID*, Ahwaz, Iran.

Sermet, Y. and Demir, I. (2020). Virtual and augmented reality applications for environmental science education and training. In Linda Daniela (ed.) *New Perspectives on Virtual and Augmented Reality*, pp. 261–275, Routledge.

Sermet, Y. and Demir, I. (2021). A semantic web framework for automated smart assistants: A case study for public health. *Big Data and Cognitive Computing*, 5(4), 57.

Sermet, Y. and Demir, I. (2022). GeospatialVR: A web-based virtual reality framework for collaborative environmental simulations. *Computers & Geosciences*, 159, 105010.

Shahrodi, A.A., Chizari, M. and Pezeshki Rad, G. (2007). Water user's cooperative effect on farmers' attitude towards the agricultural water management: A case study on Khorasan. *Journal of Agricultural Economic and Development*, 22(2), 71–85. (In Persian).

Singh, A., Kumar, G., Saha, R., Conti, M., Alazab, M. and Thomas, R. (2022). A survey and taxonomy of consensus protocols for blockchains. *Journal of Systems Architecture*, 102503, Accessed 21 November 2022, doi: 10.1016/j.sysarc.2022.102503

Szabo, N. (1998). There is no universal security architecture [online]. *Phonetic Sciences*, Amsterdam, Accessed 21 November 2022. Available from: https://www.fon.hum.uva.nl/rob/Courses/InformationInSpeech/CDROM/Literature/LOTwinterschool2006/szabo.best.vwh.net/universal.html

Vermillion, D.L. (1997). *Impacts of Irrigation Management Transfer: A Review of the Evidence* Vol. 11, International Irrigation Management Institute, Colombo, Sri Lanka.

Voogd, R., Rudberg, P.M., de Vries, J.R., Beunen, R., Espiritu, A.A., Methner, N., Larsen, R.K., Fedreheim, G.E., Goes, S. and Kruger, E. (2022). A systematic review on the role of trust in the water governance literature. *Water Research*, 100147, Accessed 21 November 2022, doi: 10.1016/j.wroa.2022.100147

Wang, X., Ni, W., Zha, X., Yu, G., Liu, R.P., Georgalas, N. and Reeves, A. (2021). Capacity analysis of public blockchain. *Computer Communications*, 177, 112–124, Accessed 21 November 2022, doi: 10.1016/j.comcom.2021.06.019.

Wang, Y., Li, Z., Tang, Z. and Zeng, G. (2011). A GIS-based spatial multi-criteria approach for flood risk assessment in the Dongting Lake Region, Hunan, Central China. *Water Resources Management*, 25(13), 3465–3484.

Xiang, Z. and Demir, I. (2022). Flood Markup Language: A standards-based exchange language for flood risk communication. *Environmental Modelling & Software*, 152, 105397.

Yildirim, E. and Demir, I. (2022). Agricultural flood vulnerability assessment and risk quantification in Iowa. *Science of the Total Environment*, 826, 154165.

Zarei Dastgerdi, Z., Iravani, H., Shabanali Fami, H. and Mokhtari Hesari, A. (2008). Analysis of the participation of farmers in the management of water supply network of Jarghouyeh city. *Journal of Rural Development*, 12(3), 72–73. (In Persian).

Zhang, Q., Gu, X., Singh, V.P., Shi, P. and Sun, P. (2018). More frequent flooding? Changes in flood frequency in the Pearl River basin, China, since 1951 and over the past 1000 years. *Hydrology Earth System Science*, 22(5), 2637–2653.

13 Blockchain and Climate Action

Pioneering Solutions for Emission Reduction and Sustainability

Esmaeil Sajadi and Saeid Eslamian

13.1 INTRODUCTION

Besides its vast contribution to cryptocurrency, blockchain has great undetected potential in many aspects of science. The nature of blockchain is a digital ledger in a distributed storehouse that can not be tampered and it usually doesn't have a central authority. The community involved can document their transactions which, at the initial level, can't be changed after publication (Yaga et al. 2019) define it as:

> Blockchains are distributed digital ledgers of cryptographically signed transactions that are grouped into blocks. Each block is cryptographically linked to the previous one (making it tamper-evident) after validation and undergoing a consensus decision. As new blocks are added, older blocks become more difficult to modify (creating tamper resistance). New blocks are replicated across copies of the ledger within the network, and any conflicts are resolved automatically using established rules.

It may be difficult to change it after the implementation and widespread adoption of a blockchain. Whenever data is stored in a blockchain, it's always there even if mistakes are made. This particular characteristic would allow us to use it in activities that have many distributed and untrusted parties and need trustworthy untampered data. With this technology abstraction, changes to working data can be made and a history of changes can be established. These characteristics are desirable to some organizations. For others, this may be a barrier to the use of blockchain technology.

Hydrological topics, due to their nature, have uncertainties and errors that need to be taken care of to have a better understanding of the environment and the potentials and dangers caused by it. Climate change is a controversial issue that has a critical role in the future state of the planet Earth, therefore, applying every possible method to gain a better understanding of this matter is only logical. It started when a difference in temperature and precipitation was detected and the major trigger was the discovery of the ozone hole in 1987. Although the main idea of climate change was formed by Svante Arrhenius, a Swedish chemist, in 1896 it was not a serious political matter until the 1990s (Luterbacher and Sprinz 2001). Countries, since then, have been trying to project and find solutions to combat the negativity that has and still is yet to affect the environment and its living things.

In this chapter, an attempt is made to review the studies that have been conducted on the subject of the application of blockchain technology in climate change. Starting with the risks of climate change, a connection with blockchain technology will be made. Then by interpreting global warming and the role of cryptocurrencies, the opportunities to battle the effects or reduce them by different methods of adaptation and mitigation are shown and also represent some activities that can be done to bring us closer to the directions of the Paris Agreement goals that was written in 2015 by the UNFCCC (United Nations Framework Convention on Climate Change) (UNFCCC 2016). At last,

DOI: 10.1201/9781003599593-20

after showing some arguments between adverse and agreeable points of view, a frame of mind for future research is demonstrated that can be done to help achieve effective processes and results.

13.2 CLIMATE CHANGE AND ITS RISKS

The use of more and more fossil fuels will make the Earth a giant greenhouse, increasing temperature and affecting natural processes such as El Niño, La Niña, and the Pacific Decadal Oscillation. Climate change is the persevering evolution of Earth's average climate patterns, which outlines its local, zonal, and international climates. Weather and climate may be interchangeably applied, but according to the National Oceanic and Atmospheric Administration, weather is defined as atmospheric conditions at a particular point or time. The climate is the average weather conditions in a given area or region at a particular time of the year, measured over about 30 years. Also, climate change projection is conducted decadal and long term, which is different from weather forecasts (less than a year) and predictions (one year to a decade) and has less accuracy, which is why climate change is reported in several scenarios (Eslamian and Eslamian 2023).

Since 1990, the Intergovernmental Panel on Climate Change has studied, presented reports, and drawn up policies in this area. By the time this chapter was written, the last assessment report was the sixth, released in 2023. Based on this report global surface temperature was 0.95°C–1.2°C higher in 2011–2020 than in 1850–1900 of which it is likely that 1.07°C is human-caused. This change can lead to damage to agriculture production, infrastructure, and key economic sectors, besides inland and coastal flooding, and also physical and mental health problems (Calvin et al. 2023). Adaptation and mitigation, which are discussed in Section 13.5, are the two main policy options to deal with climate change.

A comprehensive review by (O'Neill et al. 2017) was presented and categorized the reasons for concerns (RFC) based on the IPCC Fifth Assessment Report (AR5), including

- *Risks to unique and threatened systems (RFC1)*:
  Human systems, alongside ecological ones with geographic boundaries due to climate conditions that have many characteristics such as high endemic wildlife, are in this category. Exceptional human-driven threats are also present in many of these systems. Examples of these are tropical glaciers, coral reefs, mangrove ecosystems, biodiversity hotspots, and specialized Native communities.
- *Risks associated with extreme weather events (RFC2)*:
  Risks from climatic extreme events like heat waves, heavy precipitation, drought, and related wildfires are covered by RFC2 for human health, livelihood, assets, and ecosystems.
- *Risks associated with the distribution of impacts (RFC3)*:
  Climate change effects that have an unequal impact on certain groups. Climate change impacts that unequally affect certain groups due to the dissimilar distribution of physical climate change hazards, exposures, and vulnerabilities are reflected in this category of risks. Unevenness shall be defined as the difference between geographical location, salary, age, sex, or other physical and socioeconomic qualities.
- *Risks associated with global aggregate impacts (RFC4)*:
  These categories of risk reflect impacts on social systems that can be cumulated across the world by one metric, for example, lives affected, money lost, diversity of species in extinction or degradation risk, and loss of many ecosystems at a global level. Due to their different responses, the degradation of the ecosystem is probably caused by the broad transformation of biomes, the large-scale damage to species brought on through climatic range loss, and the disturbance of ecosystem functioning.
- *Risks associated with large-scale singular events (RFC5)*:
  These kinds of incidents, sometimes referred to as tipping points or critical thresholds, are relatively large, abrupt, and sometimes irreversible changes in the physical, conical, and social systems due to smooth variations in driving forces accompanied by natural

variability. In assessing this risk, AR5 focused on two types of events: sea level rise on a large and rapid scale led by the separation and shattering of ice sheets of Greenland and the West Antarctic, as well as coral reef collapses and Arctic systems that caused noticeable changes in ecosystems.

Understanding the scale of climate change and its risks and effects, using all technical capacity to prevent or at least minimize future catastrophes is needed.

13.3 HOW CAN THESE TWO CONNECT?

According to the research findings of Gatteschi et al. 2020, which took four main principles into account: decentralization, transparency, immutability, and authentication in particular, a wide variety of application areas for blockchain technology has been found based on popularity; the question remaining is whether blockchain can meet the expectations.

Greenhouse gases such as carbon dioxide, nitrous oxide, and methane, which come from factories and livestock farming as much as deforestation, can be helped by blockchain. This shows that this technology is not just about cryptocurrencies (Thalhammer et al. 2022).

According to the work of Arshad et al. (2023), the use of blockchain seems to have a significant and positive impact on achieving the United Nations Sustainable Development Goals (SDGs). In the last few years, governments and organizations all over the world have been focusing on adopting a variety of technological solutions to solve questions related to gas emissions, climate change, and biodiversity loss. One tool that could be very helpful in dealing with climate disasters is the use of cryptocurrencies. This technology is crucial to energy and the environment, according to experts. However, little thought has been given to how blockchains can be used in developing countries until now.

A new report from the Social Alpha Foundation (SAF) and the United Nations Environment Programme (UNEP) tried to change the situation and make new outlooks. Finding solutions to environmental challenges and improving sustainability in various ways will undoubtedly be facilitated by the use of blockchain technology. However, promoting the adoption of emerging technologies such as blockchain requires further development of digital infrastructure. Blockchain technology may be able to provide for automatic updating of the features of bonds with encryption security, which will allow them to contain important Key Performance Indicators (KPI) data. Internet of Things (IoT) tools can help monitor renewable energy production. This information could subsequently be transmitted by the Green Bonds used for financing it. In cities and surrounding areas, the use of IoT sensors may be employed for monitoring air quality. The municipality may be fined or given a reward for failing to meet the promises of improvement if it does not change its interest rates on public bonds. By connecting IoT devices to a blockchain network, online data can be collected. As the development of blockchain and other digital technologies accelerates rapidly, policymakers need to adopt legislation that can help design a future energy system.

The study showed that none of the solutions offered by researchers was adequate to deal with the issues affecting the carbon market, despite using several different techniques and technologies. Many studies have shown that blockchain technology is an important solution for obstacles such as crimes, inability, and ambiguity in the carbon emission plan, among others. Blockchain's values such as longevity, safety, clarity, and privacy-preserving properties formed it into a potential choice to improve the present carbon trading system, although, earlier frameworks have shown insufficiency and ineffectiveness. They're not taking full advantage of blockchain technology's potential to solve all the world's problems.

To cover the additional emission permits, a trading system has also been established. The trading scheme could not fulfill the purpose for which it was set up due to several difficulties. Many studies have shown that the use of decentralization technology is an important tool for environmental sustainability. In previous studies, some ways of using blockchain technology to achieve the SDGs have been identified. For example, the transparency and traceability of the supply chain can contribute to promoting sustainable production methods and conformity practices and reducing the risk of environmental damage and civil rights encroachment. To detect and manage problems such as illegal logging, forced labor, or pollution more effectively, materials can be traced safely and irretrievably through the use of blockchain technology.

Blockchain technology is useable for the development of decentralized energy systems as a way to sell and share renewable resources. If power systems based on fossil fuels become less priority to produce energy and communities move toward renewable energy production methods, blockchain can support the accessibility of energy resources and sustainability. Transactional data on climate finance alongside carbon trading can be safely and securely monitored through the use of blockchain technology. By creating a tamperproof climate finance and carbon market ledger, by streamlining the monitoring of progress toward greenhouse gas reduction targets, blockchain can build some sort of trust and reliability within climate finance and carbon trade. The development of secure, decentralized identity systems through the use of blockchain technology can meet certain SDGs. By providing people with a safe and verifiable identity, blockchain can help to reduce the problems related to financial incorporation such as lack of authorized identification or easier access to finance services and business opportunities.

Dorfleitner et al. (2021) showed the green blockchain dedication to the current worldwide state. They provided a comprehensive brief of current adaptation and mitigation measures with a blockchain technology base, taking into account the increasing interest in global reactions to climate change. They therefore present a wide range of activities and indicate possible further development. Through a variety of support opportunities, stockholders, officeholders, and society will be able to advance through knowledge about these applications and their potential to take action on climate change.

However, the experiences gained and knowledge acquired through these projects suggest that they are useful for larger-scale implementation in the future. Therefore, it is not necessary to underestimate the potential of existing applications for further climate protection measures. These results are the basis for a further part of their study, which analyses the factors contributing to application performance. To promote the development of cryptocurrencies in digital sustainability activities, it is important to identify the elements of success. Although the activity sector does not predict the success of the application, there is confirmation that the nature of the activity has a significant influence on the likelihood that the application will be operational.

It is less likely that applications in the field of energy trading, for example power network management, electricity market, and participation in peer-to-peer (P2P) transactions will advance to an operational stage. Studies show that, for a breakthrough in renewable energy development, the use of blockchain technology could be an effective alternative. In addition, developing countries have a huge influence on studies of blockchain-based energy systems, although developed countries, due to their mastery in the subject, still have full control of the basic study of energy blockchain technology. By promoting the development of renewable energy sources and contributing to the development of a more sustainable future, the harmful effects of fossil fuel combustion on the environment will be reduced using blockchain technology in the energy sector. The likely answer to these problems lies in the use of blockchain technology. A wide range of forefront applications can be seen in the concurrence of environmentally friendly blockchain applications. Many are still in development, and just a handful of real applications have been applied. For future widespread deployment, the proficiency and background gained through these efforts will be helpful. The capability of the current applications for actions to address climate change in the future should thus not be underestimated. Given the growing urgency of climate solutions, there will be a significant increase over the coming years concerning green finance and climate technology innovation. Although these are not the only means available to encounter climate change, blockchain technology and green finance will assuredly play an important role in the fight against global warming as it is more efficient in its use.

13.4 CRYPTOCURRENCY EMISSION AND GLOBAL WARMING

Cryptocurrencies have become an innovative way to carry out online financial transactions in recent years, but experts have raised concerns about their environmental impacts.

In an overview conducted by Amaka (2020) it is shown that at 15°C thermal radiation is trapped in the atmosphere, caused by greenhouse gases like water vapor, carbon dioxide, and methane; this helped create an energy balance. It is commonly referred to as a natural or beneficial greenhouse effect, which increases Earth's surface temperature by 21°C. The analysts inspected the control viability of application-specific integrated circuits (ASICs). The ASICs are customized chips to mine Bitcoin with consideration of fuels used to generate the power needed and calculation of the location where Bitcoin is probably computed and the equal CO_2 emissions.

The full effects of global warming have yet to be discovered and analyzed. On the other hand, blockchain, which is a technology that, based on data they receive, is used to power the operation of cryptocurrencies such as Bitcoin, Ethereum, and altcoins. It was expected that this activity generated 69 million tonnes of CO_2 in 2017, based on an analysis carried out by a lab group at the University of Hawaii Manoa, which measured the energy efficiency of computers used to mine bitcoins. They estimated that at the worst-case scenario in 16 years and with higher hopes within 22 years, Bitcoin would be able to raise global temperature by 2°C if it were accepted at the same rates as other advanced technology. The greenhouse gas emissions must be sustainably reduced to take control of global warming and other climate change disasters. Making the production and usage of renewable energy more efficient can help achieve this. Cryptocurrencies are a tremendous demand on the energy system, which leads to an increase in carbon dioxide emissions that is associated with the overall effects of climate change.

Bitcoin is known as a cryptocurrency that consumes so much power that it is rapidly used as a system for payment and investment. Mora et al. (2018) showed that if Bitcoin remains where it is now socioeconomic-wise, the CO_2 created from the process can warm up the earth more than 2°C in less than 30 years.

The early hypothesis that greenhouse gas emissions produced from Bitcoin activities can amount to ~231.4 GtC within 16 years and push global warming above 2°C was criticized by Dittmar and Praktiknjo (2019). To have a better understanding of the numbers: almost 63 years of generating power and emission is needed (based on 2017 observation) to create around 231.4 GtC of the carbon budget. Mora et al.'s (2018) calculation seems farfetched because it requires generating electricity three times more than now in just five years. In the last five years of their study, the capacity of electricity generation had increased by only about 17%; therefore the weak infrastructures don't let us reach that level.

The important issue, however, is to make sure that digital technologies consume less energy than they save. It might seem like low percentages but according to the European Commission, digital technologies account for between 8% and 10% of our energy consumption and 2%–4% of our greenhouse gas emissions. In this area, some changes can be made in daily digital life in order to mitigate the effect on the environment. By 2030, for example, up to 2.1 million tons of CO_2 can be saved yearly, equivalent to the removal of 1 million cars from our roads, by extending the lifetime of all mobile phones to just one year. Up to 90% reduction in energy consumption can be achieved by switching from 4G to 5G networks. For this purpose, The Cambridge Bitcoin Electricity Consumption Index (CBECI)[1] has been developed and provides up-to-date estimates of Bitcoin's daily power demand and an annualized electricity consumption estimate.

Kohli et al. (2023) divided the solutions to lower the CO_2 emissions into:

- *Alternative consensus mechanisms*: Until now Proof of Work (PoW) is the largest consensus mechanism. The possibility of exploring other consensus mechanisms that are more productive than PoW should therefore be considered as a feasible option.

- *Redundancy reduction techniques*: Among the methods to reduce storage redundancy in blockchain networks, sharding is the most important one. Transactions inside the shards are updated after creating them with a consensus mechanism.
- *Choice of mining device*: While efficient devices will help to reduce energy costs regardless of the consensus mechanism, high-efficiency equipment such as ASICs must be used if PoW continues to be employed.
- *Renewable Sources of Energy*: The consequences of their emissions and decay in the environmental footprints must be addressed, given that existing PoW blockchains consume a significant amount of electricity. The use of renewable sources of energy has been shown to reduce CO_2 emissions.

13.5 ADAPTATION AND MITIGATION

By reviewing many articles, Fawzy et al. (2020) defined three approaches for climate change mitigation. First are decarbonization technologies that lower CO_2 emissions. These technologies fall into changing fuel and using renewable energy like nuclear as much as efficiency allows, and carbon capture storage and use which are mostly traditional methods of mitigation. Second, some newer methods have been recently proposed, which are also referred to as "negative emissions technologies" (NET) and called carbon dioxide removal methods, which can be used for capturing and absorbing CO_2 from the atmosphere. These NETs can be categorized as storing and capturing carbon bioenergy and air carbon directly, biochar, weather improvement, fertilizing and enhancing alkalinity of the oceans, arboriculture and rebuilding forests, wetland development and rebuilding, mineral carbonation, and construction with biomass. Finally, it forms the basis of altering the balance of radiation through solar and Earth's radiation management. These approaches are called radiative forcing geoengineering and focus mainly on stabilizing or reducing temperature. This is achieved without affecting atmospheric concentrations of greenhouse gases, unlike negative emission technologies. Aerosol spraying, marine sky brightening, cirrus cloud thinning, space-based mirrors, surface brightening, and other radiation management techniques are among the main methods of forcing geoengineering. All of these methods are theoretical, or at best in testing stages, and have a high degree of risks and uncertainty associated with mass deployment. Current policy frameworks do not include radiative forcing geoengineering techniques.

Most common economic theories suggest that the most cost-effective way of dealing with climate change mitigation is through a global policy approach focusing on one objective. Recent data show that administrative and policy obstacles may make it more difficult to develop and implement economic policies across the board, in comparison with policies focusing on a specific sector.

Since they are more appropriate to addressing obstacles or market breakdown specific to certain parts and can be included in the package of completing policy measures, the IPCC recommends that specific sector policies be implemented. Due to the fact that almost 85% of greenhouse gas emissions are produced by electric power and heat processes, agriculture and land use, industry and corporations, and transport and transit, Bogart (2021) focused mainly on these areas.

Burning fossil fuels to get heat and power in almost all sectors causes greenhouse gas emissions, and it is included in the first category. Blockchain technologies can be a potential tool to decarbonize energy generation, by simplifying access and offering changed actions that result in renewable energy on the electrical grid. Two main ways of reducing final energy consumption can be achieved by the use of blockchain technologies: providing a verification mechanism for each product, to enable consumers to choose products based on their carbon footprint and evaluating and analyzing electricity usage according to emission allowances from one household. By expediting and promoting the usage of microgrid electrical grids operating in a small and regional area, blockchain technology can improve the efficiency of energy conversion, transmission, and distribution systems. Microgrids are meant to minimize the distance between the source and the receiver that the produced energy should travel. In another aspect, microgrids are able to supply frequency and voltage support to old power distribution systems and delay network upgrades that can be quite expensive. Therefore microgrids can reduce carbon consumption of the grid in the transport process.

The second part is responsible for around 25% of the net anthropogenic greenhouse gas emissions that can be divided into three main categories: deforestation, soil, and nutrient management that can cause agricultural emissions, and livestock and cattle. To reduce deforestation through smart contracts, blockchain technologies have the potential for usage that is operational on small scales. Rainforest, a cryptocurrency similar to SolarCoins, is using smart contracts to reward farmers in the Amazon with internationally funded financial incentives for maintaining their rainforest. In addition to reducing the associated greenhouse gas emissions, farmers can use blockchain technology for cost-effective improvements of crop and land management, grazing, or organic soil restoration by participating in a data ledger. To reduce food waste and energy inefficiency, as in many of the previously discussed economic sectors, blockchain technology can be used to simplify supply chain management, which could change consumer behavior. Using blockchain technology it is possible to trace carbon inputs of food and other products from their source to where they are bought by customers, like for renewable energy purchased from the electricity grid. It is quite ambiguous when it comes to the industry sector. The downside is that the industry sector could be emitting even more greenhouse gas emissions as additional mitigation methods are needed and produced to reduce greenhouse gas emissions in other sectors. To be able to reduce net greenhouse gas emissions, it may be needed to emit greenhouse gases like the famous phrase, "You must spend money to make money." Through a decentralized and immutable data platform, blockchain technology may help to monitor and control the food chain and supply system to reduce unnecessary greenhouse gas emissions within the supply part of the industry. On the positive side, climate change actions toward lowering greenhouse gas emissions lead to financial encouragements for the industrialists. Efficiency can lead to reduced emissions and higher profits. By incentivizing participation and the creation of circular economies, based on unlocking financial value from things that are commonly considered and usually disposed of as waste, decentralization technologies also facilitate consumer behavior to extend product life and use products more actively. This kind of action, similar to ocean plastic collection, recycling, or water conservation can be rewarded with cryptocurrencies and have financial benefits for the consumers as well.

In three key areas, blockchain technologies can be used in the transport sector: promoting and stimulating change on the demand side; helping to ensure the needed shift of traffic capacity; and streamlining road infrastructure. Studies have tried to blend transport sections with the Emissions Trading System (ETS), but it is likely to fail because the focus is only on a single stream of entities. Some studies mentioned a more promising blockchain-based emission trading system for road transport (ETS-RT) which combines upstream, midstream, and downstream entities. Trying to avoid unnecessary transport journeys, blockchain technologies may also help consumers shift their preferred mode of transport toward lower-carbon and more sustainable freight transport systems. Finally, to improve the efficiency and accessibility of public transport infrastructure in a more accessible and equitable way, blockchain technologies can be used particularly in urban areas.

Aside from adaptation and mitigation, there are other areas where blockchain technology can help mankind solve their environmental issues. These solutions can vary from social impact and smart contracts to climate-conscious coins like SolarCoin and Earth Dollar (Howson 2019), which are some examples that will be covered in the next section.

13.6 BLOCKCHAIN SOLUTIONS

To be able to reach the goals set in the Paris Agreement, Schletz et al. (2019) categorized the operation of blockchain technology in the carbon market, climate finance, and clean energy.

The potential advantages in the first aspect, which is the audit trail of the creation and transfer mitigation outcomes, are the ability to lower the management and transaction costs and boost the speed in various levels of granularity and units, trackable emissions reductions, and certification of that around the world. In climate finance, the second part, the main focus is on strengthening accountability by enabling the traceability money movement, facilitating targeted and results-based funding, and allowing micropayments to happen automatically. The third part is working on facilitating P2P energy transactions, enhancing the price for both generator and user, being able to trace and confirm the renewable energy usage, and facilitating the introduction of incentives to generate electricity tokens, for example.

Based on Nick Szabo's definition in 1994, a smart contract is a digital agreement that performs the terms written and computerized in it. The goal of a smart contract is to reduce accidental and malevolent exceptions and the presence of a trusted party while executing conditions like payments and claims, confidentiality, and prosecution. This new version of contracts would be accomplished by nodes inside the blockchain network, which will give the same results if it is executed by all nodes. These results are recorded in the blockchain memory for future needs (Yaga et al. 2019). The combination of blockchain tools with smart contracts will be a necessity in the future to ease our way of changing the phase of contracts automatically and as fast as possible. An example of this can be a change in a contract to certify the lowering of risk hazard after performing and finishing the mitigation infrastructure. It is, in fact, possible to define a smart contract as an automatic update of the terms and conditions concerning some elements that are subject to verification by blockchain tools.

Steps of this process are:

- The initial collection of data on climate phenomena and their effects, for which the blockchain can act to verify that they come from strong and dependable sources;
- The supply of the terms of the agreement, both in terms of insurance and as regards financing for mitigation activities;
- Affirming the schedule for the construction of the mitigation activities;
- Changing the path of insurance contracts after the work has been finished and confirmed, without any new contract on the actual exposure to risk after receiving all parties' signatures and approval.

Damage to private and public properties and the financial burden to restore them by individuals or the government is a great aspect of hydrological risk Vannucci et al. (2021) introduced a creative approach to benefit from combining the investment in initial risk deduction in public management with a flexible insurance system. The public administration has three different strategies to deal with the risk of flooding. The passive strategy provides for compensation to be paid as soon as this happens, to transfer the burden of compensation to the insurance market in part or in full, a standard version of strategies in which risk reduction and quantification of the premium happens after compilation, and the IIRS (Innovative Insurance Resilient Strategy) which includes a mix of an insurance plan with funding of mitigation infrastructures.

Note that in assessing some parameters of the model, a multiphase agreement may take into account certain refreshments due to taking mitigation several decades to be processed and at this time there have been possible trends in climatic phenomena. For example, those that describe the initial risk are shown by the distribution of the random harms, resulting in a change in degrees of insurance coverage.

In a study conducted by Truby (2018), some opportunities and real cases were pointed out to tackle climate change with blockchain technology. By allowing polluters to report their emissions through the use of ledger-distributed technologies, nations devoted to reducing climate change under the Paris Agreement would have been able to provide more accurate reporting and verification of progress in this regard. The transactions are based on P2P verifications, which provide transparent and verifiable records, proving the verification process. It also eliminates the need to rely on third or other parties or governments for verification and storage of transactions.

Digital currency can be traded using distributed ledger technology (DLT), recording, sharing, and validating information and events, as well as carrying out transactions or any other tasks in a more efficient, cost-effective, and dependable way. The open-access models they designed make it possible for users to share and publish data on a public or private level. Instead, the agreement requires only that countries provide information about their emissions and progress toward implementation.

To avoid disputes, the UNDP in India has been using blockchain for authenticating land records and the World Food Programme is taking advantage of it in Pakistan and Jordan so that risks relating to cash payments are eliminated.

Residents of Bangkok, Thailand use blockchain technology to create a market to sell electricity generated by renewable energy sources that are beyond their needs. To avoid pollution or profit from the sale of tokens to polluters, individual investors, environmental groups, and theorists can choose to purchase and save these tokens. This would lead to higher pollution costs and encourage the transition toward low-emissions technologies. The group proposes that the UNFCCC could watch over transactions with the help of logging them into a blockchain-based network, generate tokens of mitigation for allocating pollution rights, and have more financial focus on carbon reduction projects.

To balance the emissions of individuals and corporations through cryptocurrency trading, a group of bankers from across the world, software experts, and entrepreneurs created ClimateCoin, based in Switzerland. This kind of project allows developing countries to move investments from fossil fuel sources toward climate change businesses and create a carbon balance market and renewable resources.

Although there are numerous attempts to find the best solutions using blockchain, it is possible that this new and yet-to-be-fully-known technology can backfire. The next section describes the opposite beliefs about the utility of blockchain technology.

13.7 BLOCKCHAIN VERSUS CLIMATE CHANGE

After considering all the pros and cons of blockchain technology application in climate change, it begs the question of whether this technology can help reduce the effects of climate change or is only another way to reinforce it. In this section, the two fronts in this matter are examined.

Since the inception of the Internet, blockchain has become the most disruptive technology and it is largely supported by cryptocurrencies, which have given rise to well-known applications (Suyambu et al. 2020).

In this regard, Thalhammer et al. (2022) highlighted the overlaps between blockchain and climate change areas and also looked at blockchain's presence in our lives by analyzing the vast range of opportunities that this controversial technology has brought to face global warming. Their main focus was on emission trading certificates that include the ETS applications, private carbon balance programs, and tokenization of CO_2 or energy.

P2P energy trading and grid management, flexibility and mobility concepts, electrical vehicles and blockchain possibilities for charging stations, green finance, and sustainable investment are the main topics of their study. The conclusion drawn was that, based on 19 case studies in literature research, blockchain could support the climate protection measures. Greenhouse gas emissions are the most promising approach to support for blockchains. These categories demonstrate a correlation between the application of blockchain technology to combat the change of climate and the domains that account for the majority of emissions. In the category of green financing, sustainable investment based on the Paris Agreement is crucial for third-world countries. The links show a connection between the applications and their respective domains.

On the other hand, due to redundancies in different types of blockchain technology, it is inherently difficult to achieve a net reduction in energy consumption (Sedlmeir et al. 2020). With the rapid development of blockchain technology and its use of data-secure and distributed permissionless public ledgers, the modern Internet allows for the creation of a variety of digital goods that are very hard to find (Brekke and Fischer 2021). Rare belongings are predicted to remain in our day-to-day lives and even enter into various economic aspects to create value in society. These elements in our digital economy have both sustainability and climate concerns that require further attention.

Goodkind et al. (2020) had a rough calculation that in 2018 each $1 (US) of Bitcoin market value that was produced by mining was combined with $0.49 (US) in health and climate damages in the United States; in China the number is $0.37 (US).

In another study covering the period of 2016–2021, Jones et al. (2022) noticed that the damage caused by Bitcoin to the climate has increased. They realized that on 6.4% of the days, the damage of Bitcoin to climate was lower and on 30.6% of the days the value of this digital coin became more than 50%. The overall damage share was in the order of 35% and in a comparison it falls in between gas usage and meat production and less than coal electricity production.

Calculation of the Bitcoin emission can be done by multiplying the global mean greenhouse gas emission factor (EF) by the electricity consumption per Bitcoin. Also, a rough estimation of electricity usage for mining Bitcoin is concluded from the daily hash rate in the network of Bitcoin, which is in the scale of gigahashes per second (GH/s).

The calculations show that about 39% of PoW required for mining power is supplied by renewable energies and still most of the electricity comes from traditional fossil fuel–burning methods. Cryptocurrency mining, due to its noticeable fossil fuel energy use, has a significant carbon emission (Stoll et al. 2019) with associated environmental damages (Mora et al. 2018).

The researchers presented an empirical approach to assessing the sustainability of Bitcoin while society and individual countries are weighing benefits and costs of various digital limited products. Bitcoin's climate damages per unit market price are extremely higher than wind and solar generation. From a climate damage point of view, Bitcoin operates more like digital crude, while the promoters often offer Bitcoin as a kind of digital treasure.

As shown, blockchain can be both a tool to help and also a threat to be controlled; therefore it is still a grey subject and needs much more time to be completely developed and utilized. In the next section, some guidelines that help the world of science to progress in said issue are suggested.

13.8 PROJECTIONS AND HOPES

A large proportion of blockchain technology is still in the development stage, and only a limited number of operational applications are being developed. Therefore, it may be considered that the current contribution to protecting the environment and implementing the objectives of the Paris Agreement is insignificant. The empirical results of research like that done by Dorfleitner et al. (2021) contribute the valuable information for blockchain planners and users to improve the productivity for methods of design and development. This study took the important and permissible steps toward creating a political framework for green blockchain with the view of bringing success to political establishments.

Since the DLT is still in the first steps of development, it is suggested that recorded documents should be up to date and analyzed more. Deeper studies are needed to discover the full potential of this technology: however, because of the lack of quality information and incomprehensiveness, individual applications are more likely to reveal potential in the near future. In addition, more research should be done in the field of consensus mechanisms to be able to connect CO_2 emissions in application and implementation parts. Also, long-term studies are recommended to find out how else blockchain technology applications can be useful to protect the environment.

Wider use of blockchain technology is probable in case of more research and determining legality and governance, alongside confidence increase for users. Particularly it's going to be possible if green blockchain links with artificial intelligence, big data, and the IoT. Evidence of the future of blockchain technology to drive environmental action is presented in the empirical findings. From a comprehensive analytical point of view of success factors and present portfolios, valuable information on how to exploit the potential of existing and future usage in upcoming activities to protect climate can be gained.

Moreover, there are several weaknesses associated with the use of cryptocurrencies and climate change. First, the areas of research, where active research is being carried out, have a wide range of subareas, which are constantly generating new information. Using information from practitioners, more focus can be placed on detecting the success factors that happen in climate change areas by the application of blockchain. The focus should be narrowed and an in-depth analysis of individual applications with regard to their environmental impact and potential negative impacts is recommended. In addition, it is necessary to perform additional case studies to learn more about the feasibility of frameworks for business purposes and their bias toward this. This topic is already an area of interest in blockchain and needs to be given more attention, particularly to the aforementioned applications. The connection to outside data was a further weakness in the applications identified. Research should be directed to find a way of making sensor inputs tamper proof and detecting fraud.

Investigating the interactive relationship between electricity consumption and cryptocurrency transaction dynamics, Zheng et al. (2023) realized that there is no causal relationship between transaction activities and power usage; regardless of long periods of time and a rate network that does not increase cryptocurrency prices. Nevertheless, by indicating electricity consumption in the Common Market sign, transaction behavior contributes to investors' or regulators' ability to anticipate potential longer-term developments of transactions as a result of possible outside shocks. Furthermore, for daily and longer time periods, they have shown that transaction behavior has a causal effect on electricity consumption. Political organizations, governments, and even environmental institutions should consider the usage of electricity and the greenhouse gas emissions produced to provide that much power and then promote a level of price for cryptocurrencies. Many proofs that digital money mining can cause CO_2 emissions and consume a significant amount of energy have already been discussed. Regulators should reassess their attitudes to the development of cryptocurrencies, as transaction behavior continues to affect energy usage and lead to environmental damage that is hardly reversible. Therefore the next generation of research should focus on using new methods to comprehend the nexus of price and power usage of cryptocurrencies.

SDGs like sustainable consumption and production, and climate action where blockchain technology can be used, were detected by Arshad et al. (2023). Future studies can explore the potential of cryptocurrencies for supporting other SDGs, for instance clean water and sanitation assistance as well as energy saving in living underwater. Every test and result alongside error trials that have been conducted in different locations in a vast range of economic, environmental, and social situations can be quite useful. It might seem farfetched but even medical data is considered helpful in recognizing the success of blockchain technology in renewable energy generation and climate change. Detailed and narrow studies are needed to grasp the full understanding of contradictory effects on the environment. Also, marginal research like finding a tamper-proof sensor input and fraud detection method is going to be handy in the future.

In a more recent study, Pham et al. (2024) conducted a detailed analysis of historical developments and current integration patterns between facilitators and obstacles that could affect the behavior of smallholder farmers toward the use of blockchain technology after showing the negative consequences of climate change on agricultural uncertainty outputs and the livelihood of the interested parties. The fact that climate change is or will affect different aspects of our environment is undeniable; the trick is to find solutions and applications of upcoming technologies with consideration of their damages and backfires.

NOTE

1 https://ccaf.io/cbnsi/cbeci.

REFERENCES

Amaka M (2020) Cryptocurrency and climate change: An overview. *International Journal of Mechanical Engineering and Technology* 11:15–22

Arshad A, Shahzad F, Ur Rehman I, Sergi BS (2023) A systematic literature review of blockchain technology and environmental sustainability: Status quo and future research. *International Review of Economics & Finance* 88:1602–1622. https://doi.org/10.1016/j.iref.2023.07.044

Bogart G (2021) *Using blockchain to address the ipcc's climate change mitigation strategies*. Environmental Law Reporter, Washington, DC USA.

Brekke JK, Fischer A (2021) Digital scarcity. *Internet Policy Review* 10:1–9. https://doi.org/10.14763/2021.2.1548

Calvin K, Dasgupta D, Krinner G, et al (2023) IPCC, 2023: Climate Change 2023: Synthesis Report. *Contribution of Working Groups I, II and III to the Sixth Assessment Report of the Intergovernmental Panel on Climate Change* [Core Writing Team, H. Lee and J. Romero (eds.)]. IPCC, Geneva, Switzerland.

Dittmar L, Praktiknjo A (2019) Could Bitcoin emissions push global warming above 2°C? *Nature Climate Change* 9:656–657. https://doi.org/10.1038/s41558-019-0534-5

Dorfleitner G, Muck F, Scheckenbach I (2021) Blockchain applications for climate protection: A global empirical investigation. *Renewable and Sustainable Energy Reviews* 149:111378. doi:10.1016/j.rser.2021.111378

Eslamian S, Eslamian F (2023) *Disaster Risk Reduction for Resilience: Climate Change and Disaster Risk Adaptation*, Springer Nature Switzerland, 454.

Fawzy S, Osman AI, Doran J, Rooney DW (2020) Strategies for mitigation of climate change: A review. *Environmental Chemistry Letters* 18:2069–2094. https://doi.org/10.1007/s10311-020-01059-w

Gatteschi V, Lamberti F, Demartini C (2020) Blockchain technology use cases. In: Kim S, Deka GC (eds) *Advanced Applications of Blockchain Technology*. Springer Singapore, Singapore, pp 91–114

Goodkind AL, Jones BA, Berrens RP (2020) Cryptodamages: Monetary value estimates of the air pollution and human health impacts of cryptocurrency mining. *Energy Research & Social Science* 59:101281. https://doi.org/10.1016/j.erss.2019.101281

Howson P (2019) Tackling climate change with blockchain. *Nature Climate Change* 9:644–645

Jones BA, Goodkind AL, Berrens RP (2022) Economic estimation of Bitcoin mining's climate damages demonstrates closer resemblance to digital crude than digital gold. *Scientific Reports* 12:14512. https://doi.org/10.1038/s41598-022-18686-8

Kohli V, Chakravarty S, Chamola V, et al (2023) An analysis of energy consumption and carbon footprints of cryptocurrencies and possible solutions. *Digital Communications and Networks* 9:79–89. https://doi.org/10.1016/j.dcan.2022.06.017

Luterbacher Urs, Sprinz Detlef F (2001) *International Relations and Global Climate Change*, MIT Press, USA.

Mora C, Rollins RL, Taladay K, et al (2018) Bitcoin emissions alone could push global warming above 2°C. *Nature Climate Change* 8:931–933. https://doi.org/10.1038/s41558-018-0321-8

O'Neill BC, Oppenheimer M, Warren R, et al (2017) IPCC reasons for concern regarding climate change risks. *Nature Climate Change* 7:28–37. https://doi.org/10.1038/nclimate3179

Pham HC, Truong HQ, Thi Binh AD, Yap TL (2024) Enhancing agriculture resilience of smallholder farming to climate change through blockchain technology adoption. In: Abraham MA (ed) *Encyclopedia of Sustainable Technologies* (Second Edition). Elsevier, Oxford, UK., pp 876–884

Schletz M, Cardoso A, Salomo S, Franke L (2019) *Climate Change Challenges and Blockchain Opportunities Cop 23: Investments in Climate Action*, UNEP DTU Partnership, United Nation.

Sedlmeir J, Buhl HU, Fridgen G, Keller R (2020) The energy consumption of blockchain technology: Beyond myth. *Business & Information Systems Engineering* 62:599–608. https://doi.org/10.1007/s12599-020-00656-x

Stoll C, Klaaßen L, Gallersdörfer U (2019) The carbon footprint of bitcoin. *Joule* 3:1647–1661. https://doi.org/10.1016/j.joule.2019.05.012

Suyambu GT, Anand M, Janakirani M (2020) Blockchain: A most disruptive technology on the spotlight of world engineering education paradigm. *Procedia Computer Science* 172:152–158. https://doi.org/10.1016/j.procs.2020.05.023

Thalhammer F, Schöttle P, Janetschek M, Ploder C (2022) Blockchain use cases against climate destruction. *Cloud Computing and Data Science* 22–38. https://doi.org/10.37256/ccds.3220221277

Truby J (2018) Using Bitcoin technology to combat climate change. *Nature Middle East*. https://doi.org/10.1038/nmiddleeast.2018.111

UNFCCC (2016) *The Paris agreement, UN Climate Change Conference (COP21)*, Paris, France.

Vannucci E, Pagano AJ, Romagnoli F (2021) Climate change management: A resilience strategy for flood risk using Blockchain tools. *Decisions in Economics and Finance*. https://doi.org/10.1007/s10203-020-00315-6

Yaga D, Mell P, Roby N, Scarfone K (2019) Blockchain technology overview. https://doi.org/10.6028/NIST.IR.8202

Zheng M, Feng G-F, Zhao X, Chang C-P (2023) The transaction behavior of cryptocurrency and electricity consumption. *Financial Innovation* 9:44. https://doi.org/10.1186/s40854-023-00449-7

Part VIII

Blockchain for Hydrosystem and Industrial Transformations in Smart Cities

A Forward-Looking Approach

14 Blockchain in Hydrosystems *Research Insights, Challenges, and Future Directions*

Simon Fernandez-Vazquez, Omar Alexander León García, Julián Ramírez, and Salvatore Cannella

14.1 INTRODUCTION

The field of information technology is expanding rapidly. New technical advances have been made possible by the innovative technology known as blockchain. Since 2013, it has garnered significant interest from both domestic and international industry players (Zhang et al., 2020). Numerous sectors, including supply chain management and intelligent manufacturing, have adopted blockchain technology.

Conventional water management methods in many cities have specific problems (Eslamian, 2014). Because urban water sources and supply plants are dispersed, it is difficult to effectively exchange and provide information. Conventional water information systems that are centralized do not save their data securely (Xia et al., 2022a). Information access is slow and impacted by network congestion, and data may be manipulated or misrepresented. The process of using water is intricate and involves many connections. Because of this, it is challenging to effectively monitor this process, which drives up the cost of data verification. Several government entities are in charge of resource water management, and a lack of trust between the agencies causes information islands to form, negatively impacting corporate cooperation and lowering productivity (Shrimali and Patel, 2021).

Governments may use blockchain technology to lower fraud, increase transparency, and boost public sector confidence. Protecting client privacy and reducing information security concerns are two of the primary motivations for adopting blockchain. Blockchain technology also enables businesses to effectively trace their transactions and products back to their source (Della Valle and Oliver, 2021).

Blockchain technology can help with these issues; in particular, its dependable network that combines data and processes increases data dependability and transmission efficiency. First, each connection in the blockchain's distributed storage network can quickly query water data, guaranteeing a quick reaction and rational decision-making in the event of an accident and preventing participants from making biased judgments based only on their data. Second, data security is guaranteed by encryption technology, which also makes data tampering more difficult. Furthermore, blockchain has the capacity to boost confidence and enable low-cost, secure information sharing. Lastly, the creative decentralized information management approach increases the effectiveness of the work in managing water resources (Seven et al., 2020).

As a result, this study offers a systematic evaluation of the literature along with a detailed mapping of blockchain applications for the hydro system sector as well as guidelines, recommendations, and directions for academics working on this subject.

DOI: 10.1201/9781003599593-22

14.2 BACKGROUND

14.2.1 Blockchain

The first blockchain application was released by Nakamoto (2008) through a research white paper introducing "Bitcoin," a brand-new cryptocurrency. The rise of Bitcoin and other cryptocurrencies in recent years has been the real indicator of blockchain technology's relevance and efficacy. The technology initially piqued the interest of numerous researchers and scientists worldwide working in a variety of fields, particularly with the introduction of smart contracts with Ethereum in 2015 (Jaiswal et al., 2022).

Blockchain has the ability to make the energy sector entirely secure, transparent, unchangeable, and a reliable digital platform. The following primary attributes and features of the blockchain might be used to fulfil the needs of next hydro systems and the current ones' constraints (Khezami et al., 2022, Monrat et al., 2019).

i. Transparent: The distributed ledger contains a record of every transaction made within the business network for the various players.
ii. Immutable: A blockchain is a series of blocks connected to one another by hashes (every block holds the hash function's output from the preceding block) and arranged chronologically.
iii. Decentralized: Before a transaction is entered into the shared ledger, it must first be approved and verified by a predetermined group of participants using a consensus generated by cryptographic software. The corporate network pre-agrees on these endorsers, preventing a concentrated power for determining decisions and expediting the technique for validation.
iv. Private: In the energy industry, encryption is utilized to handle sensitive data by making sure that only pertinent portions of the ledger are visible to network participants and that all transactions are safe, verified, and authorized. Moreover, only those with authority may add new users to the network or edit the ledger.
v. Scalable: Because it is simpler to update a shared copy during a transaction than to maintain a separate ledger that necessitates separate concordance with several participants in the business network, this design encourages scalability.

14.3 METHODOLOGY

The research approach of this article is a systematic mapping investigation. A systematic mapping study aims to establish and quantify research evidence on a subject and offer an overview of the region under investigation. Gaps in the literature on a certain subject are shown by mapping studies, and these gaps may be regarded as weak points or topics for further research. The techniques portrayed by Usman (Usman et al., 2023) along with Kitchenham (Kitchenham et al., 2009) have been used to carry out this mapping project.

An extremely helpful approach to reduce time is to employ a keyword search when categorizing results and verifying that all research has been considered. A keyword search requires the execution of two stages. The first step is to read the abstract and identify the main ideas that highlight the paper's contribution. After it was finished, a mix of keywords from several articles was used to determine the general direction of the search.

A quantitative literature mapping was directed on the WoS, with the terms (Topic): "blockchain"; "water"; "hydrosystem"; "hidrosystem", "Distributed ledger technology"; "water management", "Water conservation"; "Sustainable development" Review Article (Exclude – Document Types) and Article (Document Types) and 2016–2023 (Publication Years). Dimensions were used as data visualization tools for the original bibliometric mapping, together with layout methods from VOSviewer.

After being gathered from the WoS Core Collection database, the data was organized into an Excel table.

14.4 RESULTS

This segment highlights the main results of the search previously described.

14.4.1 Year of Publication

Figure 14.1 displays the percentage of papers from the final selection corresponding to the publication year. As can be seen, there is an upward trend in the interest in publication on hydrology and blockchain in academia. Of the publications 31% correspond to 2022, and almost 25% to 2023.

14.4.2 Geographical Distribution

In regards to the geographical distribution of the chosen articles, China with 37 publications, India with 27 publications, and the United States with 24 publications are the top three nations (Figure 14.2).

For a further comprehensive interpretation of the results, Figure 14.3 represents these results in a map format.

14.4.3 Main Research Areas

Figure 14.4 summarizes the main study topics covered in the chosen articles, illuminating the wide range of subjects that researchers have investigated. With the greatest number of publications (55), the engineering discipline is the most prominent area. This strong presence highlights the importance of engineering-related research in academia and significantly emphasizes innovation, technological development, and problem-solving in this field.

With 49 articles, computer science is the second most popular research field. The significant interest in computer science reflects the increasing significance of software development, artificial intelligence, and computational techniques in solving modern problems and expanding the frontiers of knowledge.

Finally, with 24 publications, environmental science is third place and demonstrates a remarkable dedication to understanding and reducing environmental problems. The focus on environmental science in current research attempts emphasizes the urgency and worldwide awareness around ecological protection, climate change, and sustainability.

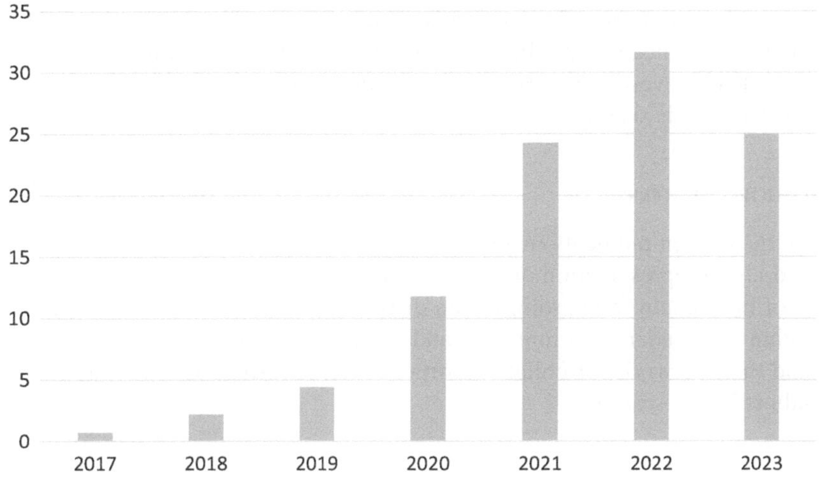

FIGURE 14.1 Year of publication of the selected papers.

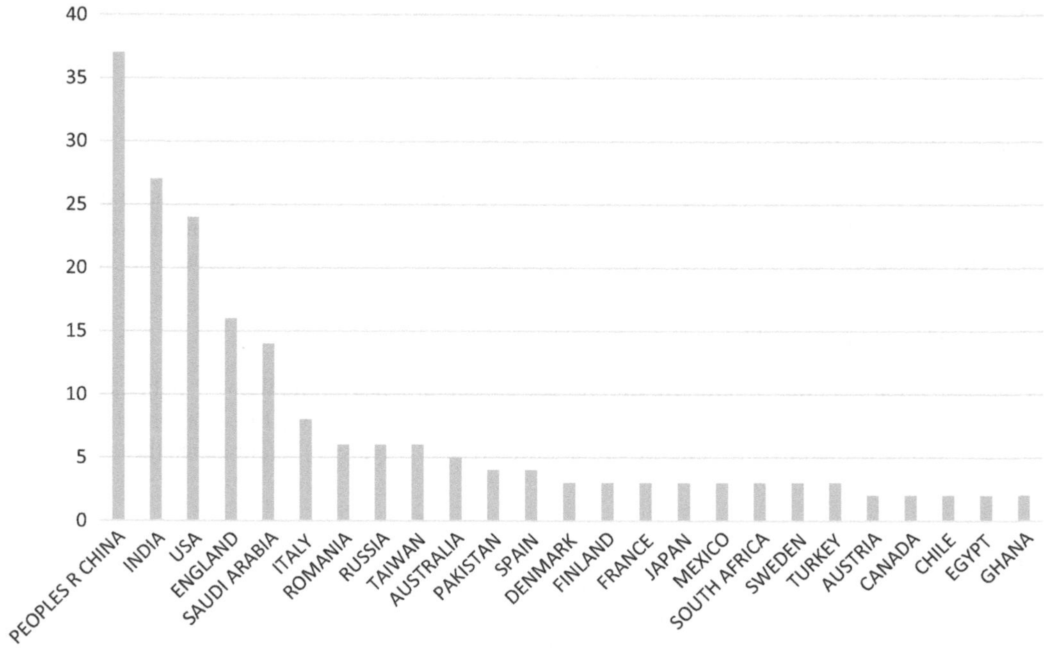

FIGURE 14.2 Geographical allocation of the chosen papers.

14.4.4 Publication Sources

Figure 14.5 offers an analysis of the sources of publication linked to the chosen works, highlighting the journals or platforms that have made a significant contribution to the spread of knowledge in this sector. Understanding the favored channels in the academic community through this examination of publication sources is crucial for providing insights into the influence and visibility of the research findings. Sustainability, IEEE Access, and Sensors, with 7, 6, and 4 publications each are the most prominent publication sources in this topic.

14.4.5 Keyword Clusters

The graphical representation of the top term clusters in the chosen publications, Figure 14.6, provides important information about the research's subject focus and interconnection. The clusters that have been found offer an insight into the fundamental ideas and overarching topics that have caught interest in the academic investigation. Blockchain technology and Internet of Things (IoT), stand out as the main clusters of this search.

14.4.6 Author Clusters

The authors of the chosen publications have created clusters, as shown in Figure 14.7, which provides a detailed visual representation of the academics who have made notable contributions to the state of study of blockchain in hydrology. The discovered author clusters highlight the patterns of authorship within the academic community and offer insightful information about the important researchers and their networks of collaboration. Sundeep,T., Shan, K. and Patel, D., are the most prominent authors in this search.

FIGURE 14.3 Map distribution of the selected papers.

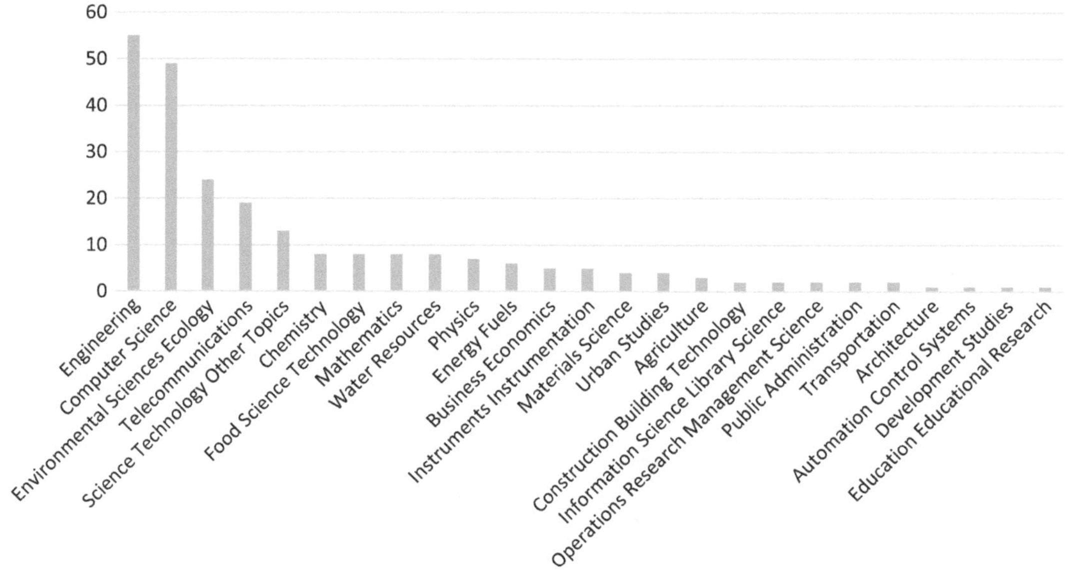

FIGURE 14.4 Main research areas for the selected papers.

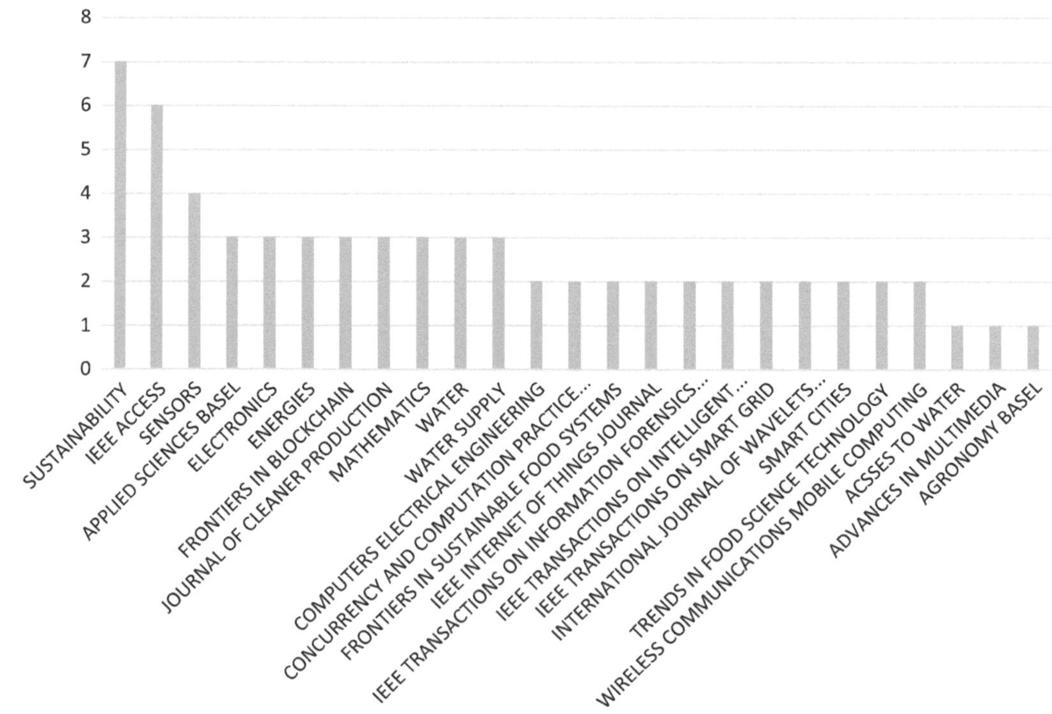

FIGURE 14.5 Publication sources for the selected papers.

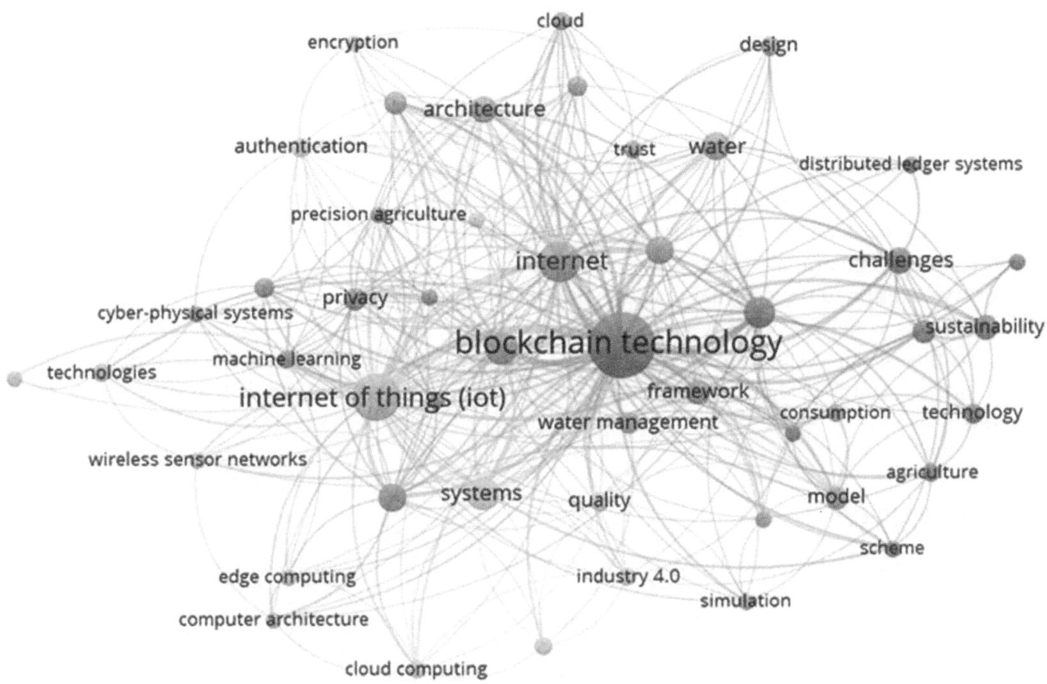

FIGURE 14.6 Keyword clusters for the selected papers.

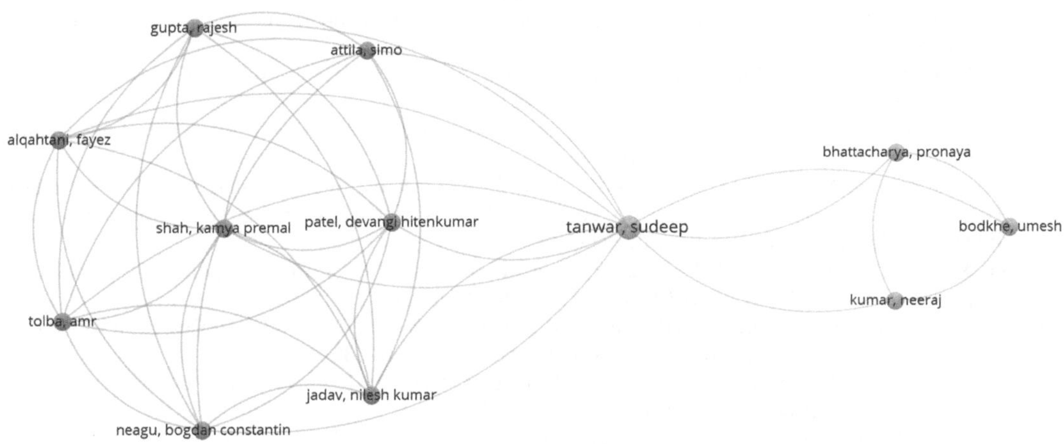

FIGURE 14.7 Author clusters.

14.5 CONCLUSIONS

This mapping investigation revealed several significant gaps in the literature. To begin with, most of the research is conducted in fields about engineering and computer science. Currently, there are relatively few studies on the usage of blockchain technology in hydrology outside these industries. In addition, a closer examination of how this distributed ledger could support the hydrology supply chain is required.

The absence of technical literature constituted a second research gap. Very few mathematical models were available for blockchain applications in hydrosystems. The majority of the articles were informative and required a methodical approach when creating practical tools to aid in overcoming the significant obstacles that were shown. More research is required to develop solutions that will assist companies to begin utilizing blockchain technology.

With a shared ledger, users may view all transactions in a decentralized network environment called blockchain. Through a collection of cryptographic methods and protocols, it offers anonymity, security, privacy, and transparency. However, these advantages also provide a number of difficulties and restrictions that need to be investigated. A comprehensive mapping analysis was conducted to understand the academic literature on blockchain that is currently available. Understanding the research issues, constraints, gaps, and possible uses of blockchain in hydrology is the goal of this chapter. We examined the original research and used maps, graphs, and charts to show the results in a way that was easier for other academics to understand.

The results of our systematic review will aid scholars and practitioners in the field in understanding the present status of the subject, organizing it, and identifying any areas that may require more research.

BIBLIOGRAPHY

Abd El-Latif, AA, Abd-El-Atty, B, Mehmood, I, Muhammad, K, Venegas-Andraca, SE, Peng, JL (2021) Quantum-Inspired Blockchain-Based Cybersecurity: Securing Smart Edge Utilities In Iot-Based Smart Cities Information Processing & Management 10.1016/J.Ipm.2021.102549

Abdo, JB, Zeadally, S (2022) Multi-Utility Framework: Blockchain Exchange Platform for Sustainable Development *International Journal of Pervasive Computing and Communications* 10.1108/Ijpcc-06-2020-0059

Abrar, A, Abdul, W, Ghouzali, S (2021) Secure Image Authentication Uuing Watermarking And Blockchain Intelligent Automation and Soft Computing 10.32604/Iasc.2021.016382

Alcarria, R, Bordel, B, Robles, T, Martín, D, Manso-Callejo, MA (2018) A Blockchain-Based Authorization System for Trustworthy Resource Monitoring and Trading in Smart Communities Sensors 10.3390/S18103561

Aldribi, A, Singh, A (2022) Blockchain Empowered Smart Home: A Scalable Architecture for Sustainable Smart Cities Mathematics 10.3390/Math10142378

Alnahari, MS, Ariaratnam, ST (2022) The Application of Blockchain Technology to Smart City Infrastructure Smart *Cities* 10.3390/Smartcities5030049

Alnowibet, K, Annuk, A, Dampage, U, Mohamed, MA (2021) Effective Energy Management Via False Data Detection Scheme for The Interconnected Smart Energy Hub-Microgrid System Under Stochastic Framework Sustainability 10.3390/Su132111836

Alrubei, SM, Ball, E, Rigelsford, JM (2022) A Secure Blockchain Platform for Supporting Ai-Enabled Iot Applications at the Edge Layer IEEE Access 10.1109/Access.2022.3151370

Alsumayt, A, El-Haggar, N, Amouri, L, Alfawaer, ZM, Aljameel, SS (2023) Smart Flood Detection with Ai and Blockchain Integration in Saudi Arabia Using Drones Sensors 10.3390/S23115148

Alzahrani, AIA, Chauhdary, SH, Alshdadi, AA (2023) Internet of Things (IoT)-Based Wastewater Management In Smart Cities Electronics 10.3390/Electronics12122590

Ameyaw, EE, Edwards, DJ, Kumar, B, Thurairajah, N, Owusu-Manu, D, Oppong, GD (2023) Critical Factors Influencing Adoption of Blockchain-Enabled Smart Contracts in Construction Projects *Journal of Construction Engineering and Management* 10.1061/Jcemd4.Coeng-12081

Angara, JSV, Saripalle, RS (2022) Towards a Virtual Water Currency for Industrial Products Using Blockchain Technology *Water Policy* 10.2166/Wp.2022.285

Anitha, R, Rai, D (2022) Internet of Things With Artificial Intelligence Detection and Blockchains of Crop Availability for Supply Chain Management *International Journal of Knowledge-Based Development* 10.1504/Ijkbd.2022.128909

Barbosa, MW (2021) Uncovering Research Streams On Agri-Food Supply Chain Management: A Bibliometric Study *Global Food Security-Agriculture Policy Economics and Environment* 10.1016/J.Gfs.2021.100517

Barinaga, E, Campos, MJZ (2023) Tinkering with Malleable Grassroots Infrastructures: Kenyan Local Currencies in Informal Settlements *Urban Geography* 10.1080/02723638.2023.2239658

Barry, S (2021) Livestock Mobility Through Integrated Beef Production-Scapes Supports Rangeland Livestock Production and Conservation *Frontiers in Sustainable Food Systems* 10.3389/Fsufs.2020.549359

Benedict, S (2020) Serverless Blockchain-Enabled Architecture for Iot Societal Applications *IEEE Transactions On Computational Social Systems* 10.1109/Tcss.2020.3008995

Berman, I, Zereik, E, Kapitonov, A, Bonsignorio, F, Khassanov, A, Oripova, A, Lonshakov, S, Bulatov, V (2020) Trustable Environmental Monitoring By Means of Sensors Networks on Swarming Autonomous Marine Vessels and Distributed Ledger Technology *Frontiers In Robotics and AI* 10.3389/Frobt.2020.00070

Bodkhe, U, Tanwar, S, Bhattacharya, P, Kumar, N (2022) Blockchain for Precision Irrigation: Opportunities and Challenges *Transactions on Emerging Telecommunications Technologies* 10.1002/Ett.4059

Bommu, S, Kumar, MA, Babburu, K, Srikanth, N, Thalluri, LN, Ganesh, G, Gopalan, A, Mallapati, PK, Guha, K, Mohammad, HR, Kiran, SS (2023) Smart City Iot System Network Level Routing Analysis and Blockchain Security Based Implementation *Journal of Electrical Engineering & Technology* 10.1007/S42835-022-01239-4

Bunge, AC, Wood, A, Halloran, A, Gordon, LJ (2022) A Systematic Scoping Review of the Sustainability of Vertical Farming, Plant-Based Alternatives, Food Delivery Services and Blockchain in Food Systems *Nature Food* 10.1038/S43016-022-00622-8

Cali, U, Çakir, O (2021) Novel Donation Sharing Mechanisms Under Smart Energy Cyber-Physical-Social System and DLT to Contend the Energy Poverty Problem *IEEE Access* 10.1109/Access.2021.3106833

Chang, YY, Xu, J, Ghafoor, KZ (2021) An IOT and Blockchain Approach for the Smart Water Management System in Agriculture *Scalable Computing-Practice and Experience* 10.12694/Scpe.V22i2.1869

Chatziamanetoglou, D, Rantos, K (2023) Blockchain-Based Security Configuration Management for ICT Systems *Electronics* 10.3390/Electronics12081879

Chaudhari, RS, Mahajan, SK, Rane, SB, Agrawal, R (2022) Modeling Barriers in Circular Economy Using Topsis: Perspective of Environmental Sustainability & Blockchain-IOT Technology *International Journal of Mathematical Engineering and Management Sciences* 10.33889/Ijmems.2022.7.6.052

Chenniappan, M, Gnanavel, D, Gunasekaran, KP, Rajalakshmi, RR, Ramya, AS, Stonier, AA, Peter, G, Ganji, V (2022) Prediction of Fault Occurrences in Smart City Water Distribution System Using Time-Series Forecasting Algorithm *Mathematical Problems in Engineering* 10.1155/2022/9678769

Chung, KHY, Li, D, Adriaens, P (2023) Technology-Enabled Financing of Sustainable Infrastructure: A Case for Blockchains and Decentralized Oracle Networks *Technological Forecasting and Social Change* 10.1016/J.Techfore.2022.122258

Clark, A, Mihailov, A, Zargham, M (2023) Complex Systems Modeling of Community Inclusion Currencies *Computational Economics* 10.1007/S10614-023-10445-9

Dang, C, Zhang, JF, Kwong, CP, Li, L (2019) Demand Side Load Management for Big Industrial Energy Users Under Blockchain-Based Peer-To-Peer Electricity Market *IEEE Transactions on Smart Grid* 10.1109/Tsg.2019.2904629

Darnell, SS, Sevilla, J (2021) 3 Stages of a Pan-African Identity Framework for Establishing Self-Sovereign Identity with Blockchain *Frontiers in Blockchain* 10.3389/Fbloc.2021.631640

Davies, FT, Garrett, B (2018) Technology for Sustainable Urban Food Ecosystems in the Developing World: Strengthening the Nexus of Food-Water-Energy-Nutrition *Frontiers in Sustainable Food Systems* 10.3389/Fsufs.2018.00084

Della Valle, F, Oliver, M (2021) Blockchain-Based Information Management for Supply Chain Data-Platforms. *Applied Sciences* 11. 8161. 10.3390/App11178161

Dhavamani, L, Priya, PP (2022) Energy-Efficient and Privacy-Preserving Approach for Internet of Things Nodes Using a Novel Hybrid Fuzzy Water Cycle and Evaporation Strategy and Matrix-Based Rivest-Shamir-Adleman Encryption Algorithm *Concurrency and Computation-Practice & Experience* 10.1002/Cpe.7336

Dmitrievsky, AN, Eremin, NA, Chernikov, AD, Zinatullina, LI (2021) Using Artificial Intelligence Methods to Prevent Complications in Well Construction *Proceedings of the Tula States University-Sciences of Earth* 10.46689/2218-5194-2021-4-1-132-144

Enescu, FM, Bizon, N, Onu, A, Raboaca, MS, Thounthong, P, Mazare, AG, Serban, G (2020) Implementing Blockchain Technology in Irrigation Systems that Integrate Photovoltaic Energy Generation Systems *Sustainability* 10.3390/Su12041540

Eslamian, S. (2014) *Handbook of engineering hydrology, Vol. 3: Environmental Hydrology and Water Management*, Taylor & Francis Group, CRC Group, USA, 606 p.

Farfan, J, Lohrmann, A (2023) Gone with the Clouds: Estimating the Electricity and Water Footprint of Digital Data Services in Europe *Energy Conversion and Management* 10.1016/J.Enconman.2023.117225

Fedorova, EP, Skobleva, EI (2020) Application of Blockchain Technology in Higher Education *European Journal of Contemporary Education* 10.13187/Ejced.2020.3.552

Frikha, T, Ktari, J, Ben Amor, N, Chaabane, F, Hamdi, M, Denguir, F, Hamam, H (2023) Low Power Blockchain in Industry 4.0 Case Study: Water Management in Tunisia *Journal of Signal Processing Systems for Signal Image and Video Technology* 10.1007/S11265-023-01880-W

Frikha, T, Ktari, J, Zalila, B, Ghorbel, O, Ben Amor, N (2023) Integrating Blockchain and Deep Learning for Intelligent Greenhouse Control and Traceability *Alexandria Engineering Journal* 10.1016/J. Aej.2023.08.027

Furones, AR, Monzón, JIT (2023) Blockchain Applicability in the Management of Urban Water Supply and Sanitation Systems in Spain *Journal of Environmental Management* 10.1016/J.Jenvman.2023.118480

Galanakis, CM, Rizou, M, Aldawoud, TMS, Ucak, I, Rowan, NJ (2021) Innovations and Technology Disruptions in the Food Sector within the Covid-19 Pandemic and Post-Lockdown Era *Trends in Food Science & Technology* 10.1016/J.Tifs.2021.02.002

Gupta, BB, Gaurav, A, Hsu, CH, Jiao, B (2023) Identity-Based Authentication Mechanism for Secure Information Sharing in the Maritime Transport System *IEEE Transactions on Intelligent Transportation Systems* 10.1109/Tits.2021.3125402

Hammedi, W, Brik, B, Senouci, SM (2023) Toward Optimal Mec-Based Collision Avoidance System for Cooperative Inland Vessels: A Federated Deep Learning Approach *IEEE Transactions on Intelligent Transportation Systems* 10.1109/Tits.2022.3154158

Hara, H, Echigo, S, Ota, T, Komatsu, R, Matsumoto, S (2018) Co-Creation with Customers to Accelerate Digital Innovation *Fujitsu Scientific & Technical Journal* 54(3). 47–54.

Holland, M, Thomas, C, Livneh, B, Tatge, S, Johnson, A, Thomas, E (2022) Development and Validation of an in Situ Groundwater Abstraction Sensor Network, Hydrologic Statistical Model, and Blockchain Trading Platform: A Demonstration in Solano County, California *ACS ES&T Water* 10.1021/Acsestwater.2c00214

Hong, YY, Alano, FI (2022) Hierarchical Energy Management in Islanded Networked Microgrids *IEEE Access* 10.1109/Access.2022.3143307

Hossain, M, Rahman, A, Chaki, S, Ahmed, H, Haque, A, Tamanna, I, Lima, S, Ferdous, J, Rahman, MS (2023) Smart-Agri: A Smart Agricultural Management with IOT-ML-Blockchain Integrated Framework *International Journal of Advanced Computer Science and Applications*, 14(7) 10.14569/IJACSA.2023. 01407107.

Huang, LY, Cai, JF, Lee, TC, Weng, MH (2020) A Study on the Development Trends of the Energy System with Blockchain Technology Using Patent Analysis *Sustainability* 10.3390/Su12052005

Huseynov, F, Mitchell, J (2023) Blockchain for Environmental Peacebuilding: Application in Water Management *Digital Policy Regulation and Governance* 10.1108/Dprg-06-2023-0080

Iftekhar, A, Cui, XH, Yang, YP (2021) Blockchain Technology for Trustworthy Operations in the Management of Strategic Grain Reserves *Foods* 10.3390/Foods10102323

Jain, G, Kamble, SS, Ndubisi, NO, Shrivastava, A, Belhadi, A, Venkatesh, M (2022) Antecedents of Blockchain-Enabled E-Commerce Platforms (Beep) Adoption by Customers-A Study of Second-Hand Small and Medium Apparel Retailers *Journal of Business Research* 10.1016/J.Jbusres.2022.05.041

Jaiswal, R, Gupta, S Tiwari, A (2022) Delineation of Blockchain Technology in Finance: A Scientometric View 17 10.1142/S2010495222500257

Jararweh, Y, Fatima, S, Jarrah, M Alzu'bi, S (2023) Smart and Sustainable Agriculture: Fundamentals, Enabling Technologies, and Future Directions *Computers & Electrical Engineering* 10.1016/J.Compeleceng. 2023.108799

Jiang, YM, Zhang, JF (2023) Distributed Detection Over Blockchain-Aided Internet of Things in the Presence of Attacks *IEEE Transactions on Information Forensics and Security* 10.1109/Tifs.2023.3279984

Kaur, E, Oza, A (2020) Blockchain-Based Multi-Organization Taxonomy for Smart Cities *Sn Applied Sciences* 10.1007/S42452-020-2187-4

Keshk, M, Turnbull, B, Sitnikova, E, Vatsalan, D, Moustafa, N (2021) Privacy-Preserving Schemes for Safeguarding Heterogeneous Data Sources in Cyber-Physical Systems *IEEE Access* 10.1109/Access. 2021.3069737

Khezami, N Gharbi, N Neji, B Braiek, N (2022). Blockchain Technology Implementation in the Energy Sector: Comprehensive Literature Review and Mapping. *Sustainability* 14. 15826. 10.3390/Su142315826

Kitchenham, B Brereton, P Budgen, D Turner, M Bailey, J Linkman, S (2009) Systematic Literature Reviews in Software Engineering: A Systematic Literature Review. *Information and Software Technology* 51. 7–15. 10.1016/J.Infsof.2008.09.009

Küfeoglu, S (2021) Sustainable Living Spaces and Open Digital Innovation Hub *The Home of the Future: Digitalization and Resource Management*. doi:10.1007/978-3-030-75093-0_1

Kumar, A, Ahuja, NJ, Thapliyal, M, Dutt, S, Kumar, T, Pacheco, DAD, Konstantinou, C, Choo, KKR (2023) Blockchain for Unmanned Underwater Drones: Research Issues, Challenges, Trends and Future Directions *Journal of Network and Computer Applications* 10.1016/J.Jnca.2023.103649

Kumar, NM, Chopra, SS (2023) Integrated Techno-Economic and Life Cycle Assessment of Shared Circular Business Model Based Blockchain-Enabled Dynamic Grapevoltaic Farm for Major Grape Growing States in India *Renewable Energy* 10.1016/J.Renene.2023.03.064

Kumar, R, Kumar, P, Aljuhani, A, Islam, AKMN, Jolfaei, A, Garg, S (2023) Deep Learning and Smart Contract-Assisted Secure Data Sharing for IOT-Based Intelligent Agriculture *IEEE Intelligent Systems* 10.1109/Mis.2022.3201553

Lavelli, V, Beccalli, MP (2022) Cheese Whey Recycling in the Perspective of the Circular Economy: Modeling Processes and the Supply Chain to Design the Involvement of the Small and Medium Enterprises *Trends in Food Science & Technology* 10.1016/J.Tifs.2022.06.013

Li, HM, Chen, XY, Guo, ZW, Xu, JL, Shen, Y, Gao, X (2021) Data-Driven Peer-to-Peer Blockchain Framework for Water Consumption Management *Peer-To-Peer Networking and Applications* 10.1007/S12083-021-01121-6

Li, J, Hu, K, Zhu, J, Bodeveix, JP, Ye, YF (2022) Formal Modelling of PBFT Consensus Algorithm in Event-B *Wireless Communications & Mobile Computing* 10.1155/2022/4467917

Li, JN, Wu, J, Li, CL, Yang, W, Bashir, AK, Li, JH, Al-Otaibi, YD (2022) Information-Centric Wireless Sensor Networking Scheme with Water-Depth-Awareness Content Caching for Underwater IOT *IEEE Internet of Things Journal* 10.1109/Jiot.2021.3058272

Li, WJ, Cheng, JQ (2021) Blockchain-Based Secure and Trustworthy Sensing for IOT-Based Smart Cities Communication Technologies for Networked Smart Cities, In book: *Communication Technologies for Networked Smart Cities*, pp 301–317, Ch13, Edited by Shree Krishna Sharma, Dushantha Nalin K. Jayakody, Symeon Chatzinotas, Alagan Anpalagan, The Institution of Engineering and Technology. https://doi.org/10.1049/PBTE090E_ch13

Li, Y, Xie, JC, Yang, JH, Ren, J, Zhai, N (2022) Application of Blockchain Technology in Water Rights Trading in the Irrigation Area Under the Internet-Of-Things Environment *Security and Communication Networks* 10.1155/2022/8700730

Lin, CY, Dai, GL, Liu, Y, Zhang, MQ, Liu, Y, Jiang, W, Fu, XM, Chen, HX (2023) Ecological Value of Mariculture Shellfish Resources in China: Assessment and Management *Marine Policy* 10.1016/J.Marpol.2022.105406

Lin, YP, Mukhtar, H, Huang, KT, Petway, JR, Lin, CM, Chou, CF, Liao, SW (2020) Real-Time Identification of Irrigation Water Pollution Sources and Pathways with a Wireless Sensor Network and Blockchain Framework *Sensors* 10.3390/S20133634

Liu, H, Wang, YQ, He, GY, Ma, RL, Fu, SL (2023) The Impact of Environmental Information Disclosure of Origin Using Blockchain Technology on Online Consumer Behaviour: A Combination of SEM and NCA Approaches *Journal of Cleaner Production* 10.1016/J.Jclepro.2023.138449

Liu, XL, Yang, Y, Jiang, YS, Fu, YL, Zhong, RY, Li, M, Huang, GQ (2023) Data-Driven ESG Assessment for Blockchain Services: A Comparative Study in Textiles and Apparel Industry *Resources Conservation and Recycling* 10.1016/J.Resconrec.2022.106837

Liu, Y, Shang, C (2022) Application of Blockchain Technology in Agricultural Water Rights Trade Management *Sustainability* 10.3390/Su14127017

Loo, BPY, Wong, RWM (2023) Towards a Conceptual Framework of Using Technology to Support Smart Construction: The Case of Modular Integrated Construction (mic) *Buildings* 10.3390/Buildings13020372

Lu, LY (2023) Evaluation on the Application Mode of Blockchain Technology in Water Transportation Engineering Project Management *Water Supply* 10.2166/Ws.2023.255

Lukic, I, Milicevic, K, Köhler, M, Vinko, D (2022) Possible Blockchain Solutions According to a Smart City Digitalization Strategy *Applied Sciences-Basel* 10.3390/App12115552

Luzzani, G, Grandis, E, Frey, M, Capri, E (2021) Blockchain Technology in Wine Chain for Collecting and Addressing Sustainable Performance: An Exploratory Study *Sustainability* 10.3390/Su132212898

Mahmoud, HH, Wu, WY, Wang, YH (2021) Wdschain: A Toolbox for Enhancing the Security Using Blockchain Technology in Water Distribution System *Water* 10.3390/W13141944

Maw, A, Adepu, S, Mathur, A (2019) ICS-Blockops: Blockchain for Operational Data Security in Industrial Control System *Pervasive and Mobile Computing* 10.1016/J.Pmcj.2019.101048

Mendi, AF (2022) Blockchain for Food Tracking *Electronics* 10.3390/Electronics11162491

Mian, IS, Twisleton, D, Timm, DA (2020) What is the Resource Footprint of a Computer Science Department? *Place, People, and Pedagogy Data & Policy* 10.1017/Dap.2020.12

Miskiewicz, R, Matan, K, Karnowski, J (2022) The Role of Crypto Trading in the Economy, Renewable Energy Consumption and Ecological Degradation *Energies* 10.3390/En15103805

Mohammadi, R (2023) A Comprehensive Blockchain-Oriented Secure Framework for SDN/FOG-based IOUT *International Journal of Information Security* 10.1007/S10207-023-00683-1

Monrat, A, Schelén, O, Andersson, K (2019) Survey of Blockchain from the Perspectives of Applications, Challenges and Opportunities *IEEE Access* 1–1 10.1109/Access.2019.2936094

Mqamelo, R (2022) Community Currencies as Crisis Response: Results from a Randomized Control Trial in Kenya *Frontiers in Blockchain* 10.3389/Fbloc.2021.739751

Munir, MS, Bajwa, IS, Cheema, SM (2019) An Intelligent and Secure Smart Watering System Using Fuzzy Logic and Blockchain *Computers & Electrical Engineering* 10.1016/J.Compeleceng.2019.05.006

Nakamoto, S. (2008). *Bitcoin: A Peer-to-Peer Electronic Cash System.* [White paper].

Nalinipriya, G Balajee, M Priya, C Rajan, C (2022) Ransomware Recognition in Blockchain Network Using Water Moth Flame Optimization-Aware DRNN *Concurrency and Computation-Practice & Experience* 10.1002/Cpe.7047

Nalinipriya, G Maram, B Vidyadhari, C Cristin, R (2021) Optimized Deep Stacked Autoencoder for Ransomware Detection Using Blockchain Network *International Journal of Wavelets Multiresolution and Information Processing* 10.1142/S0219691321500223

Ozpinar, A (2023) A Hyper-Integrated Mobility as a Service (Maas) to Gamification and Carbon Market Enterprise Architecture Framework for Sustainable Environment *Energies* 10.3390/En16052480

Pahontu, BI Petcu, A Predescu, A Arsene, DA Mocanu, M (2023) A Blockchain Approach for Migrating a Cyber-Physical Water Monitoring Solution to a Decentralized Architecture *Water* 10.3390/W15162874

Pandey, S, Twala, B, Singh, R, Gehlot, A, Singh, A, Montero, EC, Priyadarshi, N (2022) Wastewater Treatment with Technical Intervention Inclination Towards Smart Cities *Sustainability* 10.3390/Su141811563

Patel, DH, Shah, KP, Gupta, R, Jadav, NK, Tanwar, S, Neagu, BC, Attila, S, Alqahtani, F, Tolba, A (2023) Blockchain-Based Crop Recommendation System for Precision Farming In IOT Environment *Agronomy-Basel* 10.3390/Agronomy13102642

Peneti, S, Kumar, MS, Kallam, S, Patan, R, Bhaskar, V, Ramachandran, M (2021) BDN-Gwmnn: Internet of Things (iot) Enabled Secure Smart City Applications *Wireless Personal Communications* 10.1007/S11277-021-08339-W

Pervez, Z, Khan, Z, Ghafoor, A, Soomro, K (2023) Signed: Smart City Digital Twin Verifiable Data Framework *IEEE Access* 10.1109/Access.2023.3260621

Pincheira, M, Vecchio, M, Giaffreda, R, Kanhere, SS (2021) Cost-Effective IOT Devices as Trustworthy Data Sources for a Blockchain-Based Water Management System in Precision Agriculture *Computers and Electronics in Agriculture* 10.1016/J.Compag.2020.105889

Poonia, V, Goyal, MK, Gupta, BB, Gupta, AK, Jha, S, Das, J (2021) Drought Occurrence in Different River Basins of India and Blockchain Technology Based Framework for Disaster Management *Journal of Cleaner Production* 10.1016/J.Jclepro.2021.127737

Predescu, A, Arsene, D, Pahontu, B, Mocanu, M, Chiru, C (2021) A Serious Gaming Approach for Crowdsensing in Urban Water Infrastructure with Blockchain Support *Applied Sciences-Basel* 10.3390/App11041449

Raboaca, MS, Bizon, N, Trufin, C, Enescu, FM (2020) Efficient and Secure Strategy for Energy Systems of Interconnected Farmers' Associations to Meet Variable Energy Demand *Mathematics* 10.3390/Math8122182

Radu, LD (2020) Disruptive Technologies in Smart Cities: A Survey on Current Trends and Challenges *Smart Cities* 10.3390/Smartcities3030051

Ramsey, E, Pesantez, J, Fasaee, MAK, Dicarlo, M, Monroe, J, Berglund, EZ (2020) A Smart Water Grid for Micro-Trading Rainwater: Hydraulic Feasibility Analysis *Water* 10.3390/W12113075

Rottondi, C, Verticale, G (2017) A Privacy-Friendly Gaming Framework in Smart Electricity and Water Grids *IEEE Access* 10.1109/Access.2017.2727552

Samawi, VW (2021) Smcsis: An IOT Based Secure Multi-Crop Irrigation System for Smart Farming *International Journal of Innovative Computing Information and Control* 10.24507/Ijicic.17.04.1225

Seven, S, Yao, G, Soran, A, Onen, A, Muyeen, SM (2020). Peer-to-Peer Energy Trading in Virtual Power Plant Based on Blockchain Smart Contracts. *IEEE Access* 10.1109/Access.2020.3026180

Sgroi, F (2022) The Role of Blockchain for Food Safety and Market Efficiency *Journal of Agriculture and Food Research* 10.1016/J.Jafr.2022.100326

Shahidehpour, M, Yan, MY, Shikhar, P, Bahramirad, S, Paaso, A (2020) Blockchain for Peer-To-Peer Transactive Energy Trading in Networked Microgrids: Providing an Effective and Decentralized Strategy *IEEE Electrification Magazine* 10.1109/Mele.2020.3026444

Sharma, K, Sharma, C, Sharma, S, Asenso, E (2022) Broadening the Research Pathways in Smart Agriculture: Predictive Analysis Using Semiautomatic Information Modeling *Journal of Sensors* 10.1155/2022/5442865

Shou, M, Domenech, T (2022) Integrating LCA and Blockchain Technology to Promote Circular Fashion: A Case Study of Leather Handbags *Journal of Cleaner Production* 10.1016/J.Jclepro.2022.133557

Shrimali, B., Patel, B (2021) Blockchain State-of-the-Art: Architecture, Use Cases, Consensus, Challenges and Opportunities. *Journal of King Saud University: Computer and Information Sciences* 34. 10.1016/J.Jksuci.2021.08.005

Smajgl, A, Schweik, CM (2022) Advancing Sustainability with Blockchain-Based Incentives and Institutions *Frontiers in Blockchain* 10.3389/Fbloc.2022.963766

Sobrinho, RP, Garcia, R, Maia, AG, Romeiro, AR (2019) Blockchain Technology and Complex Flow Systems as Opportunities for Water Governance Innovation *Revista Brasileira de Inovacao* 10.20396/rbi.v18i1.8654757

Song, SJ, Huang, T, Ma, JM, Mao, XX, Gao, H, Zhao, Y, Jiang, WY, Lian, LL, Chen, BQ, Liu, LQ, Wang, ZX (2022) Assessing Safety of Market-Sold Fresh Fish: Tracking Fish Origins and Toxic Chemical Origins *Environmental Science & Technology* 10.1021/Acs.Est.2c00495

Sriyono, E (2020) Digitizing Water Management: Toward the Innovative Use of Blockchain Technologies to Address Sustainability *Cogent Engineering* 10.1080/23311916.2020.1769366

Su, XL, Hu, YC, Liu, W, Jiang, ZP, Qiu, C, Xiong, J, Sun, J (2023) A Blockchain-Based Smart Contract Model for Secured Energy Trading Management in Smart Microgrids *Security and Privacy* 10.1002/Spy2.341

Sumathi, M Rajkamal, M Raja, SP Venkatachalapathy, M Vijayaraj, N (2022) A Crop Yield Prediction Model Based on an Improved Artificial Neural Network and Yield Monitoring Using a Blockchain Technique *International Journal of Wavelets Multiresolution and Information Processing* 10.1142/S0219691322500308

Tang, CSS (2022) Innovative Technology and Operations for Alleviating Poverty Through Women's Economic Empowerment *Production and Operations Management* 10.1111/Poms.13349

Taranto, F, Assom, L, Chiolerio, A (2023) Energy-Based Economic Sustainability Protocols *Applied Sciences-Basel* 10.3390/App13116554

Thakur, T, Mehra, A, Hassija, V, Chamola, V, Srinivas, R, Gupta, KK, Singh, AP (2021) Smart Water Conservation Through a Machine Learning and Blockchain-Enabled Decentralized Edge Computing Network *Applied Soft Computing* 10.1016/J.Asoc.2021.107274

Tian, K, Chen, Z (2022) What Roles Do Smart Sensors Play In Citizens' Water Use? from the Perspective of Household Water-Saving *Water Supply* 10.2166/Ws.2021.385

Ting, L, Khan, M, Sharma, A, Ansari, MD (2022) A Secure Framework for IOT-Based Smart Climate Agriculture System: Toward Blockchain and Edge Computing *Journal of Intelligent Systems* 10.1515/Jisys-2022-0012.

Torabi, R, Gomes, A, Morgado-Dias, F (2023) Sure: A Smart Failover Blockchain-Based Solution for the Recycling Reuse Process *Energies* 10.3390/En16104146

Usman, M., Ali, N., Wohlin, C. (2023) A Quality Assessment Instrument for Systematic Literature Reviews in Software Engineering *E-Informatica Software Engineering Journal* 17. 230105. 10.37190/E-Inf230105

Vangala, A, Das, AK, Mitra, A, Das, SK, Park, Y (2023) Blockchain-Enabled Authenticated Key Agreement Scheme for Mobile Vehicles-Assisted Precision Agricultural IOT Networks *IEEE Transactions on Information Forensics and Security* 10.1109/Tifs.2022.3231121.

Vannucci, E, Pagano, AJ, Romagnoli, F (2021) Climate Change Management: A Resilience Strategy for Flood Risk Using Blockchain Tools *Decisions in Economics and Finance* 10.1007/S10203-020-00315-6

Vignesh, R, Prasad, KM (2023) Blockchain Assisted Ahmfa Authentication in Employee Performance Assessment System *Multimedia Tools and Applications* 10.1007/S11042-023-15846-4.

Waleed, M, Latif, R, Yakubu, BM, Khan, MI, Latif, S (2021) T-smart: Trust Model for Blockchain Based Smart Marketplace *Journal of Theoretical and Applied Electronic Commerce Research* 10.3390/Jtaer16060132

Wang, MH, Liang, XD, Li, Z (2021) Research on the Evaluation Index System of the Soil Remediation Effect Based on Blockchain *Land* 10.3390/Land10111274

Wang, YC, Fan, R, Liang, XL, Li, PG, Hei, XH (2022) Trusted Data Storage Architecture for National Infrastructure *Sensors* 10.3390/S22062318.

Wang, ZK, Wang, KN, Wang, YC, Wen, Z (2022) A Data Management Model for Intelligent Water Project Construction Based on Blockchain *Wireless Communications & Mobile Computing* 10.1155/2022/8482415.

Wu, HT, Tsai, CW (2019) An Intelligent Agriculture Network Security System Based on Private Blockchains *Journal of Communications and Networks* 10.1109/Jcn.2019.000043

Wu, JS (2022) Sustainable Development of Green Reverse Logistics Based on Blockchain *Energy Reports* 10.1016/J.Egyr.2022.08.219.

Wu, YL (2021) Cloud-Edge Orchestration for the Internet of Things: Architecture and AI-Powered Data Processing *IEEE Internet of Things Journal* 10.1109/Jiot.2020.3014845

Xia, W, Chen, X, Song, C (2022a) A Framework of Blockchain Technology in Intelligent Water Management. *Frontiers in Environmental Science* 10. 909606. 10.3389/Fenvs.2022.909606.

Xia, WJ, Chen, XH, Song, C (2022b) A Framework of Blockchain Technology in Intelligent Water Management *Frontiers in Environmental Science* 10.3389/Fenvs.2022.909606

Yakubu, BM, Khan, MI, Javaid, N, Khan, A (2021) Blockchain-Based Secure Multi-Resource Trading Model for Smart Marketplace *Computing* 10.1007/S00607-020-00886-7.

Yang, WH, Wang, RX, Li, L (2022) Method and System For Detecting and Recognizing Floating Garbage Moving Targets on Water Surface with Big Data Based on Blockchain Technology *Advances in Multimedia* 10.1155/2022/9917770

Yu, F, Zhang, XC, Liu, P, Chen, BB, Ma, J (2023) Blockchain-Like MIL-101(cr)/Carbon Black Electrodes for Unprecedented Defluorination by Capacitive Deionization *Small* 10.1002/Smll.202205619

Yuan, MH, Wang, DD, Zhang, F, Wang, SQ, Ji, S, Ren, YJ (2022) An Examination of Multi-Key Fully Homomorphic Encryption and Its Applications *Mathematics* 10.3390/Math10244678

Zeng, H, Dhiman, G, Sharma, A, Sharma, A, Tselykh, A (2023) An IOT and Blockchain-Based Approach for the Smart water management system in agriculture. *Expert Systems*, *40*(4), e12892.

Zhang, C, Zhou, G, Li, H, Cao, Y (2020). Manufacturing Blockchain of Things for the Configuration of a Data- and Knowledge-Driven Digital Twin Manufacturing Cell. *IEEE Internet of Things Journal* 7(12). 11884–11894. 10.1109/Jiot.2020.3005729

Zhang, J (2023) Application of Blockchain Technology in Water Management and New Energy Economic Development *Water Supply* 10.2166/Ws.2023.258

Zhang, RW (2022) Research on Financial Development of Water Resources Enterprises Based on Blockchain Technology *Mobile Information Systems* 10.1155/2022/3289301

Zhang, Y, Luo, WW, Yu, FF (2020) Construction of Chinese Smart Water Conservancy Platform Based on the Blockchain: Technology Integration and Innovation Application *Sustainability* 10.3390/Su12208306

Zhao, GQ, Liu, SF, Lopez, C, Lu, HY, Elgueta, S, Chen, HL, Boshkoska, BM (2019) Blockchain Technology in Agri-Food Value Chain Management: A Synthesis of Applications, Challenges and Future Research Directions *Computers in Industry* 10.1016/J.Compind.2019.04.002

Zhao, PF, Li, SQ, Hu, PJH, Gu, CH, Lu, S, Ding, SX, Cao, ZD, Xie, D, Xiang, Y (2023) Blockchain-Based Water-Energy Transactive Management with Spatial-Temporal Uncertainties *IEEE Transactions on Smart Grid* 10.1109/Tsg.2022.3230693

Zhou, BQ, Li, N, Lu, D, Xia, JY (2022) Correlation Analysis Between Economic Growth and Environmental Quality *Computer Systems Science and Engineering* 10.32604/Csse.2022.017750

Zhu, J, Wu, J, Bashir, AK, Pan, QQ, Yang, W (2023) Privacy-Preserving Federated Learning of Remote Sensing Image Classification with Dishonest Majority *IEEE Journal of Selected Topics in Applied Earth Observations and Remote Sensing* 10.1109/Jstars.2023.3276781

15 Application of Blockchain Technology in Supply Chain Management

Nguyen Thi Hang

15.1 INTRODUCTION

Faced with the strong impacts of the fourth industrial revolution, digital transformation technology on social life changes the way people are doing and performing every day. This has created great value for innovation, industrial revolution, society and especially social production and business activities (Szewczyk, 2019). One of the powerful fourth technologies that is the foundation for implementing digital transformation (Luo et al., 2021), the technologies have created solutions, methods or digital techniques to help collect, analyze and process data in daily work in the digital environment, specifically artificial intelligence, virtual reality, augmented reality, blockchain, Internet of Things (IoT),. (Cole et al., 2019). Among the typical technologies used for digital transformation platforms, each technology has its own advantages and limitations and is suitable for a group of fields. For example, Blockchain technology has high applicability: it is strongly used for product traceability applications, applications for managing business chains, supply chains or supply chain financing activities, banking (because the characteristics of this technology are safety, security, openness and transparency), secure data management. (Chang et al., 2022; Lohmer & Lasch, 2020; Dietrich et al., 2021).

According to the group of authors (Kopyto et al., 2020), the characteristics of blockchain technology are safety, data security with high reliability and consistency. Data managed and stored with blockchain technology before being exploited and used must be verified, authenticated, recorded and, especially when the data becomes immutable (i.e. cannot be modified or repaired anymore), all data are encrypted and ensure safe transactions. Blockchain technology ensures the operating mechanism of data blocks linked together and authenticated using hash function encryption algorithms. With hash function techniques, any fraud or signs of fraud, abnormal behavior in transaction activities in the data chain will not be allowed (Batwa & Norrman, 2020). Thus, Blockchain technology will prevent and limit potential risks to traditional data, ensuring security, safety and data integrity in online transactions (Rejeb et al., 2019).

In 2019, author Gurtu John researched and published a scientific paper on testing blockchain technology in supply chain operations. The study focused on tracing the origin of products, tracking input materials to production process as well as monitoring product output effectively. Each product data will be recorded, authenticated and stored in the blockchain, where it will become immutable thanks to smart and unchangeable links. Any activity to add information about the product will not change the data about the origin of the product. Based on the clear profile of the product, users can verify and authenticate product information, learn product origin easily, and prevent product counterfeiting. Thereby, it is possible to use product QR scanning systems. Thanks to this, customers can easily detect fraudulent behavior, counterfeit goods and goods of unknown origin. In addition, Blockchain technology manages real-time data, allowing users to know and monitor other updated information, which enhances publicity and transparency and increases interaction and connectivity

DOI: 10.1201/9781003599593-23

between parties participating in transactions in the supply chain. In addition, according to Alazab et al., 2021, thanks to technology, IoT sensors help data blocks in the blockchain link together, connect and create many new applications such as humidity monitoring, product temperature monitoring during transportation and storage to ensure safety and quality and avoid damage.

With Blockchain technology, according to Saberi et al., 2019, it is characterized by the ability to establish a tightly connected database of products, customers and related participants in the blockchain, ensuring publicity, transparency, accuracy and connectivity, interoperability and sharing common information. Therefore, support will closely monitor the origin of products in supply chain management activities in business.

According to Hirata et al., 2020, it is believed that relying on the characteristics of blockchain technology, which is to establish transparency, openness, accuracy and efficiency in terms of data that it can support in supply chain management through the creation of an immutable and unchangeable database. Moreover, with the characteristics of distributed data management and administration, this has created conditions for establishing data for customers and for users participating in the supply chain to be able to access it quickly, accurately and publicly. Once information has been recorded in blocks in the chain it cannot be changed, ensuring that all participants can access, exploit and use (Madumidha et al., 2019) and participants in the supply chain can view transactions and the history of transactions publicly and transparently (Di Vaio & Varriale, 2020).

This feature of blockchain technology has helped build trust between parties, components, and agents participating in transactions in the blockchain, because information and transactions are public, transparent, and data cannot be changed to hide information, and there are no acts of fraud, forgery, or data editing. This feature has helped agents and intermediaries participating in supply chain activities to verify information and communicate with each other accurately and publicly in the supply chain, allowing information to be automatically authenticated through the recording process in the chain (Akhtar & Moridpour, 2021; Boschi et al., 2018; Tokkozhina et al., 2023). And it allows to increase the processing speed and accuracy of information transmission in transaction activities because it eliminates dependence on intermediaries, that is, there is no need for third-party participation (Hader et al., 2020). Therefore, transactions will be carried out quickly, information will be continuously updated in real time and allow participating parties and agents to confirm information quickly and accurately without having to wait for verification from intermediaries (third parties) or navigate complex authentication steps (Härting et al., 2020; Zhu et al., 2022).

In 2020, the authors Çolak et al. followed in 2022 by the authors Dursun et al., studied the secure data distribution of blockchain technology, showing that data on blockchain is decentralized data and ensures data security and safety. Therefore, it is difficult for data attackers to attack, steal or illegally access data in the blockchain network (Chowdhury et al., 2023). According to the authors (Arabian et al., 2022; Park & Li, 2021), even if some nodes are compromised or illegally controlled, the system continues to operate through other nodes, thus making it difficult for malicious actors to create vulnerabilities and still ensuring data security and integrity. Attackers cannot influence or change data because information and data on the blockchain are recorded transparently and cannot be changed after being added (Dutta et al., 2020). Attackers with unauthorized access can only change historical data when they modify all copies of the blockchain on all network nodes (Gohil & Thakker, 2021; Shahzad et al., 2022), which is not possible for large blockchain networks. Furthermore, according to the authors Anand et al. (2022) and Bhadoria et al. (2020), a security issue of blockchain networks is the use of mechanisms such as proof of work (PoW) or proof of stake (PoS) to verify transactions and create new blocks to control data. This makes unauthorized attacks difficult and ineffective as both mechanisms require nodes to actively contribute to maintain the functionality of the hierarchy in protecting data and information in the supply chain (Agarwal et al., 2022; Khanfar et al., 2021).

Along with the trend of strong digital transformation, especially the power of 4.0 technology such as blockchain with outstanding features such as safety, data security, publicity, transparency, immutability, smart contracts, and so on. Enterprises have taken advantage of this to research, propose application solutions, build operational processes to support production and business activities, especially applying blockchain technology in monitoring, tracing product origin, and so forth, contributing to improving the efficiency of management and business of enterprises. To have a real view of the application of blockchain in enterprises, the article will propose solutions to build a process of applying blockchain technology to standardize processes and manage supply chain activities in enterprises in Vietnam in the trend of digital transformation.

15.2 THE IMPACT OF BLOCKCHAIN TECHNOLOGY ON SUPPLY CHAIN MANAGEMENT ACTIVITIES

15.2.1 Impact on Suppliers in the Supply Chain

In 2020, the authors (Vikaliana et al., 2020; Wamba & Queiroz, 2020) stated that the supply chain is a complex information and data system that includes organizations, activities, information, individuals, and resources that help maintain the regular operation of an organization, ensuring that goods and services are delivered from manufacturers and service providers to businesses and consumers. Therefore, individuals, components participating in the supply chain and resources to maintain the operation of the chain are all indispensable components of the supply chain (Tönnissen & Teuteberg, 2020), information related to product data, shipping and orders are required for the operation of the chain to be operated to ensure the transportation of goods and services from production to the final consumer. Tse et al. (2017) stated that each stage in the supply chain impacts the transportation process from the manufacturer to the final shipper to the buyer.

Based on the diagram in Figure 15.1, it can be seen that blockchain technology has completely revolutionized or automated the operations of manufacturers. One issue to note is that before the application of blockchain technology, suppliers of goods or services had to take on the traditional role of providing materials and components, the necessary conditions for production (Lim et al., 2021). For example: information on origin and quality can be provided through paper documents and shipping information can be communicated via email, fax or information management systems. It can also provide the ability to verify the origin and quality of goods, problems of shortages, and difficulties that may occur due to lack of transparency and the ability to change information. Blockchain technology can increase transparency in storing information related to the origin and quality of products (Wong et al., 2020). When every transaction is made and the information recorded in the blockchain becomes immutable, it cannot be changed or deleted. This has increased

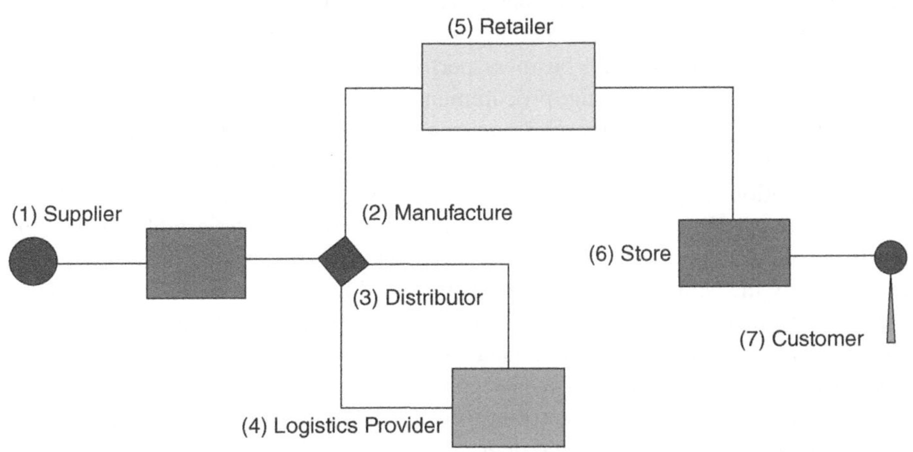

FIGURE 15.1 Entities involved in supply chain management.

Source: Designed by the author.

transparency compared to storing information about origin, quality, and shipping in immutable blocks. Therefore, participants in the supply chain can access the same copy of the data, but no single entity has the authority to independently control or change the data. This has created a transparent environment, eliminating misunderstandings or ambiguities related to information, creating trust and confidence in each other in transactions.

A prominent feature of the application of blockchain technology is that it allows suppliers to attach RFID chips to products and enter data into the blockchain to increase the authenticity of information and origin of the product. At the same time, it allows the creation of an immutable record of the origin and transportation of goods or smart contracts. Any action or transaction, or impact on data such as product quality inspection can be recorded in the blockchain, creating a reliable, transparent and public inspection history. With blockchain allowing the exploitation of information about product origin, transportation and quality closely linked in the blockchain, this has increased the level of authenticity, accuracy, and transparency when interacting and transacting with other entities in the supply chain (Kumar & Iyengar, 2017). In addition, according to Kim and Shin (2019), blockchain technology allows suppliers to conduct and update transactions in real time, allowing parties in the supply chain to update information quickly and accurately, and to be authenticated in a timely manner. This has created a transparent, public environment, increased authenticity and provided quality information and built trust in transactions on the internet environment, contributing to improving interaction and supply chain management.

15.2.2 The Influence of Blockchain Technology on Manufacturer Operations in the Supply Chain

Suppliers, as businesses participating in the supply chain, act as suppliers, producing products that are goods or services, then transporting and distributing them to end consumers through building a supply chain process for goods and services on a digital platform. Prior to the adoption of blockchain technology, managing the supply sources of goods and services, as well as tracking the origin of components throughout the production process, presented numerous challenges. Information pertaining to origin, quality, and production processes was typically recorded and stored manually through paper-based records. Documents such as certificates, delivery notes, quality inspection reports, and related materials were archived and preserved within the company (Figure 15.2).

Businesses utilize management systems to oversee their operational activities and manage information through various management software. Enterprise Resource Planning (ERP) systems are employed by businesses to integrate core business cycles, often in real-time, with the assistance of technology and software. This aids in collecting, storing, and managing information and data, facilitating strategic decision-making and business performance assessment. Businesses can carry out routine tasks such as product planning; procurement; marketing and sales; inventory management; finance and accounting; and human resource management through ERP software.

To conduct verification and confirmation activities related to products, businesses may opt for third-party organizations to perform these tasks (Srivastava & Dashora, 2022). Typically, in the food supply chain where products are sold directly to consumers by business establishments, the competent authorities conduct sampling and supervision, with testing results obtained from testing laboratories designated by the Ministry of Agriculture and Rural Development, ensuring compliance with prevailing food safety regulations and technical standards (Barua & Eslamian, 2021).

For businesses, cooperatives, and individual business owners who own and manage the entire supply chain from initial production to primary processing, processing and product commercialization, it is necessary to establish a monitoring process and mechanism for food safety across the entire product chain. These entities should undergo inspections by regulatory agencies to ensure that food safety conditions are met at every stage of production and business within the supply chain of food products (Qian and Papadonikolaki, 2021; Queiroz et al., 2020).

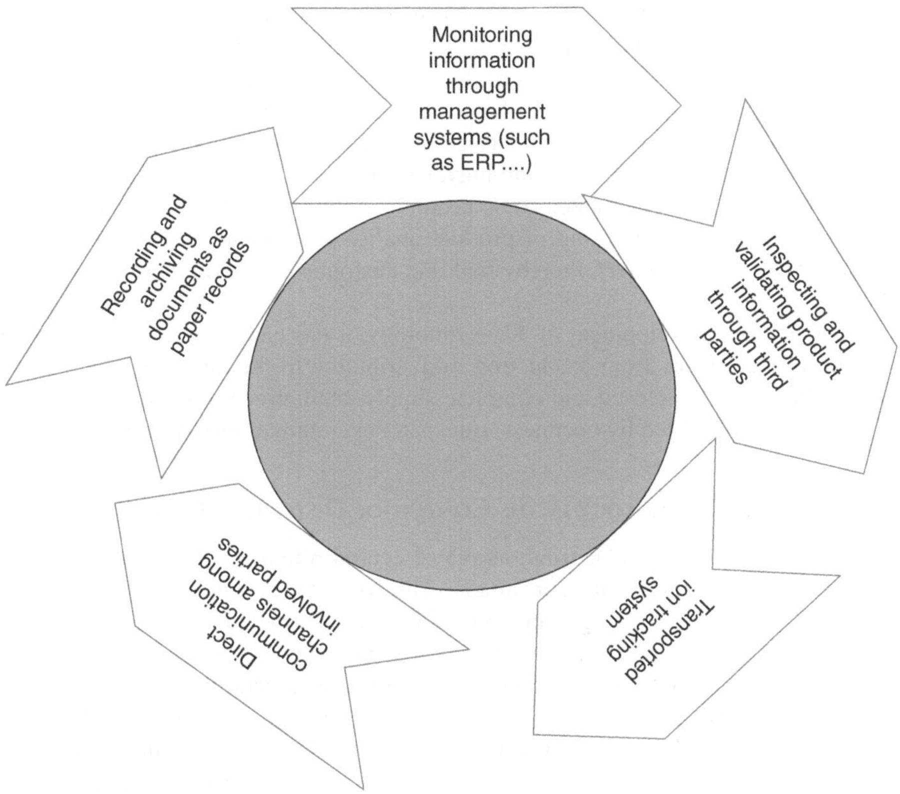

FIGURE 15.2 Traditional manufacturer operations in the supply chain.

Source: Designed by the author.

In the traditional approach to product transportation and distribution, businesses often monitor these activities through transportation management systems or other transportation information systems. In the supply chain, product distribution is a critical factor that directly impacts overall profitability as it significantly affects costs and customer experiences. In some cases, information regarding origin, quality, and transportation may be exchanged through direct communication between relevant parties. Typically, businesses choose communication methods such as email or phone calls. However, a reality that can happen in traditional supply chain operations is that in the process of production or transportation, distribution of goods may encounter limitations such as lack of publicity and transparency; information about goods can easily be changed, counterfeited, or fraudulent; and a lack of tools to accurately authenticate information, therefore leading to a lack of trust in supply chain operations. According to Pal (2020), the advent of blockchain technology has overcome the limitations of traditional supply chains, creating a breakthrough, allowing businesses to create a decentralized, secure, and anti-counterfeit data storage system.

In supply chain management, one of the current issues is to trace the origin of products to ensure that the supplied products have a clear origin and are monitored. Blockchain technology helps to deploy the application of tracing the origin of products. Thereby, it helps to effectively prevent fraudulent activities such as counterfeiting or changing unreliable information about a product. With blockchain technology, it helps to store detailed information about the origin of each component in the product as well as the production process of that product. According to Manzoor et al. (2025), blockchain technology has an important and outstanding feature, which is to create a transparent environment, with public information, where information is difficult to change illegally; prevent fraudulent activities and protect the integrity, data security, and information related to the origin of the products of the enterprise when participating in supply chain activities.

In addition, blockchain technology uses hash encryption techniques with a complex encryption system, allowing data and information to be transmitted securely during transactions. This has helped optimize the process of managing service quality as well as products in the chain. With a supply chain system using blockchain technology, it will ensure that information related to product quality and product origin is guaranteed to be accurate, complete and reliable, improving the efficiency of management and monitoring of product quality and enhancing the ability to verify the origin of products to end customers, thereby building customer trust (Mukherjee et al., 2022; Reda et al., 2020).

Thus, with the strong development of 4.0 technology, it has had a strong impact on the supply chain activities of enterprises from production to distribution to the final consumer, ensuring that there is a compliance with a closed and complete supply chain process based on the use of blockchain technology, characterized by openness, transparency, immutability, and smart contracts.

15.2.3 Blockchain Technology in the Context of Distributors' Operations

In the conventional supply chain, Distributors play a central role in the distribution and delivery of products from manufacturers to end consumers. The role of distributors is highly significant in inventory management, coordinating transportation, and ensuring timely product delivery to the final consumers. These tasks are often associated with complex processes, multiple parties, and various challenges related to transparency, traceability, and authentication (Müßigmann et al., 2021). The application of blockchain technology has brought transformative changes to the operations of distributors within the supply chain. Leveraging the decentralized and immutable nature of blockchain, distributors can enhance efficiency, transparency, and the security of essential information.

Once products are packaged, manufacturers will transport them to the distributors. Upon receiving comprehensive information about the livestock products that have been produced and packaged, distributors base their decisions on customer details and delivery dates to engage logistics services for product transportation. Depending on the nature of their business, companies will opt for suitable supply chain models. The conventional supply chain encompasses four primary stages (Figure 15.3).

The direct sales distribution channel involves distributing products directly from the manufacturer to the customers. This channel works well for products with moderate pricing, targeting customers with average income levels. Distributor intermediaries play a role as intermediaries between

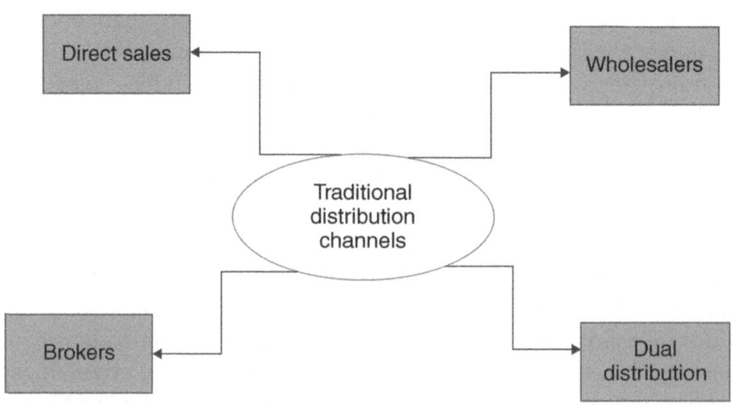

FIGURE 15.3 Traditional distribution channels.

Source: Designed by the author.

the manufacturer and the retailers; they don't directly transport products but manage sales contracts. Wholesalers purchase products in large quantities from the manufacturer and sell them at higher prices to consumers. As resellers, wholesale companies face higher risks if products remain unsold since consumers can buy directly from them. Dual distribution involves using multiple strategies to reach customers.

With the characteristics of blockchain technology in terms of transparency, publicity, immutability and enhanced security, blockchain technology acts as a technological messenger and empowers distributors in the supply chain of goods and services. Blockchain technology establishes and creates transaction records and operates in a reliable, immutable and unchangeable manner. Therefore, it supports distributors in controlling data accuracy and product authenticity, using a production and distribution process, and ultimately providing quality products of clear origin to end consumers. With the outstanding application characteristics of blockchain, it is inventory management by allowing accurate and transparent tracking of product quantities in storage. This continuously updates and shares inventory information between manufacturers, distributors and retailers through the blockchain network, thereby reducing ambiguity and inaccuracy related to business inventory.

In inventory management, businesses use blockchain technology, a system that allows distributors to maintain real-time records and accuracy of their inventory levels. In the supply chain, every transaction related to the transportation and distribution of goods and products is recorded on the blockchain, providing a public and transparent transaction history. By helping businesses minimize the risk of errors, discrepancies, and unauthorized access to audit data and, especially, anti-counterfeiting of changes in inventory, blockchain technology will help distributors and suppliers to enhance management and forecast delivery times and transportation of goods. At the same time, the system will help distributors verify the origin and quality of products, thereby building customer trust and helping distributors ensure that products meet quality standards based on the origin of the product. Blockchain technology allows tracking and monitoring of product origin, accurate information on the transportation and delivery process of products, information on the location, and the time and status of goods can be recorded transparently, publicly and stored immutably and unchangeably.

One characteristic of blockchain technology is that in the transaction activity there is no need for a third party (the intermediary is eliminated), creating a faster and more efficient process, reducing the time required to process transactions between distributors and other stakeholders in the supply chain. This thereby helps businesses bring transparency and reliability to all aspects of the supply chain and helps distributors to easily share information with partners and customers about goods, services, and distribution processes, enhancing trust, understanding, and cooperation between stakeholders involved in the supply chain.

15.3 DISCUSSIONS

The operating mechanism of the blockchain is that the parties or entities participating in the blockchain use data nodes linked together, each blockchain is considered a distributed network consisting of many nodes operating simultaneously, enhancing publicity, transparency, and ensuring reliability in the activities and data in each block. Each node in the block stores a copy of the entire data chain. With this way of storing and managing data, the transaction and distribution process will prevent excessive centralization of data and create transparency because all network members can access information for public and transparent exploitation. Each transaction is combined with the signature of the participant, creating a strong verification mechanism that helps convey reliability and authenticity to transactions. Transactions on the blockchain can be verified by encryption and digital signatures. Blockchain technology helps businesses manage their supply chains, from design and implementation to system monitoring and maintenance, promoting transparency and trust in supply chain management processes based on decentralized system design.

Figure 15.4 depicts the schematic diagram of the process of designing the structure of a blockchain, including the tasks and components of setting up specific parameters of the data to be recorded

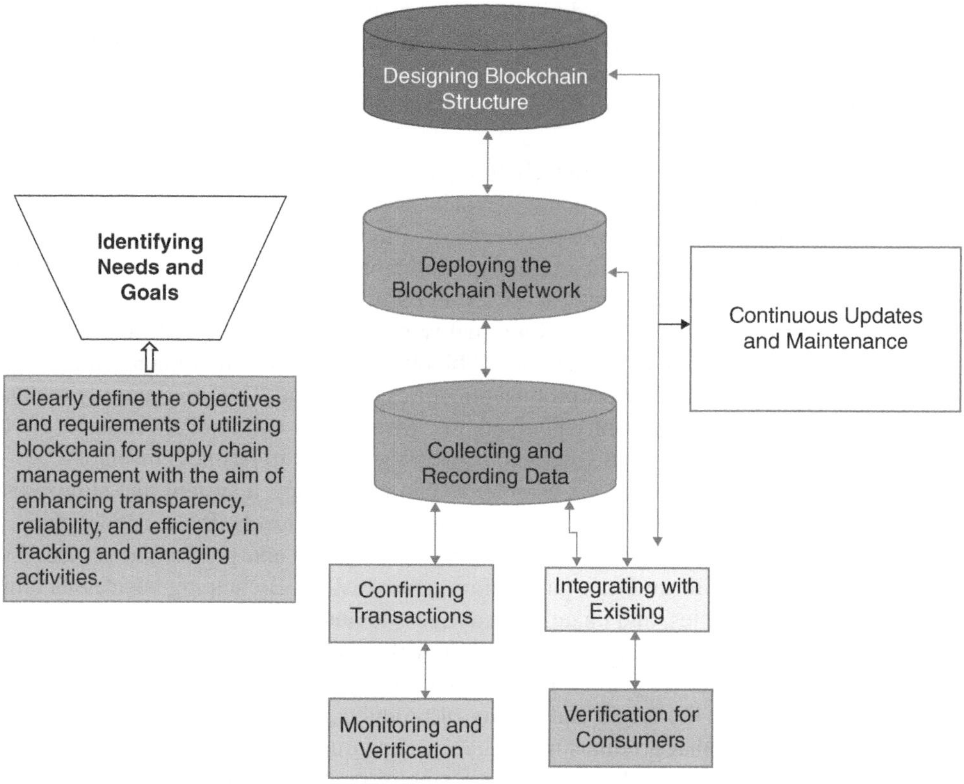

FIGURE 15.4 A decentralized data design and management scheme designed on a platform using blockchain technology.

Source: Designed by the author.

on the chain; determining the number of participating nodes and access rights to the system; determining other relevant information based on the goals and requirements of the system; determining the input data in the blockchain, including detailed information about the origin of the product; designing the production process; specific transportation parameters; quality control criteria and establishing other supply chain activities. To establish a supply chain operation using blockchain, we must also determine the type of data to be stored on the chain, as determining the input data also includes the number of nodes participating in the blockchain network. In the supply chain, it is necessary to identify nodes representing stakeholders, such as manufacturers, distributors, retailers, and raw material suppliers.

It's crucial to define access rights and permissions for different nodes within the network. Not all participants have equal access to all data. Managing access rights ensures that sensitive information is only accessible to authorized parties. Privacy concerns are addressed by deciding which data is visible to all participants and which data is encrypted or access-restricted through the implementation of robust security measures to safeguard data integrity, access rights, and other pertinent information.

The process of deploying a blockchain network is determined by establishing participating nodes, including relevant parties within the supply chain such as manufacturers, distributors, retailers, and raw material suppliers. Setting up the blockchain network with participation from nodes of supply chain stakeholders involves node configuration, data replication, ensuring data synchronization, deploying security measures, defining consensus mechanisms, and establishing guidelines for node interaction and management. Based on the characteristics of blockchain technology, which is to use

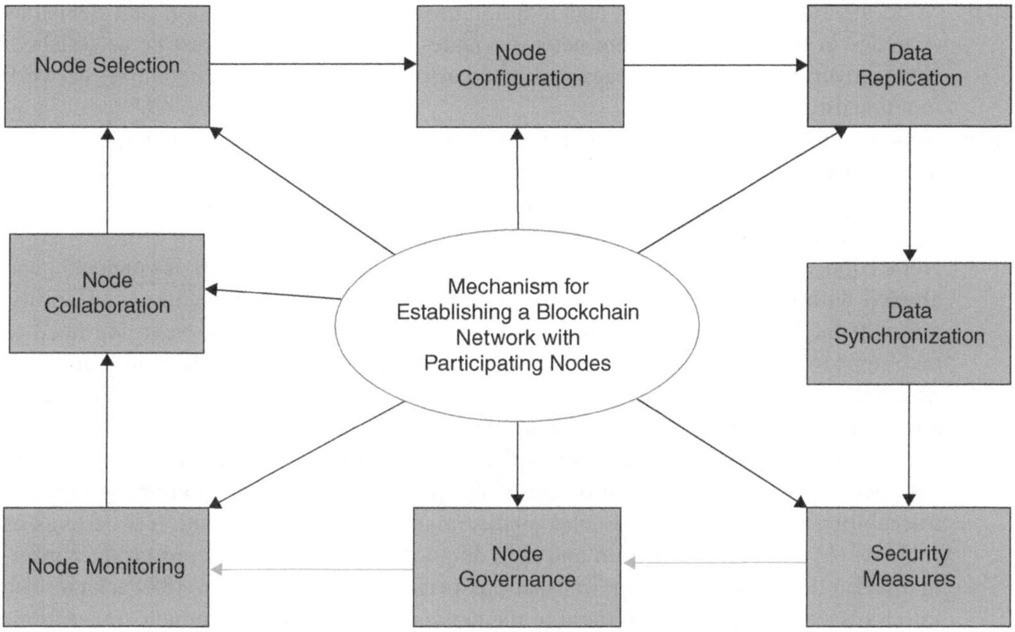

FIGURE 15.5 Illustrating the process of establishing participating nodes in a blockchain network in a supply chain.

Source: Designed by the author.

data in a decentralized approach, this method helps to increase publicity, transparency, data security, and especially to create trust and confidence in supply chain management of stakeholders such as manufacturers, distributors, retailers, and raw material suppliers, specifically including the main stages illustrated in Figure 15.5.

Based on Figure 15.5, we need to perform:

Step 1: Node selection: by identifying the participants – the main entities in the supply chain – that will participate in the Blockchain network. The stakeholders participating in the supply chain include manufacturers, distributors, and consumers (enterprises).

Step 2: Node configuration: to configure each node to ensure that they are compatible with the selected blockchain protocol. Identifying and configuring nodes in the chain depends on the scale and requirements of the supply chain. Nodes can be separate dedicated servers or individual devices.

Step 3: Data replication: to perform data at each node through copying and storing a complete copy of the data chain. This step ensures data redundancy and availability across the entire network, leveraging distributed data storage for security and fault tolerance.

Step 4: Data synchronization: to ensure that all nodes remain synchronized in real time and that any interactions with data such as updates, additions, or edits to the data chain must be transmitted to all nodes, maintaining a consistent and accurate ledger. In particular, with blockchain, nodes are used to reach consensus on the validity of transactions, ensuring that all copies of the data chain on the nodes are identical, to prevent data forgery and fraud.

Step 5: Security measures: to ensure that data must be secured by implementing security mechanisms, protecting nodes from unauthorized access and potential attacks. Uses encryption, authentication, and access control methods to protect the integrity of the blockchain network.

Step 6: Node governance: to manage and monitor the participation, behavior, and decisions of nodes in the new blockchain network, node governance rules must be established. Node governance will help ensure a standardized approach to maintaining network functionality.

Step 7: Node monitoring: to manage and monitor the status, performance, and activities of each node in the network to help identify and resolve new issues in a timely manner, it is necessary to implement monitoring tools. At that time, updating each node as well as regular maintenance will ensure optimal performance and security.

Step 8: Node cooperation: to facilitate and ensure that the blockchain network is connected and aligned with the common goals of the supply chain; maintaining continuous stability, this will facilitate cooperation and communication between nodes. Next, we need to take steps to collect data related to activities in the supply chain. At the same time, it is necessary to build a supply chain management process that includes tasks such as managing and authenticating information about product origin, transportation, and distribution activities, as well as monitoring and controlling quality. In order to collect product data, we need to determine the origin of the product through the monitoring process, traceability, production system, and quality management. The operation process is as follows: (1) Collect information and data about the product. (2) Record product information on the network through transactions performed in the system. (3) Package data into transactions and then add them to the network chain. (4) Confirm new transactions added to the chain by a number of nodes participating in the blockchain network. Note: this confirmation must ensure the accuracy and reliability of information in the transaction. In particular, it must prevent interference or modification of unreliable data.

Following is an illustrative experiment on the implementation of blockchain technology in the electricity industry, in which the electricity transaction model on the power grid is deployed according to the peer-to-peer model. The system is deployed with the priority goal of providing electricity to customers reliably, quickly, accurately, and promptly at the lowest cost. This deployment is carried out according to the supply chain management model using blockchain technology. Figure 15.6 is an illustrative image of the implementation of the blockchain technology application model in the electricity industry:

Many countries require their electricity systems to efficiently harness solar and wind energy sources. Conversely, some other nations aim to leverage the battery resources of electric vehicles (EVs) to balance peak loads. However, managing this becomes more difficult and intricate when numerous customers integrate various devices – from solar panels to energy storage systems and smart controllers – to oversee their own energy generation and grid consumption. Incorporating blockchain technology into peer-to-peer electricity trading can enhance transparency, efficiency, and security in transactions among involved parties.

The peer-to-peer trading model can be broadly applicable to businesses engaged in trading goods and services. When employing blockchain, they can optimize transactions with customers, provided they adhere well to the following requirements (Figure 15.7).

In the process of peer-to-peer electricity trading, utilizing blockchain technology ensures secure and efficient verification of identity and ownership, enhancing security measures. To achieve this, both the buyer and seller create personal accounts on the blockchain system, providing necessary information for identification and differentiation of participating entities in the transaction. To ensure accuracy and reliability, personal information and ownership rights of each user are verified, encrypted, and stored on the blockchain for permanent record-keeping and security assurance. The paramount advantage of blockchain is creating a trustworthy environment for conducting electricity trading transactions in a peer-to-peer network. Therefore, the identity and ownership verification process ensures clear identification of all transaction participants and safeguards their personal information.

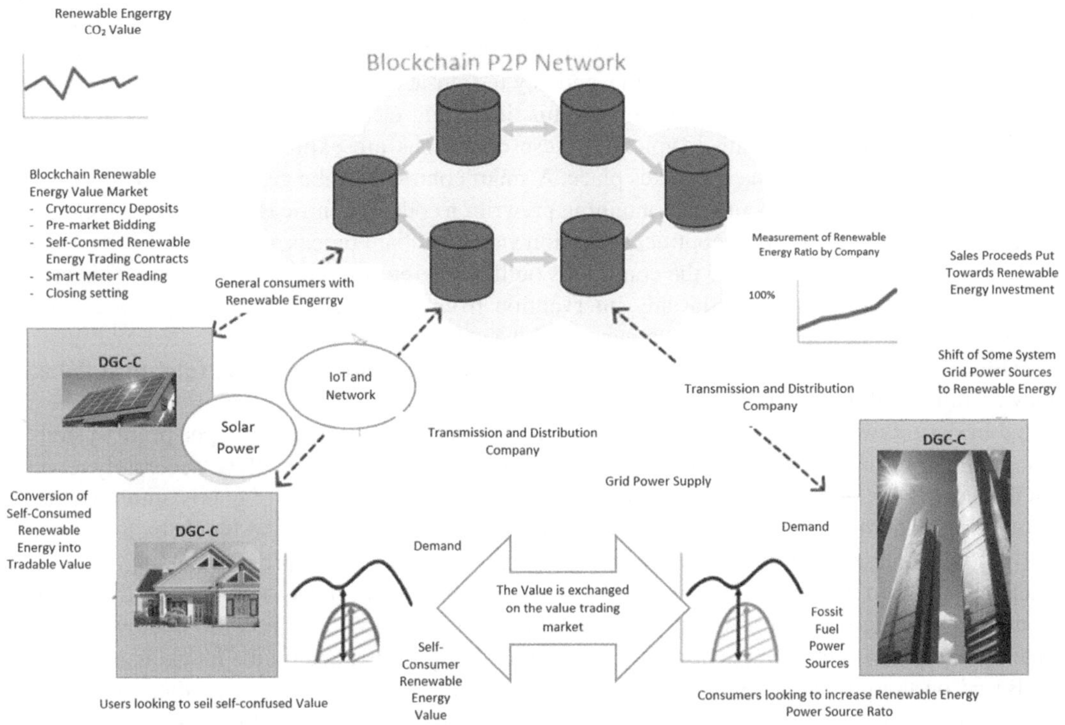

FIGURE 15.6 Application model of blockchain in peer-to-peer electricity trading.

Source: https://cpc.vn/.

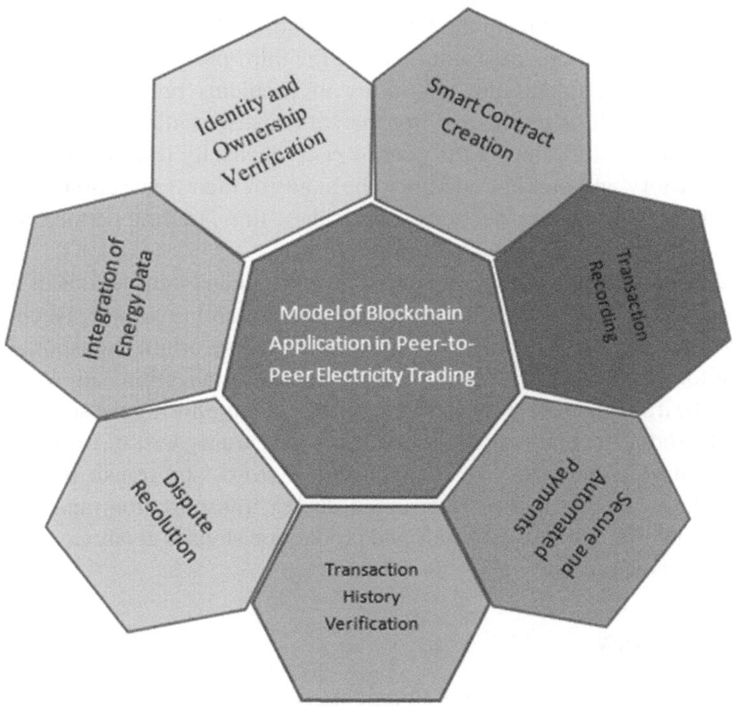

FIGURE 15.7 Model of blockchain application in peer-to-peer electricity trading.

Source: Designed by the author.

Once the accounts are verified, to execute peer-to-peer electricity trading transactions, the blockchain will support the creation of smart contracts. Applying smart contracts in peer-to-peer electricity trading introduces automation and transparency to transaction conditions. In this phase, both the buyer and seller will agree on transaction terms, including electricity price, quantity, transaction time, and other relevant details. Typically, these conditions are explicitly defined and mutually agreed upon before the transaction takes place. A smart contract will be generated on the blockchain platform. This is a digital document containing prewritten code that intricately describes the transaction conditions. Through smart contract execution via blockchain blocks, the contract will automatically trigger the transaction once the conditions outlined in the smart contract are met. This process is executed autonomously without any intervention from third parties, preventing losses and the exposure of information from both the buyer and the seller.

Throughout the process of peer-to-peer electricity trading, participants can fully trust the accuracy and immutability of data on the blockchain, as the transaction recording process will be executed in the following step. This will establish a transparent, secure, and tamper-proof system for transaction information. As a result, every electricity trading transaction will be accurately recorded on the blockchain system. All details regarding the buyer, seller, quantity of goods, unit price, transaction time, and other related specifics will be stored in a data block on the blockchain. These data will be automatically encrypted and cryptographically linked to the previous block, forming an unchangeable and irreversible chain of blocks (blockchain). Thus, the system ensures that information about trading transactions will never be altered or modified once recorded, allowing supply chain participants to trust in the accuracy and immutability of the data on the blockchain.

Based on the unique advantages of blockchain technology in decentralized data management, using a peer-to-peer network system where transactions in the system deploy applications on peer-to-peer transactions, the electricity industry simultaneously uses smart contracts to facilitate the electronic payment process to be safe and automatically performed. This has brought convenience to both businesses and consumers. The use of smart contracts requires that the conditions set out need to be met, the payment process will be automatically activated, and the corresponding amount will be automatically transferred from the buyer's account to the seller's account without the intervention of a third party. With this payment method, the third party will be eliminated, so the automatic payment process eliminates the possibility of problems related to delayed or unreliable payments. Transactions will be public and transparent because it allows both buyers and sellers to review their transaction history during the peer-to-peer electricity transaction process. Buyers and sellers can access history and monitor and track the status of their transactions on the blockchain to see information about their past transactions such as date, time, trading partner, amount of electricity, price, and any other transaction related information.

Blockchain technology helps monitor and verify the origin and transactions of parties involved in the supply chain, which has brought transparency and reliability to supply chain management, allowing stakeholders to access detailed information about the origin, transportation, distribution, and quality monitoring of products. Deploying blockchain applications allows information about product quality to be transparently recorded by tracking and monitoring the status on the blockchain, and the origin of product components can be clearly and transparently traced. Management and tracking of information about transportation schedules, stops, and transit times can be recorded, ensuring that products are not lost or interrupted during transportation, and monitoring detailed information about the origin of raw materials and production steps is always available, building trust and transparency for consumers.

15.4 CONCLUSIONS

For businesses, supply chain activities play a particularly important role. It is a complete, closed, and tight process that helps businesses manage their operations from the production and distribution of goods and services to the end consumer. Researching the advantages and typical features of blockchain technology and researching and applying them to supply chain activities will bring many advantages to businesses in the process of managing, monitoring, and supervising business activities, helping to create trust for customers to participate in transactions with smart functions such as product traceability, monitoring transaction status with smart contracts, data immutability, and so on. It's essential to identify areas in the supply chain where blockchain technology can be applied to optimization, with features such as enhancing transparency, inventory management, product quality monitoring, and enhancing traceability. To do this effectively, businesses need to clearly define objectives and functions that the blockchain network needs to fulfill within the supply chain, as well as how supply chain elements will participate in the blockchain network and how they will interact with it.

Based on the requirements and functions defined in the previous step, businesses need to develop and deploy a blockchain application to ensure accurate integration with existing systems and data. It is important to conduct testing of the blockchain application to ensure stability and proper operation as planned. To ensure understanding and optimal operation of the system, training for employees and stakeholders on how to use and interact with the new blockchain system is necessary. The system application must be widely deployed to supply chain participants and ensure they understand how to use the new technology. This helps businesses effectively implement blockchain technology into the operations of manufacturers and stakeholders within the supply chain, creating transparency, efficiency, and integration in managing and monitoring manufacturing activities and transactions.

Research on the application of blockchain technology in supply chain management contributes to a deeper understanding of the potential and benefits of blockchain in various economic sectors. It helps businesses recognize the power of blockchain in improving transparency, security, and efficiency in supply chains. Therefore, the article plays an important role in promoting digital transformation and enhancing supply chain management for businesses. It provides guidance and specific procedures for effectively applying blockchain technology in supply chain management, thereby generating significant progress in the performance and competitiveness of businesses.

COMPLIANCE WITH ETHICAL STATEMENT

This is the author's research work. The information in the article is given honestly, with full citations. The author of the article is fully responsible for the content presented in his article.

ACKNOWLEDGMENT

This paper is the result of the implementation of a ministerial-level Science and Technology project funded by the Ministry of Education and Training of Vietnam. Project title: Enhancing the sustainability of supply chains through the integration of digital transformation and supply chain finance solutions in Vietnam. Project code: B2025-TNA-15.

REFERENCES

Agarwal, U., Rishiwal, V., Tanwar, S., Chaudhary, R., Sharma, G., Bokoro, P. N., & Sharma, R. (2022). Blockchain technology for secure supply chain management: A comprehensive review. *IEEE Access*, *10*, 85493–85517.

Akhtar, M., & Moridpour, S. (2021). A review of traffic congestion prediction using artificial intelligence. *Journal of Advanced Transportation*, *2021*(1), 8878011.

Alazab, M., Alhyari, S., Awajan, A., & Abdallah, A. B. (2021). Blockchain technology in supply chain management: an empirical study of the factors affecting user adoption/acceptance. *Cluster Computing*, *24*, 83–101.

Anand, A., Seetharaman, A., & Maddulety, K. Jain, S. (2022). Implementing blockchain technology in supply chain management. In *International Conference on Data Mining and Machine Learning*, Orlando, FL, USA.

Arabian, M., Ghadiri Nejad, M., & Barenji, R. V. (2022). Blockchain technology in supply chain management: Challenge and future perspectives. In *Industry 4.0: Technologies, Applications, and Challenges* (pp. 201–220). Singapore: Springer Nature Singapore.

Barua, P., & Eslamian, S. (2021). Food Production and Consumption in the face of Climate Induced Displacement: An Exploration of Indigenous Knowledge Management Practices of Southeastern Bangladeshis. *IUP Journal of Knowledge Management*, *19*(4), 52–83.

Batwa, A., & Norrman, A. (2020). A framework for exploring blockchain technology in supply chain management. *Operations and Supply Chain Management: An International Journal*, *13*(3), 294–306.

Bhadoria, R. S., Sharma, N., & Pandey, M. K. (2020). Demystifying the power of blockchain technology in supply chain management. *International Journal of Applied Evolutionary Computation (IJAEC)*, *11*(4), 38–53.

Boschi, A. A., Borin, R., Raimundo, J. C., & Batocchio, A. (2018, September). An exploration of blockchain technology in supply chain management. In *Proceedings of the 22nd Cambridge International Manufacturing Symposium*, Cambridge, MA, USA, pp. 27–28.

Chang, A., El-Rayes, N., & Shi, J. (2022). Blockchain technology for supply chain management: A comprehensive review. *FinTech*, *1*(2), 191–205.

Chowdhury, S., Rodriguez-Espindola, O., Dey, P., & Budhwar, P. (2023). Blockchain technology adoption for managing risks in operations and supply chain management: evidence from the UK. *Annals of Operations Research*, *327*(1), 539–574.

Cole, R., Stevenson, M., & Aitken, J. (2019). Blockchain technology: Implications for operations and supply chain management. *Supply Chain Management: An International Journal*, *24*(4), 469–483.

Di Vaio, A., & Varriale, L. (2020). Blockchain technology in supply chain management for sustainable performance: Evidence from the airport industry. *International Journal of Information Management*, *52*, 102014.

Dietrich, F., Ge, Y., Turgut, A., Louw, L., & Palm, D. (2021). Review and analysis of blockchain projects in supply chain management. *Procedia Computer Science*, *180*, 724–733.

Dursun, T., Birinci, F., Alptekin, B., Sertkaya, I., Hasekioglu, O., Tunaboylu, B., & Zaim, S. (2022). Blockchain technology for supply chain management. In *Industrial Engineering in the Internet-of-Things World: Selected Papers from the Virtual Global Joint Conference on Industrial Engineering and Its Application Areas, GJCIE 2020*, August 14–15, 2020 (pp. 203–217). Springer International Publishing.

Dutta, P., Choi, T. M., Somani, S., & Butala, R. (2020). Blockchain technology in supply chain operations: Applications, challenges and research opportunities. *Transportation Research Part e: Logistics and Transportation Review*, *142*, 102067.

Gohil, D., & Thakker, S. V. (2021). Blockchain-integrated technologies for solving supply chain challenges. *Modern Supply Chain Research and Applications*, *3*(2), 78–97.

Gurtu, A., & Johny, J. (2019). Potential of blockchain technology in supply chain management: A literature review. *International Journal of Physical Distribution & Logistics Management*, *49*(9), 881–900.

Hader, M., Elmhamedi, A., & Abouabdellah, A. (2020, December). Blockchain technology in supply chain management and loyalty programs: Toward blockchain implementation in retail market. In *2020 IEEE 13th international colloquium of logistics and supply chain management (LOGISTIQUA)*, Fez, Morocco, pp. 1–6, IEEE.

Härting, R. C., Sprengel, A., Wottle, K., & Rettenmaier, J. (2020). Potentials of blockchain technologies in supply chain management-A conceptual model. *Procedia Computer Science*, *176*, 1950–1959.

Hirata, E., Lambrou, M., & Watanabe, D. (2020). Blockchain technology in supply chain management: Insights from machine learning algorithms. *Maritime Business Review*, *6*(2), 114–128.

Khanfar, A. A., Iranmanesh, M., Ghobakhloo, M., Senali, M. G., & Fathi, M. (2021). Applications of blockchain technology in sustainable manufacturing and supply chain management: A systematic review. *Sustainability*, *13*(14), 7870.

Kim, J. S., & Shin, N. (2019). The impact of blockchain technology application on supply chain partnership and performance. *Sustainability*, *11*(21), 6181.

Kopyto, M., Lechler, S., von der Gracht, H. A., & Hartmann, E. (2020). Potentials of blockchain technology in supply chain management: Long-term judgments of an international expert panel. *Technological Forecasting and Social Change*, *161*, 120330.

Kumar, M. V., & Iyengar, N. C. S. (2017). A framework for Blockchain technology in rice supply chain management. *Advanced Science and Technology Letters*, *146*, 125–130.

Lim, M. K., Li, Y., Wang, C., & Tseng, M. L. (2021). A literature review of blockchain technology applications in supply chains: A comprehensive analysis of themes, methodologies and industries. *Computers & Industrial Engineering*, *154*, 107133.

Lohmer, J., & Lasch, R. (2020). Blockchain in operations management and manufacturing: Potential and barriers. *Computers & Industrial Engineering*, *149*, 106789.

Luo, Y., Xu, S., & Zhou, Q. (2021, August). Application of Blockchain Technology in Supply Chain Management. In *The Sixth International Conference on Information Management and Technology*, Changsha, China, pp. 1–4.

Madumidha, S., Ranjani, P. S., Vandhana, U., & Venmuhilan, B. (2019, May). A theoretical implementation: Agriculture-food supply chain management using blockchain technology. In *2019 TEQIP III Sponsored International Conference on Microwave Integrated Circuits, Photonics and Wireless Networks (IMICPW)*, Tiruchirappalli, India, pp. 174–178, IEEE.

Manzoor, R., Sahay, B. S., & Singh, S. K. (2025). Blockchain technology in supply chain management: An organizational theoretic overview and research agenda. *Annals of Operations Research*, *348*(3), 1307–1354.

Mukherjee, A. A., Singh, R. K., Mishra, R., & Bag, S. (2022). Application of blockchain technology for sustainability development in agricultural supply chain: Justification framework. *Operations Management Research*, *15*(1), 46–61.

Müßigmann, B., von der Gracht, H., & Hartmann, E. (2021). Blockchain technology in logistics and supply chain management: A bibliometric literature review from 2016 to January 2020. *IEEE Transactions on Engineering Management*, *67*(4), 988–1007.

Pal, K. (2020). Information sharing for manufacturing supply chain management based on blockchain technology. In Idongesit Williams (Ed.) *Cross-Industry Use of Blockchain Technology and Opportunities for the Future*, pp. 1–17, IGI Global.

Park, A., & Li, H. (2021). The effect of blockchain technology on supply chain sustainability performances. *Sustainability*, *13*(4), 1726.

Qian, X., & Papadonikolaki, E. (2021). Shifting trust in construction supply chains through blockchain technology. *Engineering, Construction and Architectural Management*, *28*(2), 584–602.

Queiroz, M. M., Telles, R., & Bonilla, S. H. (2020). Blockchain and supply chain management integration: A systematic review of the literature. *Supply Chain Management: An International Journal*, *25*(2), 241–254.

Reda, M., Kanga, D. B., Fatima, T., & Azouazi, M. (2020). Blockchain in health supply chain management: State of art challenges and opportunities. *Procedia Computer Science*, *175*, 706–709.

Rejeb, A., Keogh, J. G., & Treiblmaier, H. (2019). Leveraging the internet of things and blockchain technology in supply chain management. *Future Internet*, *11*(7), 161.

Saberi, S., Kouhizadeh, M., Sarkis, J., & Shen, L. (2019). Blockchain technology and its relationships to sustainable supply chain management. *International Journal of Production Research*, *57*(7), 2117–2135.

Shahzad, K., Zhang, Q., Khan, M. K., Ashfaq, M., & Hafeez, M. (2022). The acceptance and continued use of blockchain technology in supply chain management: A unified model from supply chain professional's stance. *International Journal of Emerging Markets*, *18*(12), 6300–6321. https://doi.org/10.1108/IJOEM-11-2021-1714.

Srivastava, A., & Dashora, K. (2022). Application of blockchain technology for agrifood supply chain management: A systematic literature review on benefits and challenges. *Benchmarking: An International Journal*, *29*(10), 3426–3442.

Szewczyk, P. (2019). Application of blockchain technology in supply chain management. *Zeszyty Naukowe. Organizacja i Zarządzanie/Politechnika Śląska*, *136*, 591–600.

Tokkozhina, U., Lucia Martins, A., & Ferreira, J. C. (2023). Uncovering dimensions of the impact of blockchain technology in supply chain management. *Operations Management Research*, *16*(1), 99–125.

Tönnissen, S., & Teuteberg, F. (2020). Analysing the impact of blockchain-technology for operations and supply chain management: An explanatory model drawn from multiple case studies. *International Journal of Information Management*, *52*, 101953.

Tse, D., Zhang, B., Yang, Y., Cheng, C., & Mu, H. (2017, December). Blockchain application in food supply information security. In *2017 IEEE International Conference on Industrial Engineering and Engineering Management*, pp. 1357–1361. IEEE.

Vikaliana, R., Rasi, R. Z. R. M., Pujawan, I. N., & Sham, R. (2020). The application of blockchain technology in agribusiness supply chain management in Indonesia. *Solid State Technology*, *63*(6), 16522–16533.

Wamba, S. F., & Queiroz, M. M. (2020). Blockchain in the operations and supply chain management: Benefits, challenges and future research opportunities. *International Journal of Information Management*, *52*, 102064.

Wong, L. W., Tan, G. W. H., Lee, V. H., Ooi, K. B., & Sohal, A. (2020). Unearthing the determinants of Blockchain adoption in supply chain management. *International Journal of Production Research*, *58*(7), 2100–2123.

Zhu, Q., Bai, C., & Sarkis, J. (2022). Blockchain technology and supply chains: The paradox of the atheoretical research discourse. *Transportation Research Part E: Logistics and Transportation Review*, *164*, 102824.

16 Blockchain in Smart Cities *Revolutionizing Sustainability and Cyber-Physical Systems*

Mohammad Karrabi, Farkhondeh Jabari, and Asghar Akbari Foroud

16.1 INTRODUCTION

In recent years, blockchain technology has attracted worldwide attention due to the urgent need to digitalize the power and energy industry using smart contracts, reliable and secure communication platforms, producers and active prosumers, smart metering infrastructure, and so on from various points of view such as economic, environmental, energetic, social, political and legal prospects. Firstly, blockchain technology, introduced in 2008 by Bitcoin, was used in digital payment systems. Then, its applications were enhanced to smart contracts, banking and financial services in different fields such as health and treatment, power and energy systems, and other uses, especially in combination with artificial intelligence. Using blockchain technology as a distributed digital ledger in all fields of business, industry, and commerce, there is no need for a central control or trusted third authority for all peer-to-peer (P2P) transactions. Moreover, it provides immutable and time-stamped blocks of data entering a set of computers and all end-users in a cyber-physical-social secure manner (Tripathi et al., 2023).

In this way and using peer-to-peer transactions and blockchain, it stores a list of records as data and a database. With introduction of blockchain technology in distribution energy networks and smart cities without any intermediary, one of advantages of this technology in smart grids is decentralization of peer-to-peer energy and financial exchanges.

Among these advantages of blockchain technology, some important points can be mentioned:

- Immutability of data: In economic-social processes and use of blockchain, it is not possible to change data over time. This gives credit to financial transactions, asset ownership and identity management. Making these transactions on a blockchain platform makes it very difficult to change them. Also, use of a peer-to-peer network and its organization, which has a centralized architecture, and single-point attacks are prevented, unlike networks with centralized servers which are disrupted under attacks.
- Smart contracts: Smart contracts on blockchain platforms allow users to conduct exchanges in a distributed ledger between them and transfer anything of value such as assets, money and content on this platform.
- Consensus model: In this process, all members and users of the blockchain network can reach and agree on correct status and healthy data.

The main applications of blockchain technology are summarized in Figure 16.1.

There have been many technological advances in smart grid energy systems to meet energy demand. Optimization of energy networks, grid automation and energy commercialization, and

DOI: 10.1201/9781003599593-24

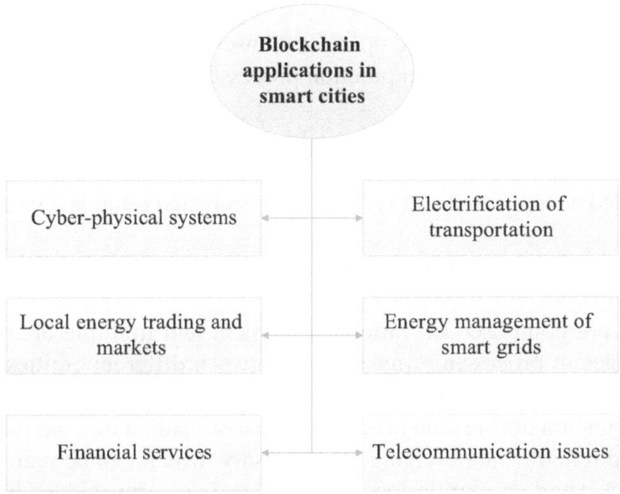

FIGURE 16.1 Different applications of blockchain technology in cyber-physical power and energy systems.

management should be considered. Researching blockchain systems and their advantages and disadvantages, and integrating them into smart energy grids, paves a way for fast and optimal energy exchanges. Some prominent features of using blockchain in energy networks can include peer-to-peer energy exchange, smart measurements, transparency and self-operation (Uddin et al., 2023).

16.2 ENERGY MANAGEMENT OF SMART GRIDS

Blockchain technology plays a key role in management of smart energy grids. There are three management strategies: Centralized platform, multi-agent structure and decentralized framework. Three categories can be stated for energy management strategies (Alladi et al., 2019; Miglani et al., 2020):

- Central energy management framework
  A center or trust third party is required for managing all generation, transportation, distribution, energy storage and consumption in a smart power network. Hence, resiliency and flexibility of these systems are low because of high risk of cyber-physical security as a result of data storage in one data center.
- Multi-agent management infrastructure
  In developing countries, the operational cost of the multi-agent energy management infrastructure is relatively high.
- Peer-to-peer energy trading platform
  In this energy management approach, data is directly exchanged between all nodes of microgrids including customers, producers, and prosumers, aggregators, retailers, owners of storage units and other emerging technologies.

In comparison with multi-agent energy management and centralized schemes, the blockchain-based P2P energy trading framework can be more reliable and secure, cost-energetic-environmentally effective, flexible, and resilient in smart energy microgrids (Wang et al., 2021). In a blockchain-based water-energy hybrid trading model, there are a series of blocks for recording all transaction information, which makes it possible to have a distributed ledger in a P2P transaction (Li et al., 2023). From technical points of view, there are three types of blockchain technology as follows (Cantillo-Luna et al., 2022; Yang et al., 2020):

- Public blockchain
  Each blockchain node is known and open for all users providing an opportunity for them to collaborate in transactions and computation processes.
- Private blockchain
  A node is open and its information is available for those authorized by developers.
- Consortium blockchain
  Each node should be registered and verified in a secure and reliable data exchange interface.

The main objective of blockchain technology is to reduce the decentralized interactions in the completely closed networks.

Smart microgrids are composed of various equipment and multiple areas, which may lead to some losses or leakages in processing transactions between different entities. Similar to advanced power and energy systems, microgrids have several components and zones. Meanwhile, any loss or leakage of node information during data processing between authenticating parties will be probable. Therefore, secure and reliable data exchange is necessary, which can be realized using blockchain. It includes permissioned and un-permissioned categories. In a permissioned blockchain framework, some secure channels are used for data exchange and safe energy transactions based on identity access. As a result, it can adopt shared consensus and improve computational efficiency of transactions. Private blockchain and some consortium ones can be considered as permissioned types. A permissioned blockchain makes it possible to procure security and privacy of transactions between multiple individuals, especially providing various advantages for effective energy management in smart grids without using a trusted third party (Wang et al., 2021). Smart contracts and cryptocurrency are two important concepts in microgrid energy management strategies, which are typically used in the data exchanging process for P2P energy-trading platforms (Khan et al., 2021). All energy transactions can be realized between producers and consumers via a decentralized structure using blockchain. These local energy exchanges can be done in microgrids and local energy hubs with a decentralized structure and transparent and immutable transactions for any participant. Data integrity and user authentication are guaranteed, encrypted and unhackable. All scenarios defined by these smart contracts are facilitated by P2P exchanges. Energy trading is used in charging vehicles, electricity exchanges in microgrids, management of distribution grids and network automation, and it allows users to trade energy in a safe and transparent platform with full authority (Yapa et al., 2021). Some key features of smart grids are stated as follows:

- Integration of renewable energy sources and storage technologies
- Encouraging and educating the community to participate in electricity production and exchange with the local grid
- Creating bills and trading energy in a transparent environment and at a reasonable price in real time
- Using fast decision-making protocols for effective network management
- Analyzing energy data in a safe environment and storing them and maintaining privacy
- Self-healing network performance

Smart grids and blockchain applications have many advantages, including reliability, transparency, immutability and distributed systems. However, there are also challenges facing researchers in integration of blockchain with smart grids (Figure 16.2). These challenges include:

- Connecting distributed energy sources on a large scale and at different voltage levels to the grid and checking imbalance between supply and demand
- Lack of reliable communication technologies and network connectivity issues

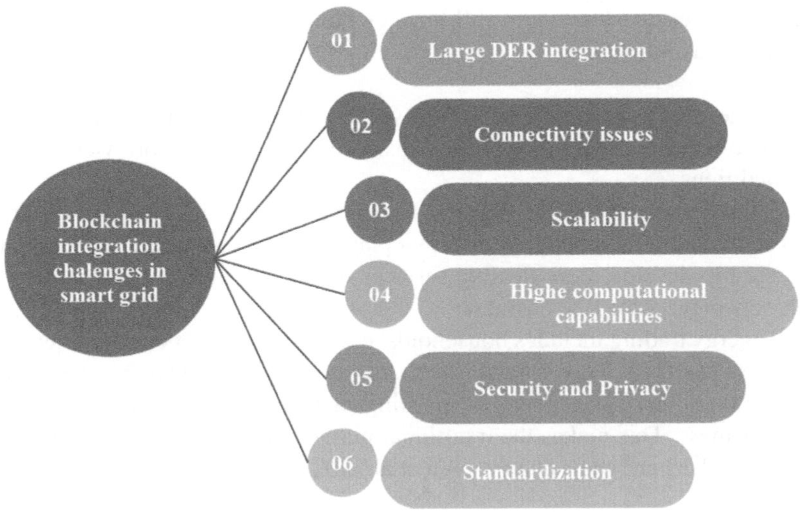

FIGURE 16.2 Blockchain integration challenges in smart grids.

- Scalability metrics
- High computing capabilities and problem solving
- Integration of blockchain into the grid raises privacy concerns
- Compatibility and cooperation of blockchain and lack of a suitable standard

16.3 LOCAL ENERGY TRADING AND MARKETS

Smart homes, load response aggregator, distribution network operator, fossil fuels and renewable energies–based distributed generation units, energy storage systems, aggregator/parking lots of pure and hybrid electric vehicles, two-way communication between consumers and data concentrators in demand aggregation center, and smart meters are some of main factors of power and energy transactions (Johanning & Bruckner, 2019). In the local energy market, buyers and sellers of electricity and other forms of energy such as freshwater, heat, cool, hydrogen, natural gas and so on can participate to increase the flexibility of interactions in managing both technical and financial constraints and balancing the demand and supply (Furze et al., 2022). Moreover, consumers and prosumers, who produce a part or all of their energy requirement, can exchange energy with microgrids or other nearby local consumers (Honarmand et al., 2021). Active users with flexible and shiftable loads, electric vehicles (EVs) and air conditioning systems may be controlled by signals from the load response programs or the entity controlling the interactions of market actors. The passive users do not engage in any consumption management strategy. When net energy consumption is positive, customers can decide to sell their surplus power to other consumers. Users participating in demand-side management programs can shift a part or all of their demand in specific time intervals by responding to energy tariff signals and the expected load of the distribution network. These transactions should be done in a secure and privacy-preserving manner, while the transactions in the local energy market are carried out in a decentralized framework (Siano et al., 2019).

Various energy sources such as solar cells, wind turbines, energy storage, heat and refrigeration pumps and controllable loads are connected to the distribution network. In traditional grids, users buy their energy from water, electricity and gas companies. But now with transition toward sustainable development, more participation of these resources and energy exchange between them is increasing. Emerging blockchain systems and peer-to-peer exchanges are a suitable platform for such markets. However, these systems still have challenges that need to be considered. Some of these challenges are as follows:

- Although the energy exchange in local resources is considered as personal property, its stability and security aspects should be considered from the network viewpoint under different conditions.
- The P2P energy trading in large scales means more flexibility among different energy sources and coordination between different players of decentralized markets. Therefore, it is necessary to develop more flexible services to support market development as well as support new players for P2P business.
- The P2P energy trading includes households, corporate managers and market operators, as well as hybrid markets. Therefore, it is necessary to consider the necessary coordination for the interaction of these players and the existing challenges to promote and expand the existing P2P business. This makes the traditional markets have to align their cooperation with these new markets and their players (Wu et al., 2022).

Due to increased penetration of renewable energy resources, especially in small-scale capacities, consumers can be both producers and consumers according to grid and market situations in different time intervals. Recently, models for buying and selling electricity in a blockchain-enabled energy market have widely been discussed. In these models, interactions between production and consumption and storage technologies in local networks have been realized (Afzal et al., 2022). In smart contracts, communication between smart meters and distribution system operators is formed and energy exchange takes place through it. In other words, smart contracts are made between two parties on a blockchain-based energy trading platform (Aloqaily et al., 2020).

16.4 SUSTAINABLE DEVELOPMENT OF SMART DISTRIBUTION NETWORKS

The comprehensive concept of sustainable development is a wide range of distinctive features. Economic development brings unwanted consequences such as environmental degradation and increasing social welfare. The hidden concepts of sustainable development require a development strategy that is more inclusive. This plan should take steps to achieve three basic goals of promoting social inclusion, sustainability and environmental health, as well as economic growth. In addition to significant growth of economic activities, the Industrial Revolution has been a motivation for numerous technological developments. Alternative energy resources have no negative effects on human health and environment. All these initiatives, involving financial transactions at different levels of energy markets and governing institutions, require the development of blockchain technology and help different sections of society (Mbaidin et al., 2023).

Findings show that scalability is an important technical challenge for blockchain deployment in developing countries. Higher costs and technical issues are two barriers in rapid growth of sustainable development. For example, time required for creating a block is 19 minutes and there is a capacity limit of 1 MB for each block. Therefore, eight transactions can be processed per second. Despite its capacity and reliability as an intermediary, lost information may not be recoverable. Hence, data users and assets owners may be affected by these privacy threats. Blockchain can be used to record information and characteristics of power networks while improving their stability, reliability, resiliency, monitoring, protection and control. In a decentralized blockchain infrastructure, transparent and accurate tracking and transactions are possible, which results in a significant reduction of greenhouse gases and increase of clean energy (Sahoo et al., 2022).

16.5 ELECTRIC VEHICLES

The development of electric vehicles means traditional electricity distribution lines may be relatively reduced. However, due to barriers such as lack of charging and discharging infrastructure, customers are less willing to participate in this field. The emergence of blockchain and the ease of transactions in charging, discharging and billing makes this industry more attractive for customers. In recent years, companies are trying to attract more customers with ease of access and reducing the costs of construction and maintenance of charging and discharging infrastructures. Using the blockchain technology, the current costs of transactions will be reduced and this will remove one of the major barriers to the its development. Electric vehicles can be charged wirelessly and inductively while stopped at a traffic light, while smart contracts record transactions quickly and automatically on this ledger. Then, smart contracts on a blockchain platform are able to charge and discharge batteries of electric vehicles based on energy demands of distribution network (Livingston et al., 2018).

Electric vehicles have special features that play a distinct role in the electricity industry. They can be used in two ways: connecting the vehicle to the grid (V2G) and connecting the grid to the vehicle (G2V) (Jabari et al., 2019). An important feature is that they are not fixed and can be connected to the network at any time and in different situations. This refers to connection of an electric vehicle to the power grid for charging during off-peak hours and discharging its battery pack in peak times. The unpredictable behavior of electric vehicles, both in terms of location and power, imposes some technical challenges to system operation and energy management, especially with their increasing growth in recent years. Blockchain technology provides an opportunity for real-time monitoring, control and management of energy microgrids considering various storage facilities such as aggregated plug-in hybrid or pure electric vehicles (Al-Saif et al., 2021). By managing energy consumption in specific locations and times, it can lead to significant improvement in energy efficiency and economic and environmental sustainability (Tonev & Nikolaev, 2020).

16.6 BLOCKCHAIN APPLICATIONS IN CYBER-PHYSICAL SYSTEMS

Smart grids have been faced with several challenges and vulnerabilities in which many parts of these systems may be manipulated or attacked. The effects and threats of the attacks are classified into two physical and cyber categories. The cyber-physical attacks are different in terms of type and impact (Zhao et al., 2020).

1. Time synchronization (or time delay) attacks
2. Global Positioning System (GPS) spying attacks
3. Denial of Services (DOS) attacks.

The false data injection (FDI) attack is one of the most important attacks in cyber-physical systems, in which the attacker manipulates or injects the false data into the real measurements set or the control signals, aiming to change the system operating state. Such attacks can be great threats for stable operation of smart energy networks, and it is essential to identify them. For secure operation of smart energy grids against any intentional FDI attacks, redundant measurement sets and signals are used in detection approaches. Protection-based schemes mainly rely on increasing data redundancy. Unguaranteed effectiveness in different operation states of these systems and high requirement for measurement redundancy are two main drawbacks in safe and reliable schemes. By creating and forming a cyber-layer, blockchain can be easily integrated into different parts of smart grids. Blockchain infrastructure as a cyber-layer facilitates operation of smart systems at a high level of security and reliability. Therefore, it has various applications (e.g. electricity market, energy trading) in the smart grid environment. For example, a recently introduced application is for energy trading in electricity markets, which connect various energy traders like consumers, producers and retailers to each other via a decentralized framework. Microgrid control, including production devices and distributed generation units and other players such as prosumers, energy storage technologies, demand response aggregator, charging stations and parking lots for pure and hybrid electric vehicles, and so on, is possible using blockchain.

16.7 PROTECTION AND CONTROL OF SMART SYSTEMS

The emergence of blockchain provides more security for the electric power system. The cryptographic security together with a consensus mechanism provides the immutability of data in the blockchain framework. Once a transaction or energy data is placed on the blockchain, since this system is robust and secure, it is very difficult to delete or change this transaction for illegal purposes. Additionally, blockchain creates more flexibility in electric power systems. Because each user of the network has a copy of the ledger, there is no point of failure, which makes them less vulnerable to malicious attacks and, as a result, the electrical power system is more flexible than a centralized data architecture. The flexibility of the electric power system is increased by decentralized computing instead of being concentrated in a central computer. The smart grid layers are managed by different blockchain aggregators. These blockchains should provide a platform to protect the smart grid by providing a safe and secure information storage platform. Furthermore, there is no direct connection between the identity of a consumer/producer and the identity in the blockchain environment. Hence, the vulnerability of the system against the malicious attacks can be reduced, which makes the power system more flexible (Musleh et al., 2019). Smart grid technologies are equipped with advanced metering infrastructure (AMI), load response, micro phasor measurement units (μPMUs), renewable energy source-based generation units, hybrid electric vehicles and energy storage systems in order to improve their monitoring, protection and control. AMI, μPMUs, smart meters and other measuring units transfer data between the control centers and the system entities such as producers, customers, prosumers, retailers, and so on via the multiple communication protocols. Attackers can perform cyberattacks on communications and computing infrastructures by manipulating network applications or physical systems. The cyberattacks vary depending on their target. Attackers can perform various intrusions by using software vulnerabilities and wrong settings. Even the secure cyber systems can be attacked by insider threats, such that a trusted individual can exploit system capabilities to steal data or affect system performance (Kabalci, 2016).

AMI is based on the development of smart meters at consumption sites and provides two-way communication between the meter and the distribution company. The development of smart meters makes it possible to send real-time electricity prices to consumers, collect information about current consumption and perform advanced analysis on distribution system failures. Since AMI is connected to the distribution system, a large number of consumer meters must be compromised to have a significant impact on the reliability of the distribution system. However, a coordinated cyberattack can seriously affect the reliability of these systems (Li et al., 2019).

Figure 16.3 shows a view of the cyberattack points on the smart distribution network. The branch currents and voltages of distribution system nodes are measured by sensors (μPMUs) and sent to the controller and protection devices through the high-speed communication link to make appropriate decisions. Controllers make decisions based on system conditions and send relevant commands to specific protection and control elements (triggers). A cyberattack can affect the delays of direct or feedback paths or affect data from sensors, protective equipment and controllers. The cyberattacks on cyber-physical systems are classified into three categories: time attack, data integrity attack and response attack.

- **Time cyberattacks**
  In dynamic systems (such as control measures and protection schemes) time is considered as a critical component. Therefore, these systems cannot tolerate communication delays and are vulnerable to temporal cyberattacks. The cyberattacks sometimes affect the communication network with packets that slow it down and may cause several communication interruptions; both are unacceptable and lead to DoS.
- **Data integrity attack**
  In a data integrity attack, the data is corrupted in the forward or reverse direction. An attack may occur that directly affects the sensor data (μPMU data and commands sent to control

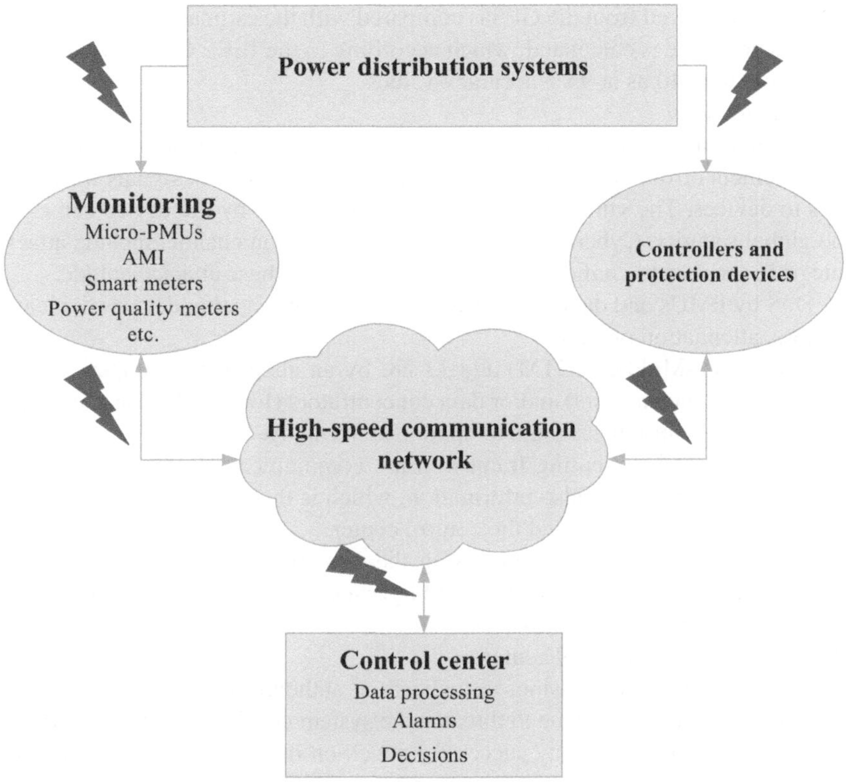

FIGURE 16.3 Possible cyberattacks to monitoring, protection and control of power distribution systems.

and protection systems). In other words, trip signals are blocked in scenarios where the controller sends a trip command to the protection elements.

- **Attack to response**
  A response attack is similar to a data integrity attack, and the attacker manipulates measurements of μPMUs or control messages by hijacking transmission packets between metering devices and the control center of the distribution network.

The cyberattacks to the smart grids are often carried out using powerful malware such as viruses, worms or Trojans, but a few attacks, such as the attack on the Pacific gas and electricity transmission lines, have pointed to the physical layers. These network attacks compromise the power and access to its control data. The cyber security of micro synchrophasors or other measuring technologies is divided into three categories (Nejabatkhah et al., 2020):

- Device and aggregator
  A device or aggregator level attack compromises data integrity and targets any device from μPMUs to data concentrators. Some of these attacks are:
  - Manipulation and change of signal measurement units through interference
  - Illegally changing the calibration of devices to report false readings
  - Data falsification to reflect incorrect measurements
  - GPS signal spoofing by broadcasting fake signals to the receiver to mis-synchronize the calculated phasors, correct satellite position shifts or broadcast GPS signals at later timestamps. GPS signal spoofing can be reduced by enabling the receiver to predict the visibility of GPS satellites at a certain and constant time position and also by using the acquisition code received from these satellites. In another strategy, the

signal measured from the GPS is compared with the estimated signal and the anomalous error rate is calculated, which according to the IEEE C37.118 standards should be less than 40 ns at 95% accuracy values

- Communications
  Smart metering is carried out based on two-way communication channels that send measurements from devices to the control center, and control signals from other channels to devices. The vulnerability of the protocols used by the device can compromise the global security. Cyberattacks on the communication channel include attacks on the integrity, access and confidentiality of data. Some of these attacks include:
  - DoS by PMUs and data concentrators or other aggregating devices that cause signal loss, attenuation or delay.
  - Man-in-the-Middle (MITM) attacks are by an attacker who impersonates μPMUs (for data concentrators) and/or data concentrators (for μPMUs) and sends false commands that cause these units to operate abnormally.
  - FDI involves intercepting frames on the communication channel and changing or replacing them with false information, which is then propagated to higher levels of monitoring, protection and the control center.
  - The delay caused by compromising the communication routers that intentionally cause a delay in the propagation will have a serious impact on the situational awareness of the network.
- Data applications in control center
  To detect and prevent intrusion, virtualization, authentication, authorization and access control, cyberattacks continue to threaten the system despite protection by organizational security tools. Obviously, any successful attack on other levels in a way that cannot be detected by enterprise security systems can be a significant threat. These attacks are at very dangerous levels because power system applications use monitoring data to perform analysis on power quality, reliability, faults and grid topology. One of the adverse effects on these calculations can compromise the self-healing nature of the network. Based on this, solutions including machine learning, game theory and data visualization can be provided.

The intensity of an attack on the target system can be understood from its impact. In blockchain-oriented smart networks, there is the ability to directly exchange information between different devices. This means that if the information of one of the elements is infected, since these systems are interconnected, the transmission of this data to other parts of the system can affect the entire network. Basically, in most attacks, the acceptability, completeness, accuracy and consistency of data are affected. There are various practical methods to reduce the impact of cyberattacks on data quality issues such as accuracy, timeliness and consistency, including the use of dedicated communication channels between smart metering devices and the data aggregation center, data encryption before communication, increasing the data exchange points, and using the firewalls and routers.

16.8 RENEWABLE ENERGY AND DEMAND RESPONSE

Blockchain technology provides the possibility of production certification from renewable energy resources, grid control and protection, coordination of distributed generation units, demand-side management programs, energy storage systems and participation of consumers in regulation and electricity markets. Some issues faced by transmission and distribution network operations are specifically related to Di Silvestre et al. (2020):

- Cooperation of active and flexible customers in demand response schemes;
- Aggregation of local energy providers to participate in energy markets;

- Ancillary services required for distributed systems; and
- Coordination of mentioned processes.

Exchanging data and various control signals and commands, which are an essential need of decentralized power systems, can threaten or maintain their security requirements. Meanwhile, diversity of distributed resources makes direct control of generators and devices used by companies almost impossible. Accordingly, customers are requested to participate in demand response strategies through their energy resource management.

To facilitate development of future power systems, blockchain is a new technology that can be offered in this field. In recent scientific papers and in work done so far by aggregators, looking at similar issues of aggregation and demand response applications from the operator's point of view, blockchain advantages can be stated as follows (Junaidi et al., 2023):

- Peer-to-peer decentralized demand response schemes;
- Demand response programs and aggregation of local distributed generation units;
- Incentive-based demand-side management strategies such as demand bidding, load reduction under emergency situations, interruptible loads and direct load control.

These three topics are mostly related in smart grid operation. Implementation of decentralized demand-side management programs in blockchain-enabled smart microgrids provides remarkable opportunities to collect information from smart meters as well as to store produced energy, and each customer expects flexible smart contracts for evaluating its reward or penalty and balancing energy between consumption sides and providers.

16.9 IMPROVING ENERGY EFFICIENCY

Growing innovation in digital technologies and consumers' participation in energy efficiency policies provide key opportunities to mitigate emissions while improving energy efficiency of power systems. Blockchain and distributed ledger technologies have gained significant popularity in security, transparency and sustainability of the energy sector, as well as its transparency and security. As mentioned in previous sections, blockchain and its various applications in the energy sector have been widely discussed in P2P energy trading field. However, it is still a concept that has not been fully explored in terms of its application in energy efficiency.

Energy services companies (ESCOs) are exploring some potential applications of blockchain for successful implementation of energy performance contracts (EPCs). EPCs have become a common way to improve energy efficiency in buildings. EPC models involve multiple stakeholders who maintain records of basic energy data, project costs, technology implementation and investment costs, as well as energy savings rates, which may cause some wastes in timing and budget of payments between stakeholders. The smart contract feature reduces transaction waste costs; the ESCOs also have the opportunity to complete their small energy efficiency schemes, which significantly reduces time and costs.

The interesting question is how blockchain can improve the added value of energy efficiency? The answer to this question is related to how to save and share energy on a blockchain platform to improve information security, service reliability, energy efficiency, and market transparency. Blockchain provides an opportunity for consumers to exchange their excess energy. Hence, consumers will have more incentive to save energy and improve the energy efficiency of their homes because they will earn more economic benefit from selling their surplus power and energy to adjacent consumers or local microgrids. In energy efficiency markets, it improves financial transactions, trust between various stakeholders, transparency and overall operations. Some of these advantages are explained below (Khatoon et al., 2019; Mao et al., 2021):

- Energy saving and its encryption
  Encryption refers to encoding information or converting it into codes to prevent unauthorized accesses. Encryption in energy saving using blockchain provides a high level of security for market transactions and improves energy efficiency of smart systems significantly. Data privacy and security are important issues in the digital world, and blockchain enhances the security of customers' energy saving data in energy efficiency markets.
- Exchange of energy saving: Blockchain technology has a high potential to increase the participation of consumers in consumption management schemes because the data related to energy saving, balanced energy bills and purchased additional energy can be encrypted and this information is stored on the blockchain platform. Blockchain in combination with information and communication technology (ICT) and automation of processes can help us to some extent in energy saving benefits.
- Higher transparency: Since blockchain is decentralized and has a distributed ledger, data can be easily and transparently shared on a secure and tamper-proof platform. Manipulating the shared data on a blockchain platform is an impossible and costly process. Blockchain can be a trustless system for sharing data with other blocks while verifying information by all of them.
- Cost saving in transactions: Blockchain can perform transactions directly through peers without any intermediaries, which reduces their complexity and costs. In smart energy contracts, this cost saving increases significantly especially when an EPC is used for energy transactions and efficiency improvement. Moreover, simplification of technical and financial processes between ESCOs, power distribution companies, customers and banks is possible using it.
- Increased reliability: During data management, trust becomes very important in the advanced world, especially when it is stored in one place or several nodes or stakeholders with no need to process automation. When data is stored in one place, it takes time to collect information, which will be difficult to track and audit. Blockchain is a decentralized platform and non-distributed ledger where data is stored in different blocks, which can significantly increase the system reliability.
- Trust, transparency and security improvement in market exchanges: When blockchain technology is used in energy efficiency markets, it can bring trust, transparency and security to energy-saving permits and payments automatically. Information and communication technology and blockchain can greatly reduce major barriers and problems of market financial flows and create a reliable path to success for higher energy efficiency.

16.10 FINANCIAL APPLICATIONS

Blockchain technology is very efficient and useful for financial services and increasing credit and security and risk management. Many institutions are using the blockchain and creating a blockchain platform for customers by improving transparency and performance as well as creating new opportunities and more income. Blockchain technology has special capabilities such as the settlement process and the identification of active people in financial institutions and banks. The development of blockchain functionality in financial fields is very prominent; the transfer of ownership of assets and communication, measurement and data analysis especially are of interest to many auditors. By clarifying the ownership of assets and liabilities, productivity increases for accountants. Blockchain-based systems will cause the faster, more customized and more cost-effective issuance of digital securities, and based on this, the market of investors will expand significantly. It will also reduce the costs of publishers, and with the customization of requests, the risks of the parties will be reduced. The transaction platform, which causes the ease of business and market management along with mutual cooperation, makes blockchain more popular (Javaid et al., 2022).

With the advent of blockchain, many studies and researches have been devoted to the structure of P2P. It integrates the decentralized paradigm and the multifactor theory as well as the two-way communication model. In this system, every user has the same privilege, and they have the same share in the electricity market. Traditional electricity users switch to electricity price predictors. This shared electricity price is determined by all projected users. In addition, there is no need for each user to send their privacy information (such as electrical loads) to the center. However, with the diversification of business, and the entry of different companies into the electricity market, problems of trust and anonymity will arise. Therefore, the reliability and safety of the energy system are not guaranteed in the preceding studies. In recent years, blockchain technology with decentralized features, transparency and anonymity has been a suitable option for creating a reliable and secure energy trading system. It can prevent the vulnerability of intensive data processing, thereby ensuring the trust of P2P energy sharing. The security and privacy of buying or selling local electricity increases P2P among electric vehicles with blockchain consortium techniques. With digital design and a multi-crypto digital identity scheme, it protects the privacy and energy sharing of the neighborhood. However, most of the work on blockchain-based P2P has been studied for energy sharing. In a building, the load price is determined by all forecast users of the building without the involvement of a trusted third party. All building users actively adjust and propose their energy purchase strategy to improve their profitability in response to the price plan. In addition, each user of the building is assigned a pseudo-digital identity in the blockchain network. Then, privacy protection can be guaranteed in process of information exchange with anonymous operations (Wen et al., 2021). The immutable, decentralized and distributed nature of smart contracts can solve many problems in the financial industry. Smart contracts execute predefined operations in a certain system state without errors (Hewa et al., 2021, Figure 16.4).

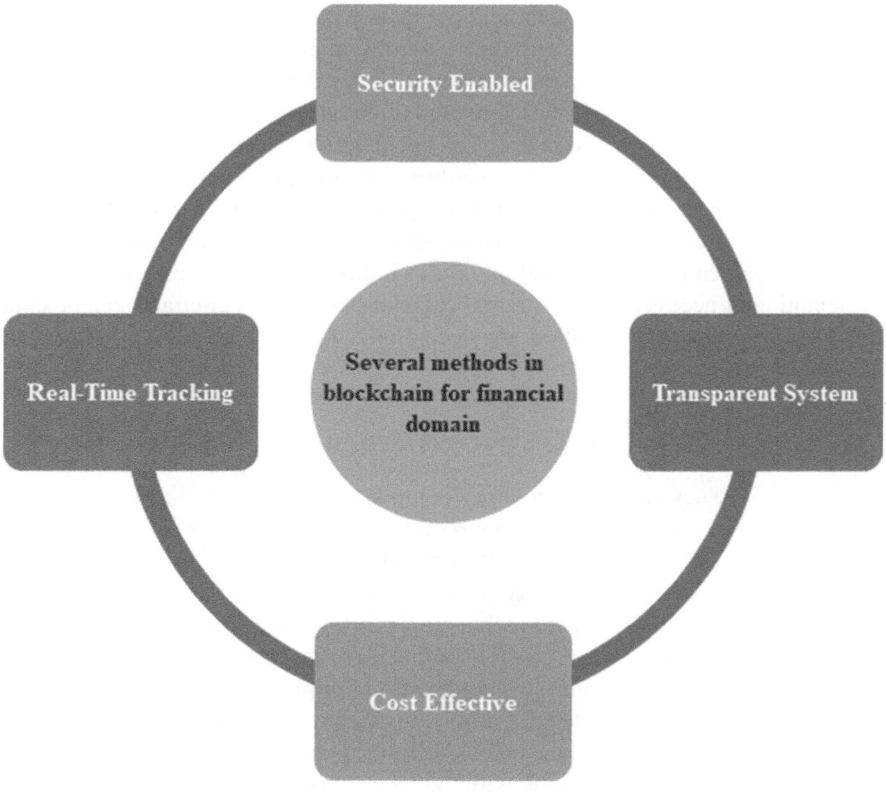

FIGURE 16.4 Several methods in blockchain for financial domain.

16.11 IDENTITY MANAGEMENT AND ACCESS CONTROL

One of the essential services in any company is access control and identity management, and these classic identity management systems are mainly centralized and equipped with expensive hardware such as smart cards and hardware security modules to connect to the system. Since centralized systems increase points of failure, they require methods such as strong backup, recovery and disaster management. These scalability limitations usually exist in centralized systems. Sometimes, due to the limited computing power of the system, there is a delay in identity management and access control when connecting to internet of things (IoT) devices. Access control and identity management systems will be the next generation of distributed ledger technologies. These blockchain-based access control systems and smart contract systems promise accuracy, fault tolerance, and availability. Personal identity information and the connection of modern technologies have a special value in human life and are increasing day by day. Much information and many specifications such as personal identity information and location are available in smart devices such as mobile phones. In most of the widely used and popular programs, including social media, they have a centralized structure in which the user can have less control over the data. Most users are still unaware of the importance and protection of their personal information. In these systems, it is possible to misuse personal data without the consent of the owner. Therefore, data owners must have the knowledge and ability to control access to their data to prevent such incidents. Therefore, to control and limit access to identity and confidential information and eliminate personal information theft, smart contracts are one of the best and most useful solutions. Relying on the decentralized nature of smart contracts, data owners are able to control access to their private data in a decentralized and transparent manner, in which a distributed ledger can be used to record access to personal data and thereby prevent unnecessary access to data by third parties (AlBadi et al., 2023; Hewa et al., 2021).

Classical identity management and access control systems such as lightweight directory access protocol, identity and access management and public key infrastructure are compatible with traditional and centralized architectures. Centralized identity management systems with development-specific access management enable administrators to meet data, security requirements and costly central vulnerabilities. If access information related to private keys is stored in a centralized server, for example in a cloud space, this information can be vulnerable. Issues such as environmental security requirements, maintenance of expensive hardware security modules and tokens with high management and maintenance costs can be imposed upon organizations. Identity theft is essential to eliminate, and secure access to services must be guaranteed. Smart contracts based on the blockchain platform have already been adopted in various industries and applications for access control. The implementation of access control using blockchain-based smart contracts can be controlled by the communications industry. In addition, the decentralized nature ensures service availability with guaranteed scalability (Pal et al., 2022).

16.12 SECURITY POLICY IN ACCESS CONTROL

Attention to security issues is a key regulatory component in any organization as well as Internet applications. The security policy includes the required security controls according to the strategic rules and regulations in every organization. The depth and scope of users' access to security resources and infrastructure depends on the security policy. These centralized security policies and governance architecture is one of the accepted architectures for most organizations. In systems with centralized security policy management, customization can be problematic and cumbersome. For example, according to the growth and promotion of an employee to the next level, the range of available resources should be expanded. Customizing a centralized security policy incurs additional costs. If the organization is located on the other side of the borders, the overhead costs will be higher. In addition to these, the risks of tampering with the security policy in the centralized system without the consent of the user can be possible (Hewa et al., 2021; Patil et al., 2021).

As soon as the predetermined conditions are met, smart contracts start the independent execution of the program, and this can be an ideal and useful solution for decentralized security policy management. Security controls can be defined and implemented programmatically in smart contracts and systems. The costs of human intervention in security policies are automatically eliminated. The distributed nature of blockchain-based smart contract systems ensures that unruly users and pool administrators cannot stop or tamper with the execution logic.

16.13 SMART CONTRACTS USING IoT

In recent years, IoT has been one of the most controversial and hot scientific and research fields in smart networks and computer and digital networks. It is expected that billions of devices in various industries will be equipped with the Internet of Things. Many new requirements and infrastructures have been identified in various dimensions to facilitate this IoT. One essential need of the future IoT devices in smart networks and digital and computer systems is to automate the system. Improvement and security are also challenging. Autonomous resource sharing requirements will be a key feature of next-generation autonomous systems. The inherently automated and decentralized nature of smart contracts based on the blockchain platform will solve many of these challenges in future IoT systems. To meet many security requirements in the field of IoT, smart contracts and networks can be used effectively and efficiently (Khan & Salah, 2018).

Smart contracts are written via the blockchain platform and contain 'if, then' conditions. Smart contracts, as their name suggests, must be smart, codes distributed and independently applied to them, and contract clauses distributed and independent for each user to be visible and enforceable. To activate and implement smart contracts, a transaction is sent to each address, all nodes must perform validation, and if all subscribers agree, this contract will be executed. Care must be taken in coding contracts because if a code is entered incorrectly, it cannot be deleted or changed (Islam, 2023).

The sharing and interconnection of different resources is an important requirement in the IoT network. A resource-constrained infrastructure requires optimal resource sharing. Resource sharing is a service that doesn't have to be an intensive process. In this case, the resource sharing process imposes another additional cost to the system. The cloud-based resource sharing mechanism causes more network traffic and computational cost. Smart networks and smart contracts enable peer-to-peer resource sharing, which improves computing and network traffic. In smart networks with smart contracts, the next generation resource sharing approach can be defined in the field of IoT. With limited hardware resources, it is very difficult to implement security with multiple cryptographic operations and with increasing key size. Devices are limited in computing capacity as well as memory. Therefore, the use of public key certificates in IoT systems and devices will be expensive and costly. In a centralized server, access control and privilege definition can be vulnerable. In general, the blockchain platform allows stakeholders to embed access control policies into smart contracts and implement them in a decentralized manner. The code in a cloud computing environment is immutable and free from any change or capture. Decentralized operation ensures that no traffic is created for the wide network when implementing security (Minoli & Occhiogrosso, 2018).

It is very important and critical for the security of IoT ecosystem trust management. Critical application areas such as smart grids and contracts, smart cities, healthcare and electrical grids with electric vehicles require high reliability for the operation of IoT nodes. Trust management includes various services such as trust creation, real-time trust evaluation, trust removal from malicious actions and so on. The future IoT infrastructure will consist of billions of connected real-time devices, and trust management frameworks are expected to grow to meet the massive connectivity needs. Managing centralized trust with expectations of future demands is challenging. In addition, computational resource limitations of design decisions drive the advanced IoT infrastructure toward decentralized architectures such as designs based on edge and fog computing. The significant added value in classical trust architecture and overcoming the limitations of classical trust ecosystems are among the prominent features of blockchain-based smart contracts in IoT. An IoT–based transaction platform consists of several layers as follows:

Layer of physical things: This layer is known as an energy layer and includes production, storage, consumption and conversion of energy on the customer side and provides all exchange and management of energy for various uses such as electricity, heating and cooling.

Perceptive and active layer: This layer is responsible for recognizing different means of identification and control. All devices placed in this layer can include various sensors, meteorological systems, smart meters and such in the production and consumption sector, or inverters, smart relays, thermostats, half-smart cables and so on in the supply and demand sector.

Connection and network layer: This layer is known as an internet-oriented layer and its task is to manage and coordinate wired and wireless communications in context of different protocols. Using IoT, data is translated and transferred from sensors to the main center for monitoring, analysis and appropriate decision-making.

Middleware layer: This is an intermediate layer and acts as a communication gap between the IoT environment and services and programs, which takes into account some complexities of lower layers as well as communication services to higher layers. In fact, this layer acts as an integrated layer and is responsible for required coordination and support between different layers and programs.

Application Layer: This layer broadly implements energy-aware applications. It provides data and information for buyers, consumers, power companies and other actors for demand-side energy management and energy exchange on a P2P platform. These programs are presented in the form of internet and also mobile phones, and in this layer, encouraging and active programs for participation of customers in energy management and also carbon reduction should be included (Wu et al., 2022).

16.14 SMART COMMUNICATIONS

Smart cities are one of the important innovations of the Internet of Things that will be used in the future infrastructure development of cities and villages. It is expected that the quality of life will be better in smart cities. Enthusiasts have many inspirations in various and important fields such as entertainment, casinos, integration with smart vehicles and so on. Since smart cities include millions of different devices and communication devices, the scalability of operating systems is an important predictor. In addition, concentration will be associated with significant overheads along with additional risks. Smart contracts and digital and smart networks in the blockchain platform will play an important role in the future of smart networks and smart cities (Aggarwal et al., 2019). The operational capability of smart contracts and the decentralized model leads to a higher value of guaranteed service availability. Peer-to-peer operational capability in smart contracts reduces network resource consumption and increases network efficiency. The communications industry is vital in all countries of the world. The volume of customers and consequently the complexity of the services provided increases. Network infrastructure and software modules need to be upgraded to match current and future needs. Smart contracts based on blockchain will strengthen the communication industry in various dimensions. Increasing trust and independent execution of smart contracts based on the blockchain platform ensures security and scalability, which is one of the mandatory requirements in modern communication systems. Subscribers of communication services, especially the widespread use of mobile phones, are increasing day by day around the world. Complexity in various aspects of the communication field causes complexity in services. Limited resources in communications need to be allocated and used in a paradigmatic manner. In addition, the user–shareholder agreement with its complexities must be compatible with future scalability and traction demands. Smart contracts based on the blockchain platform are ideal solutions for autonomously and securely implementing scalable, customized subscriber-based service level agreements. There are many applications in blockchain-based smart contracts that use these communications systems to share optimal and accurate resources (Liu et al., 2020).

16.15 CONCLUDING REMARKS AND FUTURE TRENDS

This chapter presents a comprehensive review on concept and applications of blockchain technology in power and energy systems stating their advantages and limitations. Blockchain projects have recently gained significance due to its decentralized structure, immutability, individual data control and tokenization opportunities. Higher capital investment costs, large power consumption of computers required for the mining process, environmental issues raised by high-power utilization of the mining process, money laundering using the blockchain platform due to its anonymity feature, and a time-consuming validation process due to limited size of blocks and their specific capacities to store data are some of main disadvantages of this technology in digitalization of urban development.

REFERENCES

Afzal, M., Li, J., Amin, W., Huang, Q., Umer, K., Ahmad, S. A., … Raza, A. (2022). Role of blockchain technology in transactive energy market: A review. *Sustainable Energy Technologies and Assessments, 53*, 102646.

Aggarwal, S., Chaudhary, R., Aujla, G. S., Kumar, N., Choo, K.-K. R., & Zomaya, A. Y. (2019). Blockchain for smart communities: Applications, challenges and opportunities. *Journal of Network and Computer Applications, 144*, 13–48.

AlBadi, A., Hajamohideen, F., & AlSaqri, D. (2023). *A Review on Blockchain Techniques Used for Identity Management System: Privacy and Access Control*. Paper presented at the *International Conference On Systems Engineering*, Las Vegas, NV. USA.

Alladi, T., Chamola, V., Rodrigues, J. J., & Kozlov, S. A. (2019). Blockchain in smart grids: A review on different use cases. *Sensors, 19*(22), 4862.

Aloqaily, M., Boukerche, A., Bouachir, O., Khalid, F., & Jangsher, S. (2020). An energy trade framework using smart contracts: Overview and challenges. *IEEE Network, 34*(4), 119–125.

Al-Saif, N., Ahmad, R. W., Salah, K., Yaqoob, I., Jayaraman, R., & Omar, M. (2021). Blockchain for electric vehicles energy trading: Requirements, opportunities, and challenges. *IEEE Access, 9*, 156947–156961.

Cantillo-Luna, S., Moreno-Chuquen, R., Chamorro, H. R., Sood, V. K., Badsha, S., & Konstantinou, C. (2022). Blockchain for distributed energy resources management and integration. *IEEE Access, 10*, 68598–68617.

Di Silvestre, M. L., Gallo, P., Guerrero, J. M., Musca, R., Sanseverino, E. R., Sciumè, G., … Zizzo, G. (2020). Blockchain for power systems: Current trends and future applications. *Renewable and Sustainable Energy Reviews, 119*, 109585.

Furze, J. N., Eslamian, S., Raafat S., Swing, K., 2022, *Mathematical Advances Towards Earth Systems Protection and Sustainability* Vol. 1, Springer, Switzerland, 337 p

Hewa, T., Ylianttila, M., & Liyanage, M. (2021). Survey on blockchain based smart contracts: Applications, opportunities and challenges. *Journal of Network and Computer Applications, 177*, 102857.

Honarmand, M. E., Hosseinnezhad, V., Hayes, B., & Siano, P. (2021). Local energy trading in future distribution systems. *Energies, 14*(11), 3110.

Islam, M. D. (2023). A survey on the use of blockchains to achieve supply chain security. *Information Systems, 117*, 102232.

Jabari, F., Jabari, H., Mohammadi-Ivatloo, B., & Ghafouri, J. (2019). Optimal short-term coordination of water-heat-power nexus incorporating plug-in electric vehicles and real-time demand response programs. *Energy, 174*, 708–723.

Javaid, M., Haleem, A., Singh, R. P., Suman, R., & Khan, S. (2022). A review of Blockchain Technology applications for financial services. *BenchCouncil Transactions on Benchmarks, Standards and Evaluations, 2*(3), 100073.

Johanning, S., & Bruckner, T. (2019). Blockchain-based peer-to-peer energy trade: A critical review of disruptive potential. Paper presented at the *2019 16th International Conference on the European Energy Market (EEM)*. Faculty of Electrical Engineering, University of Ljubljana (UL), Slovenia.

Junaidi, N., Abdullah, M. P., Alharbi, B., & Shaaban, M. (2023). Blockchain-based management of demand response in electric energy grids: A systematic review. *Energy Reports*, *9*, 5075–5100.

Kabalci, Y. (2016). A survey on smart metering and smart grid communication. *Renewable and Sustainable Energy Reviews*, *57*, 302–318.

Khan, M. A., & Salah, K. (2018). IoT security: Review, blockchain solutions, and open challenges. *Future Generation Computer Systems*, *82*, 395–411.

Khan, S. N., Loukil, F., Ghedira-Guegan, C., Benkhelifa, E., & Bani-Hani, A. (2021). Blockchain smart contracts: Applications, challenges, and future trends. *Peer-to-peer Networking and Applications*, *14*, 2901–2925.

Khatoon, A., Verma, P., Southernwood, J., Massey, B., & Corcoran, P. (2019). Blockchain in energy efficiency: Potential applications and benefits. *Energies*, *12*(17), 3317.

Li, F., Yan, X., Xie, Y., Sang, Z., & Yuan, X. (2019). A review of cyber-attack methods in cyber-physical power system. Paper presented at the *2019 IEEE 8th International Conference on Advanced Power System Automation and Protection (APAP)*, China.

Li, J., Herdem, M. S., Nathwani, J., & Wen, J. Z. (2023). Methods and applications for Artificial Intelligence, Big Data, Internet of Things, and Blockchain in smart energy management. *Energy and AI*, *11*, 100208.

Liu, Y., Yu, F. R., Li, X., Ji, H., & Leung, V. C. (2020). Blockchain and machine learning for communications and networking systems. *IEEE Communications Surveys & Tutorials*, *22*(2), 1392–1431.

Livingston, D., Sivaram, V., Freeman, M., & Fiege, M. (2018). *Applying blockchain technology to electric power systems*. Council on Foreign Relations, JSTOR, U.S.

Mao, W., Zhao, Z., Chang, Z., Min, G., & Gao, W. (2021). Energy-efficient industrial internet of things: Overview and open issues. *IEEE Transactions on Industrial Informatics*, *17*(11), 7225–7237.

Mbaidin, H. O., Alsmairat, M. A., & Al-Adaileh, R. (2023). Blockchain adoption for sustainable development in developing countries: Challenges and opportunities in the banking sector. *International Journal of Information Management Data Insights*, *3*(2), 100199.

Miglani, A., Kumar, N., Chamola, V., & Zeadally, S. (2020). Blockchain for Internet of Energy management: Review, solutions, and challenges. *Computer Communications*, *151*, 395–418.

Minoli, D., & Occhiogrosso, B. (2018). Blockchain mechanisms for IoT security. *Internet of Things*, *1*, 1–13.

Musleh, A. S., Yao, G., & Muyeen, S. (2019). Blockchain applications in smart grid–review and frameworks. *IEEE Access*, *7*, 86746–86757.

Nejabatkhah, F., Li, Y. W., Liang, H., & Reza Ahrabi, R. (2020). Cyber-security of smart microgrids: A survey. *Energies*, *14*(1), 27.

Pal, S., Dorri, A., & Jurdak, R. (2022). Blockchain for IoT access control: Recent trends and future research directions. *Journal of Network and Computer Applications*, *203*, 103371.

Patil, P., Sangeetha, M., & Bhaskar, V. (2021). Blockchain for IoT access control, security and privacy: A review. *Wireless Personal Communications*, *117*, 1815–1834.

Sahoo, S., Kumar, S., Sivarajah, U., Lim, W. M., Westland, J. C., & Kumar, A. (2022). Blockchain for sustainable supply chain management: Trends and ways forward. *Electronic Commerce Research*, *24*, 1–56.

Siano, P., De Marco, G., Rolán, A., & Loia, V. (2019). A survey and evaluation of the potentials of distributed ledger technology for peer-to-peer transactive energy exchanges in local energy markets. *IEEE Systems Journal*, *13*(3), 3454–3466.

Tonev, I., & Nikolaev, N. (2020). Review on blockchain applications for electric power systems. Paper presented at the *2020 21st International Symposium on Electrical Apparatus & Technologies (SIELA)*, Bulgaria.

Tripathi, G., Ahad, M. A., & Casalino, G. (2023). A comprehensive review of blockchain technology: Underlying principles and historical background with future challenges. *Decision Analytics Journal*, *9*, 100344.

Uddin, S. S., Joysoyal, R., Sarker, S. K., Muyeen, S., Ali, M. F., Hasan, M. M., … Islam, M. M. (2023). Next-generation blockchain enabled smart grid: Conceptual framework, key technologies and industry practices review. *Energy and AI*, *12*, 100228.

Wang, L., Jiao, S., Xie, Y., Mubaarak, S., Zhang, D., Liu, J., … Li, M. (2021). A permissioned blockchain-based energy management system for renewable energy microgrids. *Sustainability*, *13*(3), 1317.

Wen, S., Xiong, W., Tan, J., Chen, S., & Li, Q. (2021). Blockchain enhanced price incentive demand response for building user energy network in sustainable society. *Sustainable Cities and Society*, *68*, 102748.

Wu, Y., Wu, Y., Cimen, H., Vasquez, J. C., & Guerrero, J. M. (2022). P2P energy trading: Blockchain-enabled P2P energy society with multi-scale flexibility services. *Energy Reports*, *8*, 3614–3628.

Yang, R., Wakefield, R., Lyu, S., Jayasuriya, S., Han, F., Yi, X., … Chen, S. (2020). Public and private blockchain in construction business process and information integration. *Automation in Construction, 118*, 103276.

Yapa, C., De Alwis, C., Liyanage, M., & Ekanayake, J. (2021). Survey on blockchain for future smart grids: Technical aspects, applications, integration challenges and future research. *Energy Reports, 7*, 6530–6564.

Zhao, W., Jiang, C., Gao, H., Yang, S., & Luo, X. (2020). Blockchain-enabled cyber–physical systems: A review. *IEEE Internet of Things Journal, 8*(6), 4023–4034.

Index

Pages in *italics* refer to figures and pages in **bold** refer to tables

T

For Product Safety Concerns and Information please contact our EU representative GPSR@taylorandfrancis.com Taylor & Francis Verlag GmbH, Kaufingerstraße 24, 80331 München, Germany